Networked

Networked

The New Social Operating System

Lee Rainie and Barry Wellman

The MIT Press
Cambridge, Massachusetts
London, England

First MIT Press paperback edition, 2014
© 2012 Massachusetts Institute of Technology

MIT Press books may be purchased at special quantity discounts for business or sales promotional use. For information, please email special_sales@mitpress.mit.edu.

This book was set in Stone Sans and Stone Serif by Toppan Best-set Premedia Limited. Printed and bound in the United States of America.

Library of Congress Cataloging-in-Publication Data

Rainie, Harrison.
Networked : the new social operating system / Lee Rainie and Barry Wellman.
    p.   cm.
Includes bibliographical references and index.
ISBN 978-0-262-01719-0 (hardcover : alk. paper)—978-0-262-52616-6 (pb.)
1. Social networks. 2. Online social networks. 3. Interpersonal relations.
4. Internet–Social aspects. I. Wellman, Barry. II. Title.
HM741.R35   2012
006.7′54—dc23
2011038146

10   9   8   7   6

For Paulette and Bev

Contents

Preface

While writing our book, we had fun with the title. Should we call it "The Triple Revolution" of the turn to social networks, the personalized internet, and always-available mobile connectivity? Too complicated, we decided, although that became the title of part I, with a chapter devoted to each of the three revolutions.

Should we call our book "Networked Individualism"? This seemingly contradictory term would confuse browsers—better to explain it inside—and would downplay our interest in the internet and mobile connectivity. And so part II spends a lot of time looking at how the Triple Revolution plays out in relationships, families, work, creativity, and information.

Should we call our book "The Social Network"? Definitely not, for that term resonates too much with Facebook these days—in fact it's the title of the 2010 Oscar-winning movie about the start of Facebook—and we spend a lot of time in this book showing how social networks are much more than Facebook.

So we've called the book *Networked: The New Social Operating System*, emphasizing how networks among people have profoundly transformed how we connect, in person and electronically.

Along the way, we made a decision: Although we take the internet and mobile revolutions very seriously, this is *not* a book about the wonders of the internet and smartphones. Despite all the attention paid to new gadgets, technology does not determine human behavior; humans determine how technologies are used. Moreover, we would be instantly out of date if we took gadgets as our focal point. We are writing this in September 2011, but the book won't be published until 2012, and we are sure that many things will have changed by then and soon thereafter. At the same time, we are confident that one thing holds true: The internet and mobile phones have facilitated the reshaping of people's social networks, enabling them to be larger and more diverse. And they have reconfigured the way people use their networks to learn, solve problems, make decisions, and provide support to each other.

Who knows? By the time you read this, Facebook or tablet computers might have overtaken laptops and desktops, elevating mobile "apps" to digital supremacy over the web. But we have tried to get right the basic social processes associated with the Triple Revolution. Although we focus on the societies we know best, the United States and Canada—"North America"—the discussion should be more widely useful.

Finally, we note that all revolutions are lumpy. For example, despite all of our revolutionary talk, this book is still a traditional book—whether on paper or as ebook. We dearly hope that the next edition will have hyperlinks to all of the articles we cite and the movies we discuss. But there is still a place for a good read. We've tried to avoid jargon and write for intelligent general readers but still keep the specialists happy. We think we've succeeded, and we hope you like it even more than we've loved writing it. Please send us comments on our blog, http://networked.pewinternet.org. Thanks.

Acknowledgments

It is fitting that this book about networks comes from networked teams and via digital interactions. Although coauthors Lee Rainie and Barry Wellman met in person a few times, their basic interactions regarding the book were by email—with copious link insertions and attachments exchanging drafts—and by mobile phone.

In addition, Rainie and Wellman were at the centers of research networks: Rainie directing the Pew Research Center's Internet & American Life Project and Wellman directing NetLab at the University of Toronto.

In the Pew Internet world and the larger Pew Research Center, it is impossible to say where Lee Rainie's offerings begin and other staffers' contributions end. It is a fellowship of daily learning, creation, and kinship with Susannah Fox, Amanda Lenhart, Kristen Purcell, Aaron Smith, John Horrigan, Cornelia Carter, Kathryn Zickuhr, Deborah Fallows, Paul Taylor, Scott Keeter, and Andrew Kohut. For this book, no one's insights and careful editing skills were as important as those of Mary Madden Nesper, who was generous with her time and graceful in her interventions. Along the way, a cadre of Pew Internet researchers made important contributions: Jessica Vitak, Sydney Jones, Amy Tracy Wells, Eliza Jacobs, Lauren Scissors, Xingpu Yuan, Terrell Frazer, Margaret Griffith, and Niki Woodard.

In the wider world of Pew, the financial and substantive support provided by the board of the Pew Charitable Trusts and board of the Pew Research Center were critical. The personal interest and commitment to technology-focused research of Rebecca Rimel, Donald Kimelman, Tim Durkin, and Elizabeth Gross were essential to the work of the Pew Internet project and the contributions its work has made to this book.

Rainie has also learned at the knee of advisers and teachers whose fingerprints are all over this material: Janna Anderson, Keith Hampton, Michael Nelson, Lincoln Caplan, Adam Clayton Powell III, Esther Dyson,

Linda Stone, Clay Shirky, James Fallows, Marjorie Blumenthal, Larry Irving, Bill Tancer, Michael Delli Carpini, danah boyd, Jeffrey Eisenach, David Silver, Eileen Rudden, Jonathan Zittrain, Howard Rheingold, Jill Nishi, Joseph Turow, Paul Parsons, Ed Fouhy, Tom Rosenstiel, Amy Mitchell, Luis Lugo, and Michael Dimock.

At the center of all that matters are Paulette Rainie—shrewd reader, patient sounding board, and sublime partner—and Amanda, Christina, Abigail, and Clay—the children who tolerated with good cheer the distractions that this work created for their father.

At the University of Toronto, NetLab has been at the center of Barry Wellman's web of relationships for decades: spinning ideas, support, and community. Over the years, Professors Ronald Baecker, Dean Behrens, Bill Buxton, Dimitrina Dimitrova, Bonnie Erickson, Arent Greve, Anatoliy Gruzd, Alexandra Marin, Rhonda McEwen, Kakuko Miyata, Diana Mok, the late Judith Merril, Jason Nolan, Anabel Quan-Haase, the late Janet Salaff, Marilyn Mantei Tremaine, and Helen Hua Wang have stimulated our understanding.

They have been complemented by the NetLab diaspora of former doctoral students who have worked closely with each other—and Wellman—developing many of the ideas and research we discuss here: Susan Bastani, Kristen Berg, Jeffrey Boase, Juan-Antonio Carrasco, Wenhong Chen, Vincent Chua, Rochelle Côté, Keith Hampton, Caroline Haythornthwaite, Bernie Hogan, and Tracy Kennedy have been especially vital participants.

Preparing this book has been an intensive and extensive adventure involving many NetLabbers. Justine Abigail Yu coauthored chapters 8 and 9—"Networked Creators" and "Networked Information" plus "The Conversation Never Ends" interlude—while Xiaolin Zhuo focused on the creative use of the internet and mobile phones in the Egyptian revolt section of "Networked Creators." Maya Collum wrote the "A Day in a Connected Life" interlude; Tracy Kennedy coauthored chapter 6, "Networked Families"; Wenhong Chen coauthored chapter 7, "Networked Work"; and Christian Beermann and Zack Hayat coauthored chapter 11, "The Future of Networked Individualism."

Other NetLabbers contributed vital research support and writing: Vincent Chua for chapter 2, "The Social Network Revolution"; Mirna Ghazarian for chapter 4, "The Mobile Revolution," chapter 5, "Networked Relationships," and chapter 7, "Networked Work"; Julia Chae, Melissa Godbout, Sharanpreet Kelley, Rhonda McEwen, and Yu Janice Zhang for chapter 5, "Networked Relationships"; Julie Amoroso and Maria Majerski for chapter 6, "Networked Families"; and Anna Brady for chapter 7,

"Networked Work." Mohammad Haque researched and compiled the complex trend graphs, especially in the front end of chapter 2, "The Social Network Revolution."

Quietly supporting all of this were infrastructural heroes who made the research sing and the writing dance, most notably: Christian Beermann, Courtney Cardozo, Isabella Chiu, Sabrina Cutaia, Christine Ensslen, Maryam Fazel-Zarandi, Jennifer Kayahara, Natalia Kononenko, Chang Lin, Julia Madej, Mo Guang Ying, Barbara Neves, Ayden Scheim, Annie Shi, Lilia Smale, Sinye Tang, and Erin Weinkauf.

We have all learned a lot from each other, laughed a lot, and grown in understanding.

Most of all, extraordinarily deep thanks to Beverly Wellman who read, thought about, proposed, edited, and debated every word in every draft of every chapter—all with good humor, profound support, and uncommon intellectual sense. She has been Barry's heart, soul, and brain in their dance through life.

A series of research teams have studied East Yorkers three times since 1968. NetLab has returned for a fourth time. We have been thrilled to watch this area of Toronto change in buildings and people. We have been welcomed with generosity and insight. We have tried to repay a part of our debt here.

Our funders have created the conditions under which NetLabbers could think, communicate, and especially find and analyze evidence. Throughout, continuous grants from the Social Science and Humanities Research Council of Canada have been the foundation of NetLab's work. In addition, corporate grantors Bell Canada, Intel, Mitel, Nortel Networks, and Telus have provided generous arm's-length support. Starting in 2010, the GRAND Network of Centres of Excellence have also provided NetLab with support for research on networked organizations and interpersonal relations.

NetLab has benefited from the intellectual communities found in the University of Toronto's Centre for Urban and Community Studies and more recently in the Department of Sociology. Scholars don't work alone: They network.

We both thank our supporters and helpers at MIT Press, especially our faithful ally Marguerite Avery and our trusted pathfinders Julia Collins, Mel Goldsipe, and Katie Persons.

We also bow our heads in humble gratitude to the people whose story inspired many of our insights and whose careful reading made our work smarter: Peter and Trudy Johnson-Lenz.

I   The Triple Revolution

1 The New Social Operating System of Networked Individualism

Early on December 3, 2007, Trudy Johnson-Lenz tripped on her front steps as she was walking to her door in a rain storm.[1] She slammed her head on a rock and was knocked unconscious. Her husband Peter struggled unsuccessfully to rouse her and then called the Portland, Oregon, emergency ambulance service for help. By 8 a.m. she was on an operating table at Oregon Health & Science University. Her skull was filling with blood. To give her brain room to swell and heal, neurosurgeons removed a third of her skull, put it in the freezer for later, and removed the blood. The odds of people in her condition surviving, Peter was told, were 50–50, and of the survivors, three-quarters have some disability. Yet, beating the odds, Trudy started recovering just twelve hours later.

Before leaving her bedside, Peter used his mobile phone to snap a few digital pictures of her elaborately bandaged head and breathing tubes. He emailed the pictures and a description of the accident at midnight to some friends and was warmed by the reaction. Within 36 hours, nearly 150 people across North America had sent emails, as friends forwarded the news about Trudy to others. People sent poems, expressions of love and encouragement, and offers of help and prayers. Most were sent to Peter's computer. Urgent and logistics-related text messages came to his mobile phone.

Over the next two days, local friends stepped in to help. John Stapp came to the hospital, treated Peter to a bag lunch, and offered to manage a local meal delivery campaign for the couple. Mike Seely, director of the Pacific Northwest Transplant Bank, introduced the couple to a hospital social worker who started prepping Peter with tips about how to navigate the looming insurance, billing, and financial-aid bureaucracy. Martin Tull and Chuck Ensign ran errands and helped prepare their house for Trudy's safety once she was discharged.

More socially and physically distant acquaintances responded in other ways. Buddies who were volunteer DJs on the local jazz radio station, KMHD-FM, announced their concern about Trudy on air and dedicated shows to her. Among their many passions, Peter and Trudy co-moderate an internet forum on jazz vocalist Kurt Elling's website.[2] Several of the radio station jazz aficionados and Elling forum participants took it upon themselves to burn CDs of their favorite music to send Trudy as she recuperated.

Another recipient of a forwarded email was Lisa Kimball, a friend whom they call a "netweaver extraordinaire."[3] Lisa crafted an email in the name of the Johnson-Lenzes that did something they could not bring themselves to do on their own: ask for financial help. Lisa explained to Peter that she understood how hard and embarrassing it is to ask for money, "but I believe with all my heart that this is what networks are for." The Johnson-Lenzes are self-employed and do not have disability plans or group health insurance. Dated December 7, 2007, Kimball's email read:

Dear friends of P+T, [the online nickname Peter and Trudy have used since 1977]

If you're reading this it's because I managed to convince Peter to send it which makes me very happy even though I'm sure it makes Peter feel uncomfortable. I'm sending a check out to Oregon today. We all know about "pay it forward"—this is about "paying it backward."

P+T's work has influenced and enhanced my thinking for years and years . . . so I feel that I owe them far more than I could ever afford to pay. If we all lived in a physical village the way we're living in this global one we'd be bringing Peter healthy snacks to the hospital, shoveling their walk, filling the fridge, and doing whatever else we could to support them during a very difficult time.

Since most of us are far away, we can't do much of that but we can provide some cash to reduce the stress of figuring out how to deal with the day-to-day while they're dealing with something way more important. . . . If others have some creative ideas about more ways we can enact our network being—count me in!

lisa

Jessica Lipnack, another member of P+T's network, put Kimball's "pay it backward" email on her blog.[4] Soon, other checks arrived, including some from people who had heard of P+T through these online activities but were unknown to them.

In the following months, there were more medical ups and downs, including a harrowing period after Trudy's skull was repaired when she developed a staph infection and underwent emergency surgery. About the same time, Peter suffered a mild stroke. Local friends were indispensable in helping them get the care they required and supporting their daily needs during these periods of incapacity. For instance, it was Donna Tull, the wife of a friend, and a person Peter had never personally met, who heard about Peter's stroke symptoms and convinced him to seek help. Another "stranger," who was the spouse of a friend, is a nutritionist and recommended a probiotic that helped Trudy at a time she was on an antibiotic regimen. Many others played direct or indirect roles in the care, thanks to two websites created by Peter and Trudy and their friends. Lotsa Helping Hands is a one-stop web-based domain that allows people to set up helping communities to coordinate meal delivery, transportation, schedules of household chores and visits, and expressions of emotional support. Many of them opted for menus created by Sharon Thorne, a friend who worked at a local grocery store in the deli department and was aware of the couple's special dietary needs. Kimball set up an account on the PayPal e-commerce website to accept donations. By autumn of 2008 about ninety friends, family, and associates had made contributions, including people Peter and Trudy had never met in person and one couple who were complete strangers to them. Over thirty people, many of them at a distance, provided meal deliveries, using Lotsa Helping Hands to order from a local deli.[5] This far-flung network used a complex assemblage of email, group software, websites, regular ("landline") phones, and mobile phones to coordinate. "We're basically desktop people," Peter said, "but our cell phones came in handy when we were travelling or when I was at the hospital."

As Peter and Trudy thought about this outpouring of generosity and altruism, they reflected on the power of social networks and the amount of effort required to maintain effective support. In a series of emails to their friends, they meditated on the "art of networking" and the occasionally grueling work of making choices and tradeoffs in order to sustain a social network. Some of their emails began with their twenty-first-century update of a little-remembered quote from Shakespeare's *Timon of Athens*. Timon had said, "I am wealthy in my friends." Their rewrite and occasional email header was, "We are truly wealthy in our network." Tracy and Peter later described their experience for this book's authors:

We have been able to "get by with a little help from our friends," but we had to ask for help first, and that was a big challenge for us. . . . We have learned so much about our own resilience and that of our networks. Each relationship is a source of unique nourishment that has contributed to our healing and recovery. We thought we knew a lot about the art of networking, but this was a whole new experience.

It's been something of a challenge to manage some of the labor-intensive mechanics of networking in the current technological environment: choosing which networking tools to use and when; creating, adding to, updating, and maintaining email lists; offering opt-outs along the way; finding tools to help with scheduling food deliveries; and tracking and acknowledging contributions of money, food, and other gifts.

On the social side, we have wondered how often to send updates, with how much detail and with photos of what. What's the right balance of optimism, humor, and candidness? We didn't want to add even more to everyone's overload. We were also surprised to hear how much people appreciated getting news of our progress and being included in our circle of support. . . .

Several of you have also told us that we have isolated ourselves too much. Certified INFJs (Myers-Briggs)[6] who prefer to put things in writing and like to immerse ourselves in our projects, we unwittingly opted out of the real-time flows of talk and lots of interaction where trust grows and real work is negotiated. We realize now that we need to schmooze, circulate, and network a lot more to survive!

This is a time in our lives for radical trust, taking a leap, moving along whatever paths we take, and seeing what happens. We surrender. Heads to the floor.[7]

Networked Individualism

We read Peter and Trudy's account and we wonder about the folks who keep moaning that the internet is killing society. They sound just like those who worried generations ago that TV or automobiles would kill sociability, or sixteenth-century fears that the printing press would lead to information overload. While *oy vey*-ism—crying "the sky is falling," makes for good headlines—it isn't true. The evidence in our work is that none of these technologies are isolated—or isolating—systems. They are being incorporated into people's social lives much like their predecessors were. People are not hooked on gadgets—they are hooked on each other. When they go on the internet, they are not isolating themselves. They are conversing with others—be they emailers, bloggers, Facebookers, Wikipedians, or even organizational web posters. When people walk down the street texting on their phones, they are obviously communicating. Yet things are different now. In incorporating gadgets into their lives, people have changed the ways they interact with each other. They have become increasingly networked as individuals, rather than embedded in groups. In the world of networked individuals, it is the person who is the focus: not the family, not the work unit, not the neighborhood, and not the social group.

So Peter and Trudy's account of how they used their social networks is not only a heartwarming tale. It is also the story of the new social operating system we call "networked individualism" in contrast to the longstanding operating system formed around large hierarchical bureaucracies and small, densely knit groups such as households, communities, and work-groups. We call networked individualism an "operating system" because it describes the ways in which people connect, communicate, and exchange information. We also use the phrase because it underlines the fact that societies—like computer systems—have networked structures that provide opportunities and constraints, rules and procedures. The phrase echoes the reality of today's technology: Most people play and work using computers and mobile devices that run on operating systems. Like most computer operating systems and all mobile systems, the social network operating system is *personal*—the individual is at the autonomous center just as she is reaching out from her computer; *multiuser*—people are interacting with numerous diverse others; *multitasking*—people are doing several things; and *multithreaded*—they are doing them more or less simultaneously.

Peter and Trudy rebuilt their world through their own resourcefulness and with the help of many allies. They used varied branches of their network operating system to find support, solve problems, and improve their knowledge and skills. The actions they took to recover were different from the actions that would have been used by their parents and grand-parents. Those actions took place on a different human scale from the one that would have been available to their ancestors facing similar traumas. Those ancestors were embedded in groups and had little opportunity to navigate life by maneuvering through their networks. Yet, to networked individuals like Peter and Trudy, such art is second nature. They worked hard and thoughtfully to take advantage of the wide-ranging skills that existed in their extended social network—their closest friends, *plus* their more varied and extended system of associates, *plus* the new entrants into the network who were connected to them through personal, participatory media.

By Peter and Trudy's reckoning their network has several thinly con-nected segments. Because their friends traveled in different milieus, their friends needed contact and coordination in order to help. About twenty of those who helped were close friends and family. Beyond that inner circle was a ring of people who pitched in to help with specific issues even though they were not bosom buddies of the couple. Another ten or so were medical professionals, while another ten or so beyond that were parapro-fessionals in the health-care world, the insurance world, the social-work world, or the patient-advocate world. And there were many part-time helpers, contributors, and well-wishers. Some in the network felt tied to the couple because of their common professional interests. Others were linked through their shared passion for jazz. Still others were linked because they live in Portland—proximity still counts for something, even in the networked age. Beyond them, hundreds of others found the wherewithal to offer help from afar by sending good wishes, advice, money, and job contacts. Collaboratively, this far-flung network made contributions to the couple's emotional, financial, and logistical well-being.

The networked life epitomized by Peter and Trudy's story is different from the all-embracing village that is usually held up as the model of community. In Peter and Trudy's case, the people who were most useful in providing advice on medical decisions often did not know the people who provided emotional and social nourishment. Nor did all network members work closely in sync in providing assistance. Nevertheless, they found ways to work together in helping the couple wrestle with their daily—and future—lives.

So, what's new about this social reality? Haven't many communities pitched in before to help their members? Of course. Yet the way in which Peter and Trudy's network did this is quite different from the way their forebears' communities would have. In generations past, people usually had small, tight social networks—in rural areas or urban villages—where a few important family members, close friends, neighbors, leaders and community groups (churches and the like) constituted the safety net and support system for individuals.

This new world of networked individualism is oriented around looser, more fragmented networks that provide succor. Such networks had already formed before the coming of the internet. Still, the revolutionary social change from small groups to broader personal networks has been powerfully advanced by the widespread use of the internet and mobile phones. However, some analysts fear that people's lesser involvement in local community organizations—such as church groups and bowling leagues[8]—means that we live in a socially diminished world where trust is lower, societal cohesion is reduced, loneliness is widespread, and people's collective capacity to help one another is at risk. While such fears go back at least one hundred fifty years, the coming of the internet has increased them and added new issues: Are people huddling alone in front of their screens? If they are connecting with someone online, is it a vague simulacrum of real community with people they could have seen, smelled, heard, and touched in the "good old days"?

The evidence suggests that those with such fears have been looking at the new world through a cloudy lens. Our research supports the notion that small, densely knit groups like families, villages, and small organizations have receded in recent generations. A different social order has emerged around social networks that are more diverse and less overlapping than those previous groups. The *networked operating system* gives people new ways to solve problems and meet social needs. It offers more freedom to individuals than people experienced in the past because now they have more room to maneuver and more capacity to act on their own.

At the same time, the networked individualism operating system requires that people develop new strategies and skills for handling problems. Like Peter and Trudy, they must devote more time and energy to practicing the art of networking than their ancestors did. Except in emergencies, they can no longer passively let the village take care of them and control them. They must actively network. They need to expend effort and sometimes money to maintain their ties near and far; choose whether to phone, visit, or electronically connect with others; remember which members of their network are useful for what sorts of things (including just hanging out); and forge useful alliances among network members who might not previously have known each other. In short, networked individualism is both socially liberating and socially taxing.

Paradoxically, the technology that promises to connect people also threatens to overload them with extra work. The Johnson-Lenzes told us how it takes them just as much effort and even more time to conduct deeply satisfying electronic communications as it does to conduct person-to-person encounters. They noted that while the internet put more potential relationships at their fingertips and made relationships easy to start, it also made relationships harder to sustain because it brought so many distractions and fleeting interactions into their lives. After making a good connection via email or texting, they wrote, "we want to hear the music of each other's voices and we want to see and touch each other."

Our research supports this. An environment that spawns more social liberation also demands more social effort when people have desires or problems they want solved. This is where technology is especially useful. A major difference between the past and now is that the social ties people enjoy today are more abundant and more easily nourished by contact through new technologies. We will show throughout this book how the internet and other forms of information and communication technologies—what scholars call "ICTs"—actually aid community.

One way to look at the changed environment is to compare the Johnson-Lenzes' social network operating system to the social and media environment of their parents. As Peter and Trudy recall, their parents had a few close friends who literally meant the world to them where they grew up—Portland for Trudy and Denver for Peter. Their mothers did not work until their children were teenagers. Their parents' milieus revolved around their children, work, volunteer activities like scouting and PTA, regular bridge games, and church.

Peter and Trudy learned to read with the *Fun with Dick and Jane* primer. As children, they got information and diversions from television shows like *The Mickey Mouse Club*, local newspapers and newscasts, magazines such as *Life*, and books checked out of the local library. However, Peter says their parents rarely treated these media sources as resources or tools they could use to tackle problems. Although family members used their home encyclopedia when they needed technical information or material to help with schoolwork, they did not see it as a "go to" information trove that could answer all questions or help solve all problems. Peter's parents would talk about the news with their friends, but they never wrote letters or made phone calls to talk back to the news organizations or newsmakers. Except for gossiping with friends and family, they never created their own version of "news" to share with their acquaintances. The only personal news that they sent around was the occasional letter or card with family updates.

Nowadays, Peter and Trudy use a variety of tools to make sense of their environment and to plot their next steps: the internet, phones, books, magazines, newsletters, and interactions with friends. At the same time, the internet and their phones (landline and mobile) allow them to stay in touch with more people in their social networks, more often, and under more circumstances. They multitask with multiple devices. They find themselves sending emails to those helping them to coordinate household chores while on the same day they process contributions from strangers and do research and consulting work.

"All this technology makes it easier for me to take care of lots of things quickly," Trudy says. "It's a juggling act with all the things I need to do, but I don't know how I'd be able to work with so many people on so many different issues if I didn't have this technology." Rather than being overwhelming, the internet extends her reach—and the reach of people to her. While the internet itself is not overwhelming, Trudy notes that it introduces more demands on her life about how to allocate her attention and manage her personal interactions.

Still, the technology and the social network are not the sole solutions to Peter and Trudy's problems. Despite their wonderful network support, they have been hard-hit financially as a result of their health problems. They have gotten back on their feet with a lot of help from their friends and are slowly rebuilding their lives.

In thinking about Peter and Trudy, we have wondered if their story is unusual because they have been active networkers and community builders since the 1970s—both in person and via ICTs. To be sure, Peter and Trudy have more experience and expertise networking than legions of other Americans. They have been developing social networking concepts, software, and virtual communities since the 1970s. Fittingly, they coined the term "groupware" in 1978 to describe and construct the then-revolutionary software that allowed two or more people to work together online —even before the internet itself had been widely embraced. Today, they realize that they work in social networks, not groups.

Yet, the more we have examined the research that is the heart of this book, the more we see that while Peter and Trudy have been pioneers, many people are actively networking in similar ways. We describe this new social operating system in the rest of this chapter, and we show throughout the book how social networks—combined with personal and mobile ICTs— are shaping how people relate to others, work, play, learn together, and seek out helpful information.

Although we focus on North America, our home and the source of most of our evidence, we believe that our conclusions generally hold true for the entire developed world. These insights also have implications for the developing world, where internet and mobile phone use is mushrooming.

The Triple Revolution's Impact on How Networked Individuals Live Their Lives

Peter and Trudy Johnson-Lenz's story highlights how the Social Network, Internet, and Mobile Revolutions are coming together to shift people's social lives away from densely knit family, neighborhood, and group relationships toward more far-flung, less tight, more diverse personal networks. In their story, we see the changing realities of this new social operating system.

First, the Social Network Revolution has provided the opportunities— and stresses—for people to reach beyond the world of tight groups. It has afforded more diversity in relationships and social worlds—as well as bridges to reach these worlds and maneuverability to move among them. At the same, it has introduced the stress of not having a single home base and of reconciling the conflicting demands of multiple social worlds.

Second, the Internet Revolution has given people communications power and information-gathering capacities that dwarf those of the past. It has also allowed people to become their own publishers and broadcasters and created new methods for social networking. This has changed the point of contact from the household (and work group) to the individual. Each person also creates her own internet experiences, tailored to her needs.

Third, the Mobile Revolution has allowed ICTs to become body append-ages allowing people to access friends and information at will, wherever they go. In return, ICTs are always accessible. There is the possibility of a continuous presence and pervasive awareness of others in the network. People's physical separation by time and space are less important.

Together, these three revolutions have made possible the new social operating system we call "networked individualism." The hallmark of networked individualism is that people function more as connected individuals and less as embedded group members. For example, house-hold members now act at times more like individuals in networks and less like members of a family. Their homes are no longer their castles but bases for networking with the outside world, with each family member keeping a separate personal computer, address book, calendar, and mobile phone.

Yet people are not rugged individualists—even when they think they are. *Many meet their social, emotional, and economic needs by tapping into sparsely knit networks of diverse associates rather than relying on tight connections to a relatively small number of core associates.* This means that net-worked individuals can have a variety of social ties to count on, but are less likely to have one sure-fire "home" community. Looser and more diverse social networks require more choreography and exertion to manage. Often, individuals rely on many specialized relationships to meet their needs. For example, a typical social network might have some members who are good at meeting local, logistical needs (pet sitting, watering the plants), while others are especially useful when medical needs arise. Yet others (often sisters) provide emotional support. Still others are the ones whose political opinions carry more weight, while others give financial advice, restaurant recommendations, or suggest music and books to enjoy.

Networked individuals have partial membership in multiple networks and rely less on permanent memberships in settled groups. They must calculate where they can turn for different kinds of help—and what kind of help to offer others as they occupy nodes in others' extended networks. They have an easier time reattaching to those from their past even after extended periods of noncontact. With a social environment in flux, people must deal with frequent turnover and change in their networks.

A key reason why these kinds of networks function effectively is that social networks are large and diversified thanks to the way people use technology. To some critics, this seems to be a problem. They express concern that technology creates social isolation, as people rely on tech-based communication rather than richer face-to-face encounters.[9] We find a different story. Technologies such as the internet and mobile phones help people manage a larger, more diverse set of relationships. Consider the many people—and the many kinds of people—that Peter and Trudy could call on. The lesson is this: Rather than the internet or mobile phones luring people away from in-person contact, extensive internet use is associated with larger, more diverse, and growing networks. For example, one study of internet users shows that between 2002 and 2007, there was an increase of more than one-third in the number of friends seen in person weekly.[10]

The changing social environment is adding to people's capacity and willingness to exploit more "remote" relationships—in both the physical and emotional senses of the word. The internet especially helps to maintain contact with *weaker ties*: friends, relatives, neighbors, and workmates with whom people are not very close. While weaker, these ties often provide—as in Peter's and Trudy's case—crucial elements of information, sociability, and support as they seek jobs, cope with health issues, make purchase decisions, and deal with bureaucracies. Most importantly, they are the broader milieus that give people their places in life by providing them a means of connecting to the broader fabric of society. They can function better in a complex environment because the Triple Revolution provides them diversity in several ways, including more access to a greater variety of people and to more information from a greater variety of sources.

The new media is the new neighborhood. The internet plays a special role for networked individuals because it is a participatory medium. To be sure, people still value some neighbors, because living nearby remains important for everyday socializing and for dealing with emergencies large and small. Yet, neighbors are only about 10 percent of people's significant ties. As a result, people's social routines are different from their parents or grandparents. While people see their coworkers and neighbors often, most of their important contacts are with people who live elsewhere in the city, region, nation—and abroad. The internet is especially valuable for those kinds of connections.

Networked individuals have new powers to create media and project their voices to more extended audiences that become part of their social worlds. Their connections can ripen in important ways because the internet offers so many new options for interaction through social media such as emailing (still the most popular overall), blogging, posting Twitter messages, and Facebook activities. Social media allow people to tell their stories, draw an audience, and often gain social assistance when they are in need. Pew Internet surveys find that more than two-thirds of adults and three-quarters of teenagers have created content online. The act of creating with new media is often a social—and networking—activity, where people work together or engage in short- and long-term dialogues.

The lines between information, communication, and action have blurred: Networked individuals use the internet, mobile phones, and social networks to get information at their fingertips and act on it, empowering their claims to expertise (whether valid or not). They use social media and the web as a vast information store that can help them gather information, find and contact others who have faced similar experiences, compare options when they are making decisions, locate new experts to consult, and get second, third, and fourth opinions when they are assessing the advice they are given. Peter and Trudy had good doctors, but they used the internet and their networks to take charge of their own health care, searching on the web and asking knowledgeable friends elsewhere for advice and comfort.

Such empowerment is not limited to health crises. For example, after people have bought a product, they can turn themselves into broadcasters as they comment on the experience they have just had, rate the product they have just bought, apply their own "tags" to label it in ways that are meaningful to them, and comment about the product on the blog or news site that may have originally led them to the product. Their participation then assists those who come later and can read their comments. The interactive Web 2.0 environment provides innumerable opportunities for expanding one's reach for new relationships, even among the most remote strangers. In this world, a new layer is added to a person's social network—the audience layer sits beyond the weak ties layer. It is made up of strangers, but as Peter and Trudy discovered, even those strangers can play constructive roles when they are activated. The role of experts and information gatekeepers can be radically altered as empowered amateurs and dissidents find new ways to raise their voices and challenge authority.

In this world of expanded opportunity, community building can take new forms. Hobbyists, civic actors, caregivers, spiritual pathfinders, and many others have the option of plugging into existing communities or building their own from scratch. Networked individuals can create new communities around themselves, their interests, even their illnesses—online, in person, and mixes of both. They can also use social media such as Twitter to discover and make connections to others with whom they share something in common.

Although they do not use Twitter, Peter and Trudy relied on their communities built around futures research, sustainability, social media, virtual communities, and jazz musician Kurt Elling. Not only do they moderate a forum about Elling's work and post news, articles, reviews, and personal information; they also write supportive and informative comments on others' online blogs. They work hard to keep the Elling network vibrant. Similarly, the internet became the environment where a distinct new community formed around dealing with Peter and Trudy's medical and daily living needs, containing both new members and old friends.

Not only do networked individuals participate in social networks, they also take on specialized roles inside those groups. Many interpersonal ties are based on the particular attributes—not the full personality of the whole person. Peter and Trudy's personal health network is typical. It includes family members, neighbors, work colleagues, members of online and offline support groups, expert hunters for medical information in professional literature or reports of clinical trials, and acquaintances coming into the picture because of the particular help they can provide.

Moving among relationships and milieus, networked individuals can fashion their own complex identities depending on their passions, beliefs, lifestyles, professional associations, work interests, hobbies, or any number of other personal characteristics. These relationships often depend on context, which provides networked individuals an opportunity to present different faces in different circumstances, especially online. For example, Peter and Trudy are jazz buffs, organizational consultants, futures researchers, sustainability advocates, software designers, and friends—in multiple environments that only overlap somewhat. Yet, despite their involvement in different milieus, they are still Peter and Trudy wherever they participate. They have a networked self, a core being that emphasizes different identities as they connect with each milieu.

At work, less formal, fluctuating, and specialized peer-to-peer relationships are more easily sustained now compared with the past, and the benefits of boss/subordinate hierarchical relationships are less obvious. Pew Internet surveys show that about three-quarters of all American workers use all the basic tools of internet browsing, emailing, and messaging, and mobile phoning/texting. But that is just the starting point. Many of the most technologically connected workers have jobs built around creative effort rather than manufacturing or standardized paper pushing. This thrusts more autonomy and authority onto individual workers. Flexible arrangements with bosses, peers, and subordinates encourage independent thinking—and perhaps even creativity.

Networked workers frequently operate in multiple teams, rather than with the same colleagues every day, so their organizational life becomes more horizontal and less vertical. Peter and Trudy have always been a two-person consulting partnership, but through the years they have developed a diverse set of trust relationships to get their projects done. Sometimes such networks develop within organizations with people shifting their work relationships throughout the week. They rely heavily on the internet, within-organization intranets, and mobile phones to obtain and share information and complete tasks.

The organization of work is more spatially distributed. The classic picture of the Industrial-Bureaucratic era of the nineteenth and twentieth centuries has been of people commuting to large factories and offices. Not only was it more convenient to produce goods in one place, but it also was easier to coordinate and control operations. Yet, the Internet and Mobile Revolutions enhance the ability to coordinate and control at a distance, so that goods and services can come from multiple locations. Documents and drawings are now internet attachments or are stored in internet "clouds" where they can be accessed from anywhere. Mobile phones and wireless computers allow dispersed workers—at home, on the road, and in coffee shops—to connect with each other. Air travel—of people and goods —has joined with traditional land and sea transport to facilitate distributed operations.

Home and work have become more intertwined than at any time since hordes of farmers went out into their fields. The interpenetration of home and work goes in both directions. In one direction, workers bring work home from the office to finish off jobs or they may stay home full or part time. For example, Peter and Trudy have always lived and worked in the same place: Their home is their workplace. For others, the new media tethers them to their jobs—they cannot leave work behind when they head out the office door. On the one hand, many feel so burdened by time pressures and the constant threat of demands that they respond and complete tasks even when they are away from their place of work. On the other hand, many feel liberated by being able to avoid long, tedious, and tense commuting. They enjoy the prospect of being able to do "home" activities such as personal browsing of the web, sharing Facebook updates, shopping, and emailing family and friends while they are at work. In short, "home" activities have invaded work while "work" activities have invaded homes.

While ICTs have shattered the work-home dividing line, they have also breached the line between the private and public spheres of life. Mobile phones have made conversations more private than they were in the era when the household phone sat in the middle of the house so that everyone at home could hear at least one end of a phone conversation. Texting has brought another dimension to person-to-person contact by helping it become more private, even in close quarters. Blogs often become quasi-public diaries, and social media such as Facebook, Twitter, and foursquare enable people to inform others of their whereabouts and to announce their momentary thoughts and doings. For example, Peter and Trudy shared widely many details of their operations—including pictures of Trudy with a long row of surgical staples winding around her head. At the same time, heretofore private activity invades public spaces as people speak openly of intimate affairs on their mobiles in public spaces and work on their laptops in coffee shops (hoping that others won't peek too much).

New expectations and realities about the transparency, availability, and privacy of people and institutions are emerging. Reputation management—the selective exposure of personal information and activities—is an important element to how people function in networks as they establish credentials, build trust with others, and gather information to deal with problems or make decisions. In the really old days of wandering tribes and agricultural villages, people knew most things about each other—for better or worse. They felt both comforted by the availability of others and concerned about the surveillance of others.

The turn from groups to networks changed this as people expressed different parts of their behavior in different milieus. At first, the Internet and Mobile Revolutions aided this segmentation: Email, texting, and mobile calling are usually one-to-one media. But the rise of social media has brought people back into one network—happily or not. The most popular social media such as Facebook have offered limited ability—so far—to deal with the subtleties of how people really function in different segments of their networks. Rather, the sites tend to treat each person's network as a monolithic entity that functions in a let-it-all-hang-out milieu. To be sure, it is intoxicating at times for people to share a lot. Many social network participants, especially young adults, say that the advantages of disclosure—for instance by building friendships, enhancing status, and connecting to friends of friends—outweigh the problems they might encounter with too much sharing.

Yet with this re-emergent transparency comes a loss of privacy and the perhaps unwanted commercialization of personal information. Not only do all Facebook friends learn a lot about the person who they have "friended," but the social media companies can also aggregate and analyze this information and find out what twenty-year-old American students—and their forty-five-year old parents—are interested in. As former Google CEO Eric Schmidt boasted: "We know where you are. We know where you have been. We can more or less know what you're thinking about."[11]

In the less hierarchical and less bounded networked environment—where expertise is more in dispute than in the past and where relationships are more tenuous—there is more uncertainty about whom and what information sources to trust. The explosion of information and information sources has had the paradoxical impact of pushing people on the path of greater reliance on their networks. It might seem that the abundance of information that organizations provide on the internet would prompt people to rely less on their friends and colleagues for facts and advice. Yet it turns out that the increasing amount of information pouring into people's lives leads them to turn to their social networks to make sense of it. The result is that as people gather information to help them make choices, they cycle back and forth between internet searches and discussion with the members of their social networks, using in-person conversations, phone chats, and emails to exchange opinions and weigh options. In short, as the internet and mobile phones proliferate, people behave even more as networked individuals.

Is the Triple Revolution Having a Good or Bad Impact on Society?

The simple answer is: both and more. The research we shall present shows that networked individualism is the reality of many everyday lives. We believe that there is clear evidence that the shift to networked individualism is widespread and is changing the rules of the game. Networked individuals live in an environment that tests their capacities to deal with each other and with information. In their world, the volume of information is growing; the velocity of news (personal and formal) is increasing; the places where people can encounter others and information are proliferating; the ability of users to search for and find information is greater than ever; the tools allowing people to customize, filter, and assess information are more powerful; the capacity to create and share information is in more hands; and the potential for people to reach out to each other is unprecedented. Rather than snuggling in—or being trapped in—their groups, people must actively maneuver in their networks. Some people are more likely to be network mavens than others, better able to navigate and operate the system.

Different networks operate in different ways. Many provide *havens:* a sense of belonging and being helped. Many provide *bandages:* emotional aid and services that help people cope with the stresses and strains of their situations. Still others provide *safety nets* that lessen the effects of acute crises and chronic difficulties. They all provide *social capital:* interpersonal resources not only to survive and thrive, but also to change situations (houses, jobs, spouses) or to change the world or at least their neighborhood (organizing major political activity, local school board politics). Not only must people choose which parts of their networks to access, the proliferation of communication devices means they must also choose how to connect with others: meet in person, phone, email, text, tweet, or post on Facebook.

This is the era of free agents and the spirit of personal agency. But it is not the World According to Me—it is not a world of autonomous and increasingly isolated individualists. Rather, it is the World According to the *Connected* Me, where people armed with potent technology tools can extend their networks far beyond what was possible in the past and where they face new constraints and challenges that are outgrowths of networked life. Those primed to take advantage of this reality are the ones who are motivated to share their stories and ideas and then invite conversation and feedback. Much of the activity by networked individuals is aimed at gaining and building trust, the primary currency of social networks. There are new ways to offer trust and procure it online, and its basic value is growing because networks are so essential to people's social success. In a world of networked individuals, those who engage in the mutual exchange of intangible or mundane resources have the potential to thrive. These individuals will seek support and seek to provide support. Further, those individuals who are able to balance relationships with people in the various sectors of their social networks—kin, friends, neighbors, associates, and workmates— are better positioned to receive both broad and specialized support.

The changes wrought by the Triple Revolution—in social networks, the rise of the internet, and the advent of mobile connectivity—are not all for the good or all for the bad. Rather, some of the changes created by networked individualism are beneficial to people and make society better while others are challenging to personal fulfillment and make society harsher. Some of the changes just make it different in neither a positive nor a negative way. Moreover, the effects of networked individualism often depend on personality traits and environmental contexts.

   This book explores how this world came into being, the impact these changes have produced already, and where they are leading. In part I, we describe how the Triple Revolution—the Social Network, Internet, and Mobile Revolutions—affect networked individualism. Chapter 2 examines how the social network perspective differs from the two traditional approaches to understanding human behavior: in groups or as individuals. Chapter 3 looks at the rise of the internet in the United States, how patterns of its adoption changed over time, and the current activities of people online. It notes the special contribution that high-speed, always-on broadband has made to how people connect with each other and information. Chapter 4 shows how mobile phones have moved beyond ways in which people talk on the fly to become key means of always-available accessibility.

   Part II shows how the Triple Revolution of social networks, the internet, and mobile access play out in communities, households, and work. Chapter 5 considers the ways in which interpersonal relationships have moved beyond neighborhood communities. Chapter 6 looks inside and outside households to see how the everyday rhythms of traditional household-centered families have moved out of homes as families become networked. Chapter 7 shows the partial transformation of work, with people working in multiple teams rather than hierarchies and work organizations becoming geographically distributed. Chapter 8 describes how individuals can easily create, manipulate, share, and broadcast their ideas. Chapter 9 looks at the special features that digital technology and social networks have brought to how people obtain information.

   The two concluding chapters in part III sum things up and look to the future. They try to answer the questions, "So what?" and "Now what?" Chapter 10 organizes what we have learned about how people and organizations can perform well in the world of networked individuals, while Chapter 11 explores the technological and social trends that might affect networked individualism in the coming decade. This world will create greater opportunities for people to build networks of kindred spirits and to amass information and social support to have their needs met. Yet, this world will also offer greater uncertainties, insecurities, and opportunities for surveillance. As the Triple Revolution unfolds, the move to networked individualism will continue.

2 The Social Network Revolution

When we tell people that we are thinking about social networks, they often say *"Oh, Facebook."*[1] Many believe that the Social Network Revolution started with Facebook's emergence in 2004. To be sure, Facebook *is* somewhat of a network. But social networks are bigger than Facebook, and they have been around since the beginning of time when Cain hung out with Abel. Even computer-based networks have been around for decades before Facebook.

In the story of the rise of networked individualism, it is important to realize that the Social Network Revolution came first—before the Internet Revolution or the Mobile Revolution. It is the least obtrusive because it is not a shift in technology, but a shift in how people relate to each other. The Social Network Revolution, which throughout this book we often shorten to the Network Revolution, has been less noticed and commented upon than the technology revolutions partly because the conceptual idea of a social network is simple, yet intangible.

A social network is a set of relations among network members—be they people, organizations, or nations. From a network perspective, several things matter: Society is not the sum of individuals or of two-person ties. Rather, everyone is embedded in structures of relationships that provide opportunities, constraints, coalitions, and work-arounds. Nor is society built out of solidary, tightly bounded groups—like a stacked series of building blocks. Rather, it is made out of a tangle of networked individuals who operate in specialized, fragmented, sparsely interconnected, and permeable networks. Social network analysts focus more on the characteristics of these relationships than on the characteristics of the individual members.

If social networks have long been with us, why have we become more networked recently? The answer is that new technologies and major social changes have resonated with the footlooseness of North Americans and their desire for personal autonomy. North Americans move to new homes a lot: both locally and long-distance. North Americans also move around a lot on a daily basis: hopping in their cars for work, play, and socializing. And North Americans have especially extolled the value of personal autonomy—doing things on their own. Moreover, the turn to social networks has been a gradual movement that has recently accelerated, rather than a short-term cataclysmic event.

Yet this is not just a North American story, for similar changes have happened throughout the developed world and in many parts of other societies. People today are less bound to their national allegiance, village, and neighborhood moorings. Around the developed world flexible, maneuverable connectivity has increased, group boundaries have weakened, and information has become more directly available—all driving the shift to networked individualism. It is not that these changes have *caused* the Social Network Revolution. They have created technological, social, and economic circumstances that helped make the network operating system possible. The next three sections describe nine key changes that have facilitated the change to networked individualism.

Toward the Social Network Revolution

Widespread Connectivity

1. *Automobile and airplane trips have made travel wider-ranging and broadly affordable, helping spread social networks worldwide.*

Longer drives became more feasible with the construction of the 37,000-mile (75,000 km) U.S. Interstate expressway system between the 1950s and 1980s. Middle-range trips became routine, automobile reliability improved, and rising affluence made cars more broadly affordable. High-speed cruising on I-40 replaced getting kicks on Route 66.[2]

The percentage of households without any cars dropped from 21 percent in 1969 to 8 percent in 2001, and the percentage of one-car households dropped from 48 percent to 31 percent in the same time period (figure 2.1). Three-fifths (60 percent) of all households have two or more vehicles. The upshot is that the number of passenger-car miles driven annually by Americans rose 60 percent from 900 billion in 1970 to 1.5 trillion in 2001, and the amount of miles driven by an average car rose 20 percent during the same period, from 10,000 to 12,000. More people drove on highways in 2010 than a decade earlier, even while highway death rates went down. In short, Americans are driving more and driving farther in more cars. For better or worse, they are less bound up in their localities and their local habits.[3]

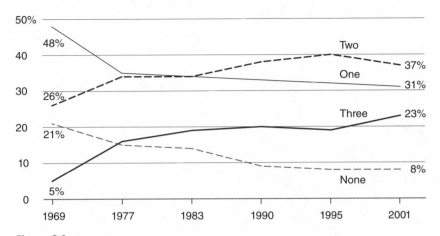

Figure 2.1
Percentage of U.S. households by number of vehicles.
Source: NHTS Summary of Travel Trends (2004); Vehicle Availability and Utilization.

In the air, the proliferation of jet airliners made long-distance flight a regular practice, starting with the advent of the Boeing 707 in 1959. Deregulation of airlines in the late 1970s, first in the United States and then in Europe, made flying more affordable. No longer were distant relatives and friends hugged good-bye and rarely seen again. Figure 2.2 shows that the average American boarded a plane once every four years in 1954, but 2.5 times *each* year in 2005. The proliferation of car and air travel has enabled social networks to become more far-flung.[4]

2. *The rapid growth of affordable telecommunications and computing has made communicating and gaining information more powerful and more personal.*

Two major changes were the automation of the telephone system locally, starting in the 1930s, and the spread of automated direct distance dialing, with regional area codes replacing operators within countries in the 1960s and between countries in the 1970s. The number of phones and the volume of calls went up rapidly domestically and internationally as per-minute costs dwindled and ease of use increased (figures 2.3 and 2.4). Distance still matters, but it has become much less of a deterrent to phone contact. At the same time, multiple phones proliferated inside homes: Even though they used the same phone number, the separate phone in each room provided more personal service and privacy.[5]

After rapid growth between 1950 and 2000, telephone use changed dramatically. The number of traditional landlines peaked at 68 per 100

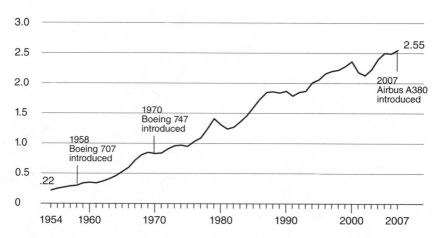

Figure 2.2
Per capita airline boardings in the United States.
Source: U.S. Bureau of Transportation Statistics.

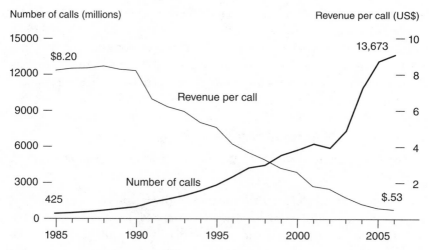

Figure 2.3
U.S. billed international calls.
Source: U.S. Federal Communications Commission.

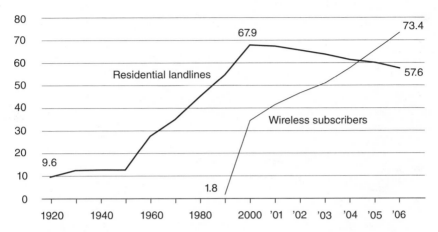

Figure 2.4
Number of landlines and mobile phones in the United States (per 100 inhabitants).
Source: U.S. Federal Communications Commission.

Americans in 2000 and declined 15 percent by 2006 to 58 per 100. The obvious causes were the skyrocketing growth of more personal, flexible, and powerful mobile phones, plus the reduced need for second landlines to access the "dial-up" internet connections as broadband became widespread. By 2006, there were more mobile lines than landlines in the United States and almost as many in Canada (figure 2.4; Canada not shown). By the end of 2010, the number of people only using mobile phones had grown to nearly 30 percent of the population and nearly half of those under age 30.[6]

Starting with the opening of the internet to the public in the late 1980s and early 1990s, the rapid growth of computer-supported information and communication technologies (ICTs) increased personal connectivity: first on the internet—itself expanding from email to ever-elaborating websites and social media—and then via mobile phones (figure 2.5). Because all countries used the same communication protocols, the growth was worldwide. The combination of proliferating ICTs with easier air and car travel shifted cities into functioning more as hubs of social networks and less as heaps of people and industries.[7]

3. *The general outbreak of peace and the spread of trade have driven commercial and social interconnectedness.*

There are fewer impediments to personal interactions across the globe. Interstate conflicts between nations are low, with the U.S.-led invasion of

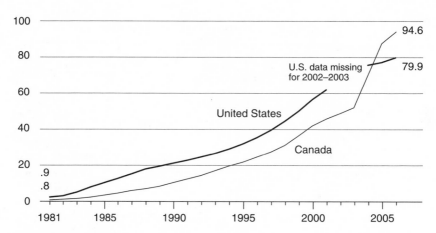

Figure 2.5
Number of personal computers in the United States and Canada (per 100 inhabitants).
Source: International Telecommunication Union, Table 1090, Utilization of Selected Media.

Iraq in 2003 the last to occur before 2011 (see figure 2.6). Within Western and Central Europe, the growth of the European Union to twenty-seven countries and the subsequent twenty-five-nation Schengen Agreement in 1985 almost erased borders among many European countries for trade, work, and travel. Although conflicts within nations persist in places such as Serbia, Sudan, Myanmar, Libya, and the Congo, conflicts have rarely taken place in the developed countries to which almost all residents of developed countries travel.[8]

A high level of globalized production and consumption has accompanied peace and more permeable borders. The growth of North American and European manufacturing bases in Asia and Latin America since the 1970s, two-way exports of resources between the developed and the less-developed worlds, the economic reforms of China in the 1980s and 1990s, and the shredding of the Iron Curtain in 1989–1990 allowed East and West, North and South to open up travel and trade. U.S. exports have become more important to the overall economy, more than doubling from 6 percent of the U.S. gross domestic product (GDP) in 1970 to 13 percent in 2008. More strikingly, imports have become even more important to the economy, more than tripling from 5 percent of the GDP in 1970 to more than 17 percent in 2008. Imports have been appreciably greater than exports since 1982 and especially since 1998 (figure 2.7).[9]

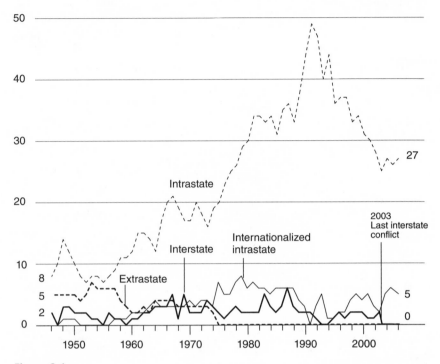

Figure 2.6
Number of world conflicts.
Source: Uppsala Conflict Data Program (UCDP)/Oslo Peace Research Institute (PRIO);
UCDP/Human Security Report Project Database.

Products now come from everywhere; indeed, from many everywheres
as pieces from different countries get assembled into the same product.
Consumer options have expanded, so that people can choose among a
wide array of goods and services from around the world, resonating with
the transformation from groups to networks. For example, a Toronto supermarket carries fruit from more than ten countries in three continents (a
sampling is shown in figure 2.8).[10]

Weaker Group Boundaries

4. *Family composition, roles, and responsibilities have transformed households
from groups to networks.*

Fewer marriages, smaller families, and more women doing paid work have
transformed traditional homemaker households into networked families.
Homes have become less of a castle and more of a base for sallying forth.

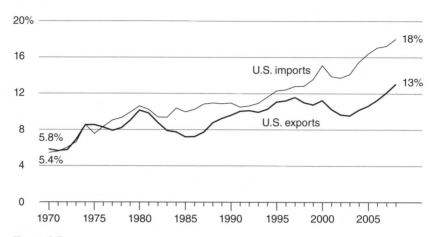

Figure 2.7
U.S. imports and exports as percentage of GDP.
Source: The World Bank Group, the Organisation for Economic Co-operation and Development, and the U.S. Department of Commerce Bureau of Economic Analysis.

Figure 2.8
Globalized fruit prices in Toronto.
Source: Photos from Fiesta Farms Supermarket, Toronto. © 2010 Gregory Jancelewicz and Barry Wellman. Used with permission.

For example, the percentage of American households comprising married couples with children declined by one-quarter between 1980 and 2005, from 31 percent to 23 percent (figure 2.9). Moreover, people, especially women, are spending less time at home. For example, Canadian women were at home and awake 36 minutes fewer per day in 2005 (8.5 hours) than they were in 1992 (9.1 hours). Those women with children were the most likely to stay home, but also experienced the largest reduction in time at home. They were home 44 minutes fewer in 2005 (8.7 hours) than in

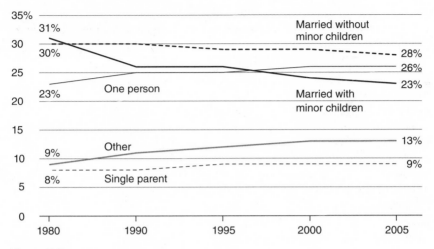

Figure 2.9
Distribution of households in the United States as percentage.
Source: U.S. Bureau of Labor Statistics.

1992 (9.5 hours). Moreover, 27 percent of Canadians were eating at least one meal alone in 2005 as compared to 17 percent in 1986.[11]

5. *Structured and bounded voluntary organizations are becoming supplanted by more ad hoc, open, and informal networks of civic involvement and religious practice.*

Political scientist Robert Putnam has shown in *Bowling Alone* that the average membership rate in thirty-two large American organizations dropped by half between 1960 and 1997. Similarly, the percentage of those Americans actively involved in such organizations also dropped by half, from 16 percent in 1973 to 8 percent in 1994.[12]

This trend toward independent lives is also reflected in the shift away from institutionalized religions where people go frequently to church. According to Putnam's *American Grace*, the 1960s were the beginning of a mass wave of interest in religious experimentation. As focus changed to exploring "spirituality," there was a corresponding drop-off in church attendance. Today, fewer Americans have a religious affiliation. Many Americans have moved to a more flexible and individualized way of engaging with religion. Traditional religions such as Baptist, Methodist, and especially Catholic have declined in childhood-to-adult retention rates, while identifications such as "non-denominational" and "unaffiliated" have increased (see table 2.1). Moreover, many Americans are not

Table 2.1

Percentage Changes in Childhood versus Current Religious Affiliation of U.S. Adults (2008)

	Childhood	Current	Net
Baptist	20.9	17.2	−3.7
Methodist	8.3	6.2	−2.1
Nondenominational	1.5	4.5	3.0
Lutheran	5.5	4.6	− .9
Presbyterian	3.4	2.7	− .7
Pentecostal	3.9	4.4	.5
Anglican/Episcopal	1.8	1.5	− .3
Catholic	31.4	23.9	−7.5
Mormon	1.8	1.7	− .1
Jehovah's Witness	.6	.7	.1
Jewish	1.9	1.7	− .2
Muslim	.3	.4	.1
Buddhist	.4	.7	.3
Hindu	.4	.4	0
Other faiths	.3	1.2	.9
Unaffiliated: atheist, agnostic, "nothing in particular"	7.3	16.1	8.8

Source: Pew Forum on Religion and Public Life/U.S. Religious Landscape Survey (2009).

keeping the religion of their birth, but switching (or jettisoning) religion later in their lives. The greatest losses are in highly institutionalized religions such as Catholicism (-8 percent net change), Baptist (-4 percent) and Methodist (-2 percent). By contrast, the greatest rates of retention and growth have been in non-denominational Christianity (+3 percent) and in the unaffiliated category (+9 percent). It is not that Americans are becoming less religious: As many as 92 percent of Americans say they believe in "God," according to a June 2008 survey by the Pew Forum on Religion and Public Life.[13]

6. *Common culture passed along through a small number of mass media firms has shifted to fragmented culture dispensed through more channels to more hardware.*

Even before the internet gave people billions of web pages and videos to peruse, people saw their media options growing and that gave them more chances to sample specialized programming. The number of television

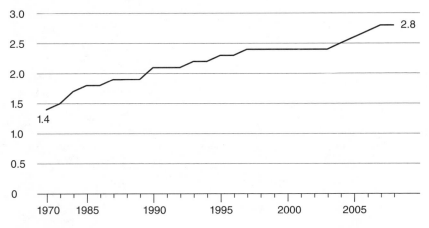

Figure 2.10
Mean number of televisions per U.S. household.
Source: U.S. Census Bureau.

channels went from a handful of options that were accessed by getting up and turning a dial (so program announcers would say: "Do not turn your dial: There is more to come") to hundreds that could be surfed via remote control. Now, there are many more outlets, but fewer shows are widely watched. Changes in hardware both reflected and encouraged the personalization of broadcast information. Television sets went from household consoles made for worshipful watching by the entire family to multiple sets for parents and children, doubling in number from 1.4 per household in 1970 to 2.8 in 2008 (figure 2.10).[14]

Increased Personal Autonomy

7. Work has become flexible in the developed world, especially the shift from pushing atoms in manufacturing to pushing bits in white-collar "creative" work.

As more workers started using personal computers rather than big machinery, employers became more willing to allow flexible schedules and work sites. This fit with a trend away from the close supervision of employees on blue- and pink-collar assembly lines and toward giving workers loosely supervised, goal-oriented directions. The decline of working-class jobs and the rise of creative jobs have changed the face of many workplaces from industrial-age hierarchies to networks of collaboration (figure 2.11 and chapter 7).[15]

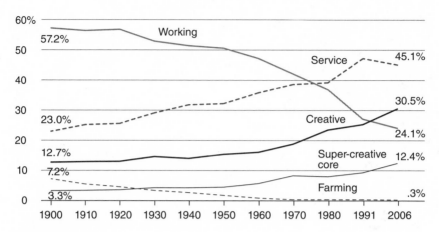

Figure 2.11
Percentage of creative class "bit workers" in the United States as percentage of total workforce.
Source: Martin Prosperity Institute, University of Toronto.

8. *American society has become less bounded by ethnicity, gender, religion, and sexual orientation.*

In 1967, racial intermarriage was illegal in seventeen states and frowned on by most Americans. It was only in the same year (on June 12, 1967) that the Supreme Court legalized such marriages nationally in the aptly named *Loving v. Virginia* case. Nowadays, interracial couples are common and more widely accepted, and the United States has elected biracial Barack Obama as president. There is a consistent trend toward interracial acceptance, with younger generations being more tolerant than their elders. The percentage of Americans *opposed to* laws banning intermarriage has risen dramatically from 63 percent in 1972 to 90 percent in 2002 (figure 2.12).[16] Indeed, the percentage became so consistently high that the General Social Survey stopped asking this question after 2002. A record 15 percent of all new American marriages in 2008 were between spouses of a different race (including Hispanic ethnicity), double the 7 percent intermarriage rate of 1980 and six times the rate in 1960.[17]

This is not a naïve hymn to American integration because there still is far to go. For example, the percentage of Americans who do not want immigrants to be their neighbors nearly doubled from 10 percent in 1995 to 19 percent in 2006. Similarly, the percentage of those who do not want Muslim neighbors jumped from 12 percent in 1995 to 22 percent in 2006. Nevertheless, ethnic, racial, gender, and sexual orientation prejudice

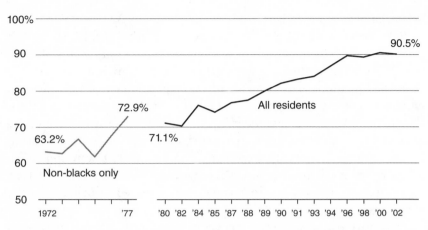

Figure 2.12
Percentage of adults aged 18 and older not in favor of a law banning racial intermarriage.
Source: James Allan Davis and Tom W. Smith, General Social Surveys 1972–2002.

(attitudes) and discrimination (behavior) have decreased. Intermarriage has increased along with occupational and residential integration. Other dividing lines also have lessened: To take just one example, Will Herberg wrote in 1955 that white America was rigidly divided along Protestant-Catholic-Jewish lines, and this book's coauthor Barry Wellman was barred in 1960 from most undergraduate fraternities at Lafayette College because he was Jewish. Such distinctions have paled even more than race and now often become conversation points rather than reasons for discrimination. Weekly attendance at U.S. church services has declined, and those services themselves have become more participatory and informal. Even homosexuality—which dared "not speak its name" until recently—has become more broadly accepted, with a majority of Americans willing to accept homosexual military members in 2010, and a trend to accepting same-sex marriages and civil unions.[18]

9. *The decline of defined benefit pensions and the rise of independent retirement accounts.*

American society is shifting away from an institutionalized employee welfare system to one where people must play a more active role managing their personal wealth, work lives, and retirement. Not long ago, many workers had "defined benefit" pension plans that motivated them to stay with their company until retirement. But medium and large companies and governments are scaling back substantially on such pension plans:

Where 84 percent of workers had them in 1980, only 32 percent had them in 2009. This downward trend is less obvious in small establishments, but still important: Where 20 percent of employees in small firms had defined benefit pensions in 1990, only 9 percent had them in 2009. The decline of defined pensions is associated with more people taking individualized retirement plans, in the form of 401Ks and individual retirement accounts (IRAs). These plans allow workers to change companies without hurting their pensions, and they mean that people must take a more active role in managing their retirement investments. A similar push toward individual initiative and away from corporate oversight has been the rise of "cafeteria plans" that require workers to become more engaged in managing their health insurance and medical costs.[19]

Taken together, the nine key changes in affordances that we have identified indicate the greater social flexibility of North American and European societies. They suggest more personalization and the weakening of traditional boundaries of neighborhood, region, nation, race, and gender. People can range more widely in their travel, communication, and information seeking. These are moves toward flexible, mobile, somewhat fragmented social systems, and they have helped to set the stage for the Triple Revolution of social networks, the internet, and mobile availability. In a leitmotif that will run through this book, such changes are both causes and results of the turn away from bounded, insulated, and homogeneous groups, and toward networked individualism.[20]

People Think They Are in Groups but They Really Are in Networks

The move toward networked individualism was largely missed because there were understandable human tendencies for observers to see the world in an either/or way: Either people related in groups or they functioned as individuals. Those trying to examine groups often took as their basic framework the 1960s countercultural rejection of group-based conformity, summed up in Malvina Reynolds's 1962 song "Little Boxes" (sung most famously then by Pete Seeger and resurrected in 2005 as the theme song of the *Weeds* TV show). The song expressed the prevailing concern that stifling groups were preventing them from "doing their own thing." In this view, members of little-box societies deal principally with fellow members of the few groups they inhabit—at home, in the neighborhood, at work, or in voluntary organizations. They:

• live in a household in which everybody gathers for family time on nights and weekends.

• live in a neighborhood like the TV show *Desperate Housewives'* Wisteria Lane
• hang out in pubs, bars, or coffee houses where "everybody knows your name" (the theme refrain from the 1980s TV show *Cheers*)
• belong to discrete, group-like organizations such as churches, bowling leagues, union locals and school associations
• work in a discrete, bounded work unit in an office or factory organized as a traditional tree-like hierarchy

All of these are groups with precise boundaries for inclusion and hence, exclusion. Each has an internal organization that is often hierarchically structured: military officers and soldiers, supervisors and employees, parents and children, pastors and churchgoers, union executives and members. In such a little-box society, observers believe each interaction usually happens one group at a time. Yet, as devotees of the TV show *Mash* know, in practice, Corporal Radar O'Reilly, the clerk, is the individual who really holds the unit together.

Still, people like to think that they operate in groups. It's cognitively easier. "I belong to the Putnam Bowling League" is easier to say—and remember—than "I belong to a shifting network of people that go bowling once in a while, although some of these people plus some others like to go to a dance club one Saturday night a month." In short, a group is often a stereotype—a shortcut for how we think about our relationships.[21]

There are understandable reasons for taking this shortcut. First, groups are governed by a culture of generalized exchange where favors given to one person are not repaid directly by that person, but by other group members. In a closed group setting, mutual help becomes a norm, as many group members are both givers and takers.

Second, prominent people may emphasize "groupiness" to bolster their power and compliance to group norms in what sociologist Émile Durkheim called "mechanical solidarity."[22] For instance, ethnic entrepreneurs who have sought to venture into mainstream markets in America have sometimes been accused of disloyalty by members of their own ethnic groups. Clearly, groups, while providing social support, may also lock people into limited opportunities.

Third, it is comforting for people who crave stability to think of themselves as belonging to a small set of groups rather than as maneuvering through murky, shifting sets of relationships at home, work, and in the community. In our research, when we ask people what groups they belong to we get stable answers most of the time. Yet, when we ask people who their close relationships are, we often get different answers each time, depending on the social circles and situations the people are involved in at that moment.

Most people in developed countries do not operate in tightly bounded, densely knit, group-centered worlds, and much of the less-developed world also is networked. People have separate agendas and schedules in their homes (chapter 6), communities (chapter 5), and at work (chapter 7). They socialize with shifting sets of friends rather than being regulars at bars or bowling leagues; many work in multiple teams at work—and increasingly away from the office. They not only receive information (chapter 9), they also create it (chapter 8). In short, people live in fluid and changing networks that go well beyond groups and Facebook.[23]

The mistake of thinking that most people belong to a single group became resoundingly clear on February 10, 2010, when Google sprung Buzz on the unsuspecting world of its Gmail (email) users. Wanting an internet application that blended Facebook's connectivity with Twitter's microblogs, Buzz sought to make things easy by putting each user's frequently contacted people into a personalized Buzz network.[24]

Google executives made two big mistakes: (1) They assumed everybody was a member of a single happy group; (2) even worse, they made that information public.

All hell broke loose. People in discreet multiple love relationships were outed; psychiatric care relationships became visible. One American, Harriet Jacobs, pointed out she was a marital rape survivor, and Buzz was linking her with her abusive ex-husband.[25] Unlike a village, where everyone knows (almost) everything about everyone else, modern people live segmented lives in which they cycle among different social networks. They handle things by a combination of compartmentalizing their relationships and overlapping their networks.

Outrage spread with the speed of the internet, and Buzz became a top-ten American "Trending Topic" on the Twitter microblogging site. Google, having tested Buzz only on its own employees, had not realized the potential problems and were apparently surprised by the widespread upset. They changed Buzz two days later to allow people to choose with whom they wanted to connect.

Buzz product manager Todd Jackson said Google was "very, very sorry, and that users were rightfully upset." Yet, damage had already been done, and Google had to pay $8.5 million to settle a class action lawsuit for the invasion of privacy. Perhaps the threat of lawsuits is why Google CEO Eric Schmidt said on February 16, 2010, that "no really bad stuff" happened when Buzz had made private relationships public. Ignoring the fact that people do different things in different parts of their networks, Schmidt also asserted, "If you have something that you don't want anyone to know, maybe you shouldn't be doing it in the first place." By contrast, U.S. Federal Trade Commissioner Pamela Jones Harbour said Google had engaged in "irresponsible conduct."[26] Ultimately, Google settled with the FTC in an agreement that bars the company from future privacy misrepresentations, requires it to implement a comprehensive privacy program, and calls for regular, independent privacy audits for the next twenty years. This was the first time an FTC settlement order has required a company to implement such a sweeping privacy-protection regime to try to prevent future privacy breaches.[27] Google discontinued Buzz in December 2011 as it implemented the more flexible Google+.

Google leaders, like many others, viewed the world in terms of groups, even though they mostly function in networks. In networked societies, boundaries are more permeable, interactions are with diverse others, connections shift between multiple networks, and hierarchies tend to be flatter and more recursive. The change from groups to networks can be seen at many levels. Trading and political blocs have lost their monolithic character in the world system. Organizations form complex alliances rather than uniting in cartels, and many workers—especially managers and professionals—report to multiple peers and superiors. We summarize some key contrasts in table 2.2—somewhat exaggerated for effect.

The new focus on the role of networks is part of a broadly based movement away from Aristotelian thinking that the world is structured in groups and Linnaean thinking that these groups can be neatly subdivided.[28] One visible example of this shift has been in modern art since Picasso, where the same objects are simultaneously shown in multiple perspectives. The change to relational thinking has been fundamental in the sciences. When the two authors were growing up, atoms were pictured as mini solar systems: electrons circling a nucleus of protons and neutrons. In the influential shift by scientists toward thinking in networks—quantum physics—the very properties of parts are defined by the interactions among them. We now picture atoms as complex interactions among a variety of particles, bonding together and connecting to others. Indeed, astronomers even describe the solar system less rigidly now: Instead of nine planets primly circling the Sun, they have demoted Pluto to a planetoid and discovered that other pieces swoop in and out of the system. They even define the universe itself relationally, linking perhaps to other multiverses.[29] Biologists, such as Richard Leowontin, now assume that "the properties of organisms are consequences of the particular interactions that

Table 2.2
Groups to Networks: Comparative Analysis

Group-Centered Society	Networked Individualism
Contact within and between groups	Contact between individuals
Group contact	One-to-one contact
Neighborhood community	Multiple communities
Local ties	Local and distant ties
Bowling leagues	Shifting networks of friends who bowl
Homogeneous ties	Diversified ties
Somewhat involuntary kin and neighbor ties	Voluntary friendship ties
Strong social control	Weak social control/shift to another network
Broad spectrum of social capital within group	Diversified search for specialized social capital
Tight boundaries with other groups	Permeable boundaries with other networks
Organized recreational groups	Shifting networks of recreational friends
Public spaces	Private spaces and online
Bulletin boards	Facebook, Twitter
Focused work unit	Networked organization
Autarky	Globalization, outsourcing

Source: Barry Wellman. © 2011, used with permission.

occur between bits and pieces of matter."[30] Scientists are realizing that many complex systems—such as highways, power grids, and the internet itself—are best understood as networks.

People Think They Act Independently but They Really Are in Networks

On the other side of the group/individual divide is the tendency of people to think they make choices independently of others.[31] We may think we are free agents, but there are others whose presence in our networks and broader environments shapes the decisions we make.[32] Neuropsychologist Craig Kinsley has even argued that the brain inherently craves social interaction, although he does not distinguish between groups and networks.[33]

While networks are often invisible, they nevertheless are important sources of our sociability, information, and social capital. For example, people are often put on corporate boards as individuals, but their connections to others are one of their key assets. And when we take a step back, we can see that these boards are interlocked so that different corporations in the same industry are indirectly connected through "old boy networks" (with a few token old girls). One study highlighted a small set of 16 men who interconnect many major European corporations, forming a total of 216 links between firms. They carry information back and forth, help coordinate, and provide access to financial capital.[34]

But even the most rugged individualists cannot stand alone. That's what golf superstar Tiger Woods found out the hard way when the public learned about his extramarital affairs. When he finally made a statement on February 19, 2010, he said: "I thought only of myself. . . . I thought I could get away with whatever I wanted to. I felt that I had worked hard my entire life and deserved to enjoy all the temptations around me. . . . I was wrong. . . . The same boundaries that apply to everyone apply to me. . . . I hurt my wife, my kids, my mother, my wife's family, my friends, my foundation, and kids all around the world who admired me."[35]

In other words: (1) Woods used to think that he was a self-contained individual, but (2) he realized that he is connected and constrained by his membership in multiple social networks.

Health care, too, has become networked. Not only is it important to have family doctors who know good specialists, but the more empowered networked individuals also seek to manage their own health care. Beverly Wellman studied why some people used complementary health care providers—like chiropractors—to deal with low back pain instead of relying only on medical doctors. Her research ruled out individual factors such as age, sex, education, and income. Instead, the key was networks: The larger and more diversified people's networks are, the more likely they are to learn about alternative health care providers. But information is not enough: People only go to chiropractors when they trust friends and relatives who have successfully used specific ones.[36]

In sum, social reality is relational. We are not a set of individuals moving past each other as disconnected grains of sand or two lovers with eyes only for each other. Instead, these relationships are parts of fragmented partial networks rather than embedded in solidary groups. Therefore, analysts focus on how these relationships and networks interconnect to provide resources and meaning.

Thinking Networked

Social scientists have been using the metaphor of "the social network" for more than a century to connote complex sets of relationships at all scales, from inter-personal to inter-national. Yet it was not until the 1950s that they started using the term systematically and self-consciously to describe patterns of ties that cut across the traditional concepts of *bounded groups* (such as villages and families) and *social categories* that treat people as discrete individuals (such as gender and ethnicity). Starting in the mid-1960s, a detailed lore and body of research has emerged to help us understand how people connect in networks.[37] Figure 2.13 tracks this, courtesy of Google's nGram program. That shows the growing use of the terms "social network" and "social networks" in books from 1950 to 2005 and after. Although the percentages of references to the terms are tiny, the rate of growth is high.[38]

"Social networks" took off as a concept because it was an apt description of social interactions. For example, people who hang out together—at work, in a café, or on the internet—can be thought of as a group, a bunch of individuals, or a social network. Those who study them as a group assume they know the membership and boundaries of the group. That is the mistake that politicians make when they talk about "the community": They wrongly assume that all the people in Ward 5—or all the Italian-Americans on the New Jersey Shore—know each other and belong to the same group. In all but desert islands and laboratory situations, people are constantly entering and leaving networks, and these networks are complex

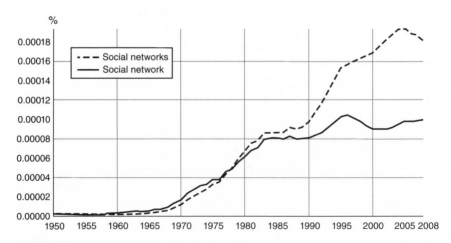

Figure 2.13
Growth in use of the terms "social network" and "social networks" in books.
Source: Googlelabs Books Ngram.

structures with clusters, cleavages, and separate ties. In addition, many important ties extend beyond a network's boundaries.

We discuss later how families and working arrangements are better understood these days as networks. The ultimate examples are national societies. While people often discuss "American society" or French, or Libyan, their interactional and cultural unity is more complex and dynamically changing. Consider how the southern United States split from the rest of America in the Civil War. Consider how Quebec and the "Rest of Canada" (a term often used) are continually readjusting their relationship. Consider how French kings, presidents, and cardinals worked for centuries to knit together the disparate regions we now consider to be part of a unitary French society. Consider how most outsiders saw the Libyan society as monolithic before the revolt starting in March 2011 made palpable the split between Cyrenaica and Tripolitania. Yet "society" is such a convenient shorthand that we all continue to use the term. The trick is not to take it—or other manifestations of groupiness—too seriously, but to use the network perspective to delve into the actual clusters, cleavages, and connections in societies.

The switch to the social network perspective raises questions about how people really operate, such as:

• Who is in what network as people make their way through a bunch of loosely connected networks?
• What kinds of relationships do people have? Are they narrowly defined relationships of love *or* money, or are they more broadly supportive, providing love *and* money and perhaps some babysitting?
• How do these relationships enable people to survive and thrive? How are these relationships interconnected? What are the clusters and cleavages? Who is central or peripheral? Who is attached to whom?
• How permeable are the boundaries of these networks? Are the networks closed shops or welcome wagons open to newcomers?
• Does networked individualism foster *critical commitment* as people have a real, but limited, engagement with each of their social circles?

Understanding the answers to these questions helps solve some interesting mysteries. For instance, Wellman's work with Ken Frank has shown that each tie between a parent and adult child tends to be supportive when the entire family has a culture of supportiveness. This is what social scientists call an "emergent property." You cannot just predict helpfulness from the characteristics of the tie between two persons, because the character of the network in which the ties are embedded also affects the likelihood that people will be helpful—or not.[39]

The social network perspective also matters because the nature of the networks may also be associated with happiness and other emotional states. Social scientists Nicholas Christakis and James Fowler's research reported that among participants in the Framingham (Massachusetts) Heart Study, people are about 15 percent more likely to be happy if they have close ties with happy people. While this may seem as if it is just happy people finding each other, the emergent property kicks in because if the friend of the happy friend is also happy, there is an additional 10 percent likelihood of that second person being happy. Even friends of friends of friends are happier.[40] By contrast, if a person is depressed, then their friends are likelier to be depressed.[41]

Christakis and Fowler have also reported other network phenomena: Friends are similar in their smoking, obesity, and alcohol drinking.[42] While these similarities might lead you to think that misery (and happiness) loves company, their analysis of changes over time suggests that company loves misery (and also happiness). So happiness, misery, and obesity may spread by contagion, by homophily (similar people befriending each other), and also by shared exposure to similar environmental factors such as poverty or a general happy environment. As Christakis's colleague Damon Centola found, this may be because multiple neighbors reinforce health behavior, which then spreads across *bridges* between network clusters.[43]

The group's most startling research report is that people who are networked tend to have similar genetic markers. Obviously, the genes came before the people connected in networks. This finding fits with Wellman's earlier speculation about a gregariousness gene—as his NetLab's research had shown that each member of a large network is more likely to be supportive than each member of a small network. We caution that some analysts fear that Cristakis and Fowler's results have been overstated because of the nature of their statistical models.[44]

The Social Network Perspective Develops

The shift to networked individualism has been accompanied by a shift in thinking about how people behave socially. Rather than seeing people as driven by individual norms or by the collective activities of solidary groups, social network analysts focus on how people's connections affect possibilities and constraints in their behavior. The network perspective has gone beyond being a suggestive metaphor to a systematic way of looking at how societies and people interrelate, with its own theoretical statements, methods, and research findings. The field of social network analysis has developed from diverse sources, including anthropological accounts of rural–urban migrants, surveys of people's long-distance communities, political upheavals, international trade relations, and of course, Facebook.

Social network analysis studies the larger patterns of what people and organizations do and how these patterns fit into society. It is a new perspective for how to look at the world. As such, it has a good deal in common with revolutionary changes in fields as diverse as astrophysics, genetics, atomic physics, and English literature—when such disciplines emphasize how patterns of relationships affect the behavior of individuals: be they stars, words, genes, atoms, people, organizations, or nations. As sociologist Bernie Hogan puts it:

Key to the idea is the emphasis on the specific patterned relationships between individuals, rather than emphasis on inner forces, such as personality, or categorical differences such as race and gender. . . . Networks are . . . ways in which individuals negotiate both opportunities and constraints, thereby demonstrating they are not independent cases, but structured actors. It is very difficult, for example, to be at the forefront of culture if one does not have access to culture's leading figures. Similarly, it is difficult to rise through the political ranks without a specific set of useful relationships that will promote the individual.[45]

The Founding Grandfather—Georg Simmel

The Social Network Revolution took many years to become an overnight success. The story begins more than a century ago:

Working in Germany, Georg Simmel (1858–1918) was the first to present a consistent network perspective and link it to the cataclysmic changes of the Industrial Revolution. Simmel argued that life—especially in cities—was a fluid form of networks. His main intellectual opponent, Ferdinand Tönnies (1855–1936), nostalgically lamented the loss of group solidarity in German villages as industrialization, bureaucratization, and urbanization took over in the late nineteenth century. Tönnies argued that the German sense of community solidarity was gone, as impersonal relations in big organizations and cities had come to dominate German life.[46]

But Tönnies had made the terrible mistake of binary thinking: If villages—*groups*—were falling apart, then contractual relations—*individualism*—had to be their replacement. Simmel thought otherwise, and wrote the first appreciation of *networked* cities in his article, "The Metropolis and Mental Life."[47] Simmel's network thinking comes in part from his position as a Jew in Germany at a time of anti-Semitism. Despite the widespread appreciation of his brilliance, Simmel never received a permanent university position. He was never part of an ingroup, and his focus was on marginal people who link different groups rather than those who inhabit a single group. He wrestled with how people maintained their individuality in modern life while the division of labor in society caused them to rely on the complementary activity of others.

Simmel was the first to see contemporary humans as networked individuals. He elaborated on how their networks function by showing how interactions among three persons are fundamentally different from interactions between two persons. It is only with three persons that you can have coalitions—two against one.

Furthermore, interactions between two persons are constrained—even shaped—when a third person is there. Think of how two guys compete for the attention of a woman. With only two persons present, if either one leaves, the interaction dies. But with a third person present, the interaction continues. None of the three has the leverage to destroy it.

The Cold War Begets Social Network Thinking

Despite Simmel's brilliance, the social network perspective did not develop until the 1960s. Its flourishing had much to do with the 1960s' *zeitgeist* of more openness—and less groupthink—and with increased funding for the social sciences, especially after the Cold War set in. The loyalties of those in less-developed countries were fluid in the 1950s and 1960s, with colonialism on its way out and huge numbers of migrants moving from rural villages to explosively growing cities. U.S. and British officials were worried about the political leanings of populations in Africa, Asia, and Latin America. They believed that while villagers had been calm in their traditional hierarchical villages, they would be up for grabs in the cities—available for Communist activists to gain their allegiance. In short, Western authorities feared a political upheaval would occur during the transition from socially controlling groups to unconstrained individualism. So they encouraged social scientists to study the situation.[48]

The researchers all came back with the same story: Things had *not* fallen apart in the less-developed countries. Community was not dead as the result of massive migration to cities. There were few isolated, alienated individuals. But the new social reality did not resemble the traditional bounded hierarchy of prewar rural life, when the village chiefs supposedly told folks what to do and all the villagers knew each other. Rather, migrants to the cities were forming what analysts soon came to call "social networks"—where they connected both with former village mates and with the people they worked with, lived near, and did business with. A more fluid form of social structure emerged.

Pioneering network analysis was also done in Western countries. Anthropologist Elizabeth Bott became the most famous social network analyst of the 1950s when she unlocked the key to why some English working-class husbands and wives spent their leisure time together while other husbands and wives went their separate ways. The answer had nothing to do with individual characteristics such as age or income. Rather, it was the strength of the women's kinship networks. Where a woman's Mum was close with her adult daughters and sisters, those women were pulled away from their husbands. Where Mum and daughters were not close, those wives spent more leisure time with their husbands.[49]

A similar fear of things falling apart arose domestically in the United States during the urban discontent crescendo of the 1960s. Three things coalesced: the Civil Rights Movement, inner-city poverty among African-Americans, and a student movement against "little boxes" conformity. The working belief of government officials was that individual rioters were disconnected and alienated from American society—a similar belief to what was driving Cold War fears. Social scientists Joe Feagin and Harlan Hahn found evidence to challenge this belief. During jailhouse interviews, the researchers found that rioters were *not* the disconnected or alienated urban poor. This meant the individualistic explanation was wrong. Rather, rioters were more likely to be the settled poor who had more of a stake in their neighborhoods, larger friendship networks, and a greater concern for protesting social inequities. Indeed, it was a combination of friendship networks and local involvement that encouraged individuals to riot.[50]

Meanwhile, American urban researchers, such as Herbert Gans, Elliot Liebow, and Carol Stack, were showing how supportive inner-city life could be: When people did not have much money, they relied on each other—although Liebow cautioned that with so much interpersonal reliance, relationships could easily explode under the strain. This is a theme that continues to the present. Sudhir Venkatesh's *Gang Leader for a Day* graphically shows how the complex webs of support and social control pervade Chicago housing projects, with so-called "urban renewal" tearing down their homes in order to erect wealthier high-rise buildings.[51]

The Cold War and tumultuous 1960s yielded a third entry point into the social network perspective. Not only were people flooding into the cities of less-developed countries, they were also having many babies.

There were more mouths to feed—and potentially, more discontented citizens emerged from rural uprooting, poverty, and overcrowding. Consequently, many Western policymakers wanted to influence people in the less-developed world to use birth control—condoms, IUDs, and the pill—to reduce the number of pregnancies. The standard marketing approach to this and other public health issues had been to use a two-step flow of communication: The mass media—newspapers, magazines, radio, and TV— were used to try to induce people to chat about and reinforce mass media messages. Yet, traditional marketing methods did not work well in societies where most did not read, listen to the radio, or watch television. The messaging had to be by word of mouth. So starting in the 1950s, scholars began to study the "diffusion of innovation." They discovered that they could model the spread of influence throughout a society and identify those who either spread the information the fastest among their own networks or who were bridges connecting different networks.[52] Although population control has been the best-known issue among policymakers, the same ideas have been applied to many things: medical innovations, the adoption of new seed types, the spread of AIDS, uncovering clandestine insurgent networks, and getting tastemakers to influence others to buy a particular brand of running shoes.[53]

The tale of internet adoption matches long-observed patterns that go through several phases. The first phase is *innovators*: The smallest percentage, they strive to create new products and services, and constantly prowl for new things to embrace and enhance. They are the people who first started using personal computers and connecting to early online communication services—some even before the internet. The second is *early adopters*: They try out new ideas after they become available to the general public and are often motivated by being at the cutting edge of the culture. The third is the *early majority*: These adopters learn about a new product or service and are happy to adopt it once they see it has some value. They came thundering in when the internet developed point-and-click graphical interfaces, powerful browsers, and search engines. The fourth is the *late majority* adopters: They are skeptical about buying into something new until many of those around them have already become adopters and the innovation has been widely embraced. Now that the internet and mobile phones are routine and easy for beginners to use, they're in. Fifth and finally is the *laggards*: They are comfortable with the status quo and not much interested in change. For example, a minority of North Americans still do not use the internet or mobile phones.

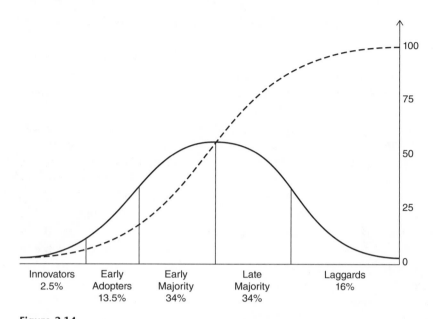

Figure 2.14

Diffusion of innovation as a percentage of the population.

Source: Tungsten, Wikipedia article "Diffusion of Innovations" under Creative Commons License.

Communications scholar Everett Rogers did the pioneering work on the diffusion of innovations. As shown in figure 2.14, the bell-shaped curve shows the proportion of each category of adopters. The S-shaped curve shows the rate at which innovations spread—slow at first, picking up pace as adoption increases and tapering off as laggards slowly stream in. However, information—and innovations—do not spread evenly.[54] Heterogeneous networks slow things down as different types of people steer clear of each other. For example, primary school boys rarely play with primary school girls: The gossips—and flu germs—of the boys and of the girls flow in different channels, although enough friendships slowly bridge the gender gap.[55]

Where would the key influencers be located in the network? On the one hand, common sense and some evidence suggest that heavily connected people in the core of a network would be most apt to spread things—be they ideas or infectious diseases. On the other hand, those on the periphery have little impact within the network, but they are often important for sending and receiving things from other social milieus.[56]

The Structure of Social Networks

Social network analysts map networks of these relations, tease out the prominent patterns in such networks, trace the flow of resources—such as information, love, or money—through them, and discover what effects they have on individuals. The analysts explore network ties, which can consist of one or more relations, such as friendship, hate, financial exchange, trade, web links or airline routes. Ties vary in (1) *quality*—for example, whether the relation provides emotional aid or companionship; (2) *quantity*—how much emotional aid and how frequent the companionship; (3) *multiplexity*—the bundling of relationships in a tie, such as friends who provide emotional aid *and* frequent companionship; and (4) *symmetry*—for example, which types of people who get emotional aid do not give it back. In addition, social networks vary in size and scale. They could be as large as the many-billion-page World Wide Web, as small as a classroom, as insignificant as a sandlot baseball game, or as world changing as the old-boy network of bankers, financiers, and government officials who brought about the Great Recession that began in 2007. Additionally, different nodes in networks have different roles. For example, the web has a central, thickly interconnected core, surrounded by thinly connected clusters, and even some totally isolated websites.

Google searches this web network with automated bots (spiders) that crawl from website to website as they follow the ties among them. But how does Google—or anyone—go from one website to another, when there are billions and billions of them? For one thing, web ties cluster, making navigation easier. It is easy, for example, to go from pewinternet.org to any other Pew site.

For another thing, there are superconnectors—like *Star Trek*'s wormholes or neighbors on the block—who know everybody, or in the case of the web, connect to lots of others. Although these highly connected hubs shorten the distance ("path length" in network language) that information must travel, recent research on Twitter by mathematical sociologist Duncan Watts has found that messages reach their destinations even without superconnectors.[57] High network density—many interconnections—is especially useful for rallying support and crowdsourcing information.

There is cumulative advantage: The more connections you have, the more you get. The distribution often follows what mathematicians call a "power law."[58] This is a variation on what sociologist Robert K. Merton called the "Matthew effect": The rich get richer and the poor get poorer.

The name comes from the Biblical verse of Matthew 25:29: "For to all those who have, more will be given, and they will have an abundance; but from those who have nothing, even what they have will be taken away."[59] In short, networks are not random assortments of disconnected ties, but complex sets of relations that channel resources such as money, information, and love toward some people and away from others.[60]

Hubs link people and places as well as websites. Hubs that bridge two otherwise separated social networks can be useful for gaining new information about jobs and housing.[61] Weak ties between acquaintances are sometimes more useful than strong ties between close friends: Weak ties are more likely than strong ties to connect to different social circles, providing access to more diverse information. For example, sociologist Mark Granovetter found that weak ties were more helpful than strong ties for Boston-area professionals and managers looking for new jobs.[62]

As organizational sociologist Ronald Burt has pointed out, "network brokerage" is about building connections across different social circles that provide more exposure to variations in opinions and behavior. Burt argues that networks are most important for finding people who are interesting and who think differently, which exposes people to more diverse information. With such diversity, corporate managers—and ordinary human beings—can perceive alternate ways of doing things.[63] Burt has extended his analysis to the many denizens of Second Life, an online virtual world. He finds that network brokers in Second Life are the high achievers who provide the social infrastructure that makes this virtual world valuable. Such brokers are more likely to found communities that survive and attract more people as members. There is a cost, though, because those living in closed Second Life networks are socially closer and more trusting than those in broker-ridden worlds.[64] We can make a similar argument about brokerage in finding homes, jobs, dates, or partners: The best strategy is to tell everyone with whom you are connected—weak or strong—that you are looking.

In short, *bridging ties* are great for getting information in and out of a cluster of relationships. But *bonding ties* that stay within a cluster are often necessary for internal trust, efficiency, and solidarity. Merton called these different structural roles "cosmopolitans" and "locals." Both are useful for societies.[65] As Uncle Jack used to tell his nephew, coauthor Wellman, about his small coat factory: "We have an 'outside man' and an 'inside man.' The outside man to schmooze with the customers, the suppliers, and the bankers; the inside man to make sure the factory makes good-quality clothes on time."

If Uncle Jack is not persuasive enough, there is now evidence from Britain showing that the spatial and social diversity of networks is related to their locality's economic development. The more prosperous regions have more far-flung networks that connect to both high- and low-status people.[66]

Analyzing Social Networks

Small networks are easily visualized in a graph as a bunch of network members connected by a bunch of lines. But once the number of network members gets above twenty or so, the graph's clutter makes it hard to visualize. The standard trick among analysts is to transform the letters and lines into a matrix, where each row and column represent a network member and each cell shows if there is a connection between the two network members (table 2.3). While we are not going to show matrices again in this book, we wanted to provide a glimpse of the matrices behind the apparent magic of social network analysis.

With matrices, analysts can easily find such things as:

• *Clusters* of people, with many interconnections among them. These are the true groups.
• How *densely knit* these clusters are. For example, relatives are usually more interconnected than friends. We would worry about families where people don't talk with each other.

Table 2.3
Strong and Weak Ties: Matrix of Who Is Connected to Whom

	A	B	C	D	E	F	G	H	I	J
A	-	2	2	2	1	0	0	0	0	0
B	2	-	0	2	0	0	0	0	0	0
C	2	0	-	2	0	0	0	0	0	0
D	2	2	2	-	0	0	0	0	0	0
E	1	0	0	0	-	1	0	0	0	0
F	0	0	0	0	1	-	2	0	0	0
G	0	0	0	0	0	2	-	0	1	0
H	0	0	0	0	0	0	0	-	1	2
I	0	0	0	0	0	0	1	1	-	2
J	0	0	0	0	0	0	0	2	2	-

Notes: 0 = no tie; 1 = weak tie; 2 = strong tie.
Source: Table by Barry Wellman. © 2011, used with permission.

• Where the *bridges* are that connect different clusters, and where the structural holes are when there are no bridges. Such holes can create opportunities for people to become connectors and brokers between two networks. What kinds of people in what kinds of positions are bridgers?

• *Networks of networks*. If two people in different clusters are connected, then the two clusters are potentially connected. Corporations connect in this way, putting members of other corporations on their boards of directors. This is also how much of the web is structured, with links to major sites (internet service providers, organizations, governments) that provide links to the smaller sites they service.

• *Indirect ties* in the network, such as from A to E to F in table 2.3. This is how information and infectious disease spread.

• People who are *structurally equivalent* to one another. For example, all of the children in a class have a structurally equivalent relationship to their teacher. Less obviously, Burton Pasternak and Janet Salaff used network analysis to identify a class of high-status communist-party members in Inner Mongolia who were routinely receiving gifts from peasants and merchants.[67]

Clusters of Networked Individuals

Networked individuals rarely are dancers in multiple, but separate, duets. Rather, many of their ties are in densely knit clusters. For example, A-B-C-D in table 2.3 is a cluster, even though B and C are not connected. While this is visually clear, finding clusters in larger networks requires manipulating the matrix using standard network-analysis software programs such as UCINet, Pajek, and NodeXL.

Are such clusters communities? For better or worse, the clustering approach focuses purely on relationships and does not take into account sentimental feelings of belonging to a community. This can be useful because many people wrongly assert the existence of communities when they do not actually exist. Despite the precision of the clustering approach, it can underestimate people's multiple attachments. This can lead to confusion. When Beverly Wellman moved to Toronto more than forty years ago, she taught English-language subjects at a Jewish day school. One day the principal asked her:

"When are you moving to be among your own kind?"

"Where do the intellectuals live?" Beverly replied.

The principal asked the question of a Jew; Beverly answered as an intellectual. Theirs was an exchange that wrongly assumed that networked individuals belong to one community rather than to multiple communities that command only some of their allegiance and attention.

Take, for example, the network of ties among those people who mutually exchange Twitter messages ("tweets") with coauthor Wellman. Within the tangled spaghetti-like mass shown in figure 2.15, there are six clusters of people who are especially apt to tweet with each other. They are clustered around some of Wellman's scholarly interests, because his other friends, relatives, and neighbors do not tweet. So this Twitterverse only contains fragments of his social network life.[68]

Where do these communities come from? Sociologist Scott Feld has come up with the useful concept of "foci" to identify some bases of community, these can be shared places and institutions, both online and offline. One focus might be people with similar characteristics who come into contact, such as an ethnic minority at work or people who go to the same church. A focus is not community: It is the basis for community by

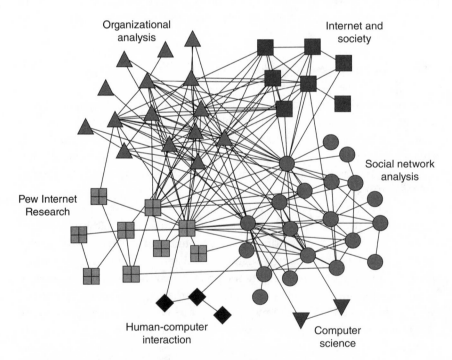

Figure 2.15
Clustering in Wellman's network of mutual Tweeters.
Source: Anatoly Gruzd. © 2010, used with permission.

providing a shared context that creates the possibility for interconnections among people.[69]

Clusters are usually regions of particularly heavy interconnection, but they rarely are mutually isolated. Look at the interconnections among the six clusters in Wellman's network (figure 2.15). In the lower right corner, information and mutual awareness is flowing among social network analysts, organization analysts, and internet and society analysts. As people retweet the messages they have received, information spreads and those receiving the forwarded messages can become aware of people in other clusters.

Fortunately for society, clusters rarely are like moated castles. The duality of persons and clusters creates both within-cluster bonding and between-cluster bridging. When two people are in the same cluster, they are linked. But if they are also members of other clusters, the tie between these two persons also connects these other clusters. Their ties simultaneously bond them within one cluster and bridge between different clusters.[70]

In addition to finding interconnected people, analyzing clusters can also identify communities of ideas—or mental models—when certain words are linked or disconnected. Organizational analyst Valdis Krebs discovered this when he analyzed the network of purchasers of American political books just before the 2008 presidential election. His results foreshadowed the problems that President Obama has had in working with Congress. Although the diagram shows a network of books—which really means a network of ideas—rather than of people, it is people who buy and read these books.

Figure 2.16 shows that ideas did not flow between the three clusters. The big Democratic cluster has a low density, reflecting the diversity of Democratic opinion: Only 14 percent of all the potential interconnections among its books actually exist. By contrast, the Republican books are twice as densely connected at 30 percent. While this is due in part to the smaller number of the Republican books—it is easier to interconnect a small number—the more densely knit Republican books portend the greater cohesiveness of Republican partisans since the presidential election. The third cluster of six books in the upper left indicates people who are more interested in Barack Obama personally than in the Republican or Democratic themes.[71]

The arrows in the book graph show the directionality of purchases. For example, people who first bought *Patriotic Grace* (top center in the Democratic cluster) sometimes went on to buy *Snowball*, and some of those

Figure 2.16
Links between book purchases before 2008 U.S. presidential election.
Source: Valdis E. Krebs, http://orgnet.com/divided.html. © 2008, used with permission.

purchasing *Snowball* also bought *When Markets Collide*. In this way, ideas can flow from *Patriotic Grace* readers to *Snowball* readers to *When Markets Collide* readers.

There are no bridges in these networks. The individual books are networked, but their idea networks do not spread widely. The Republicans, Democrats, and Obama readers are only reading their own self-supporting stuff, with little evidence of ideological crossover. Not even one book attracted the attention of more than one crowd; there is not a Democratic-Obama link. So the personal attraction of Obama to some is separate from their political leanings. The three-way pre-election split gives a good hint of the problems that Republicans, Democrats, and Obama had in working together after the election. They were talking different languages, with no bridges.

Personal Networks

So far in this chapter, we have been looking at *whole networks* from a God's-eye view: looking from above at the entire structure of these networks and what flows through them, be it information, money, disease, or love. But most people are more earthbound and self-centered, viewing social networks from their own personal perspective: the relationships that they have, directly and indirectly, via friends of friends.

People are becoming more aware that each individual is at the center of his or her own *personal network*: a solar system of one to two thousand and more people orbiting around us. Each person has become a communication and information switchboard connecting persons, networks, and institutions. At the same time, each person has become a portal to the rest of the world, providing bridges for their friends to other social circles. With the size and complexity of these networks, each networked individual has to balance out collective and interpersonal commitments in unique ways. Facebook is a good example: It consists of millions of interlinked personal networks, each "home page" connecting to "friends" and interests.

Indeed, barring the odd hermit, just about everyone is connected these days—at most by six links of interpersonal connection and often by less. This is the "six degrees of separation" that social psychologist Stanley Milgram made famous in the late 1960s. Although the concept was made into a Broadway show and a subsequent movie, the "six degrees" concept has been more celebrated than proven, because there has been limited evidence.[72] But retweeted messages in Twitter provide confirmation. The many bridges between Twitter clusters means that chains of information from one Twitter follower to a follower of that follower, and so on, encompass about 83 percent of all Twitter users within five steps of interconnection.

On average, about half of the people on Twitter are only four links away from each other. Of course, not all followers retweet each message—or even look at them—but the news spreads quickly. While messages usually get read and retweeted by those who speak the same language, they cross substantial distances, a mean of about 1,540 kilometers (955 miles). Indeed, a map of Twitter links looks much like a map of worldwide airline links: Tweets especially travel between big cities.[73]

The study of personal networks highlights the new realities of the network operating system. People are not alone, but connected with many others in a variety of social circles that provide them with diversified portfolios of social capital. The shift from groups to networks affects people's behavior and calculations about their social strategies: They must understand and navigate the interpersonal solar systems of which they are a part.

Networked Societies

Remember Moliere's *Bourgeois Gentleman* who was speaking in prose without knowing it?[74] Similarly, we all have been using networks throughout our lives without knowing it. Sometimes we need a new perspective to see things differently, just as it took fractal discoverer Benoit Mandelbrot to point out "clouds are not spheres, mountains are not cones, coastlines are not circles and bark is not smooth."[75]

Networks have always been with us, although they are more prevalent now and we are certainly paying more attention to them. So what does the discovery of social networks give us, besides a new label for an old habit? For one thing, network awareness provides new insights into the structure and functioning of our societies, and how we should operate in them. For another thing, we get away from the groupthink that leads people to cry "*oy vey*, things are falling apart" when in fact, they are just changing the world into a more diversified, more complex, and more interesting place.

We also become more aware that the structure of networks matter. For example, many social networks are sparsely knit: The fact that A knows B and C does not mean that B knows C. This provides maneuverability, some privacy, and some insecurity. But the shape of the networks makes a big difference in how we communicate and get information. Sparsely knit networks often provide bridges to multiple social worlds. By contrast, densely knit networks—like small-town villages—often provide bonding, solidarity, and security but at the probable cost of insularity and social control.[76]

We have talked principally in this chapter about some of the implications of the network operating system for social life. But there are things we have lost with the diminishing of groups: The weakened comfort of group identity comes with the gain in maneuverability; the ease of organizing group-based activities has given way to more strenuous microcoordination of networks. As sociologist Bernie Hogan puts it: With the turn to social networks, "we're not bowling alone, but texting our friends, seeing who's available, sending the electronic invitation, and waiting for people to show up, scheduling another time because someone can't make it and maybe, if we're lucky, actually getting to bowl."[77]

The turn toward a network operating system has been built on flexible connectivity between individuals and the ability to trust one another across distances and groups without requiring the cohesive force of the tribe to punish transgressions. We believe that this turn toward networks will continue—barring the loss of flexible connectivity and the loss of trust over large spatial and social distances.

3   The Internet Revolution

The pioneers of the internet did not act as if they knew they were creating anything special.[1] Internet communication began with a computer crash and meaningless nonsense. The first attempted transfer of information packets over a wire strung between two computers took place in 1969. It was not accompanied by the same kind of thundering declaration as Samuel Morse tapped out in the first telegraph message in 1844: "What hath God wrought?" Rather, Charley Kline, an engineering student at UCLA, froze his computer on October 29 when he began typing "L-O-G." (He was on his way to keying in "L-O-G-I-N" to start a file transfer program he had coded.) Programmers fixed the glitch quickly, and after a reboot, the login worked and file sharing between computers began.[2]

Email came into being two years later in 1971, and, again, there was no throat-clearing puffery or even an effort to match the practical tone of Alexander Graham Bell's initial phone call in 1876: "Mr. Watson, come here; I want you." Instead, an engineer for a U.S. Defense Department contractor named Ray Tomlinson sent a meaningless test message from one computer to another that was sitting less than five feet away. Recalling the episode later, Tomlinson wrote on his website that "the test messages were entirely forgettable and I have, therefore, forgotten them."[3] He suspects his text was probably "QWERTYUIOP," the top text line of a standard keyboard. Indeed, the most notable thing for which Tomlinson is remembered is that he picked "@" as the locator symbol in electronic addresses.

From those humble beginnings, a socially disruptive technology emerged. Though they did not proclaim it, the internet's creators built something that would enhance the broader social, economic, and political forces that were already pushing people in the direction of networked individualism in a network operating system. The internet did so by enabling people to act more effectively on their own and function more easily in large, dispersed networks. It empowered individuals and extended their reach by giving them tools to create media, search for information that mattered to them, project their voices, form groups that served their needs, and reach out to their strong and weak ties. The internet also helped people broadcast to and receive material from more sources. It helped users change the size and shape of their social networks. It even changed some peoples' communication patterns inside those networks.

In fact, the ascendance of the internet and innovations in the online environment in the last decade have strongly shaped public understanding of what it means to be part of a social network. Before the internet became a powerful cultural force, not many people thought of themselves as actors in social networks. Now they see it easily because their world is abuzz with references to social networking and because they are explicitly invited to participate in activities that are described as social networking and networking opportunities. Once it became widely domesticated, the internet made it easier for people to see that they themselves personally functioned in networks—both the technological and the social kind. In helping people understand what networks are and what networks do, the internet has become a useful metaphor of the new network operating system as well as one of its primary driving forces.

In the case of the internet, the diffusion of innovation model, shown in chapter 2, fits well. Innovators in government, technology firms, and universities, as well as individual super-geeks pretty much had the online world to themselves for two decades. No good data exist about the size of the internet population before the early 1980s, but it probably was several thousand. In September 1983, a telephone company's survey titled "Road after 1984: The Impact of Technology on Society" (after George Orwell's *1984* novel) asked a representative sample of Americans how many had computers at home: Ten percent said they did. This group was asked a follow-up question, "Do you transmit or receive information on your computer over your telephone lines or not?" Some 14 percent of those home-computer users said they did, so the internet population in the fall of 1983 in America was 14 percent of 10 percent— 1.4 percent.[4]

The computer users were then asked a daringly futuristic question: "Would your being able to send and receive messages from other people . . . on your own home computer be very useful to you personally?" Some 23 percent of the computer owners said it would be very useful, 31 percent said it would be somewhat useful, and 45 percent of those early computer users said it would not be very useful. However, 74 percent of the early adopters agreed with an even more fanciful statement: "The trouble with purchasing and bill-paying by computer is that it will be too easy to buy too many things that aren't in the family budget." So much for anticipating their future desires.

Things grew slowly in the early years as the internet—mostly email—was largely confined to universities. By the time Tim Berners-Lee wrote the code for the World Wide Web in 1989 and made it publicly available in 1991, there were probably fewer than five million "early adopters" of the internet in the world. Only those with specialized knowledge could find what later came to be called "web sites," and only real specialists could build them. So when the first browsers enabled somewhat easier search and display of data in the early 1990s, the U.S. user population was barely hovering around fifteen million—but was ready to blast off.[5]

While all of these building blocks were critical to the development of the internet we know today, 1993 is the birth year when the "early majority" cohort of adopters began to use the internet. That is when the first easy-to-use web browser, Mosaic, became available and various email services had largely come together. Where the 1980s were the decade of stand-alone computers, the 1990s became the decade of connected computers. Over the course of 1993, various test versions of the Mosaic browser were released and by November, Marc Andreessen and Eric Bina of the National Center for Supercomputing Applications had unveiled the official 1.0 version. Their browser was so popular it overwhelmed the competition and quickly established the World Wide Web as the dominant part of the internet's traffic. Why? Because bits stored on computers finally became graphically pleasing, easy to navigate, and understandable by a broad audience. Mosaic and its offspring Netscape Navigator were to the web what paper is to ink. The browser creators used a simple format called hypertext markup language (HTML) to display the data stored on other computers in a graphical format that came to be known as "web pages."[6] The revolution was underway, and the "late majority" of Americans became internet users in record-breaking time.

Over the next decade, the internet became one of the most rapidly adopted mass consumer technologies in history. It took radio thirty-eight years to attract a comparably sized audience of fifty million Americans; television took thirteen years. But it took the web just four years to amass that many, if you take as the starting date the day Mosaic was made available.

Why Was Internet Adoption So Rapid and Widespread?

The internet just didn't happen by itself. A variety of factors encouraged people to embrace it.

First, the U.S. federal government and adventurous individuals were the forerunners. After the government supported the creation of the internet, it took a light-handed regulatory role once its use became popular. Policies adopted by the Clinton administration and the Republican-controlled Congress emphasized support for internet growth and minimum regulation of what was happening online. The government transferred control of the internet infrastructure to the private sector in 1995, encouraged expansion of the internet backbone, kept new taxes off internet activity, and supported online commercial growth.[7] The Federal Communications Commission (FCC) buttressed these policies by embracing the dictate of the 1996 Telecommunications Act "to preserve the vibrant and competitive free market that presently exists for the internet and other interactive computer services, unfettered by Federal or State regulation."[8] The FCC determined that the internet was an "information" service in little need of regulation rather than a "telecommunications" service that would require more regulation. This hands-off policy enabled a spectacular amount of technological and commercial innovation and allowed user demand and feedback to play major roles in shaping the environment of the internet.

Second, the technology improved rapidly and dramatically, which increased its usability and attractiveness—and hence its potential profitability to hardware and software makers. Connected computers became cheaper and better to use over time. Gordon Moore, the cofounder of Intel Corporation, famously asserted in April 1965 that the maximum processing power of a microchip, at a given price, doubles roughly every eighteen to twenty-four months.[9] That observation—called "Moore's Law"—has held since 1958. Moore said the ability to pack more transistors onto an integrated circuit allows computers to become faster at an astounding rate and the price of a given level of computing power to decrease at an equal pace. Intel press statements put this level of change in practical terms by noting that in 1978, a commercial airline flight from New York to Paris cost around $900 and took about seven hours. If progress akin to Moore's Law had been applied to the airline industry the way they have to the semiconductor industry since 1978, that flight today would cost a fraction of a penny and take less than one second.[10]

Third, at the same time computing power was growing and processor prices were dropping, a similar trend was taking hold in communications capacity—or bandwidth. Slow "dial-up" access that limited users to text and competed with phone calls soon was supplanted by "broadband" technology that could push more bits more quickly through the copper wires—and, eventually, fiber optic cables—that tied computers together. As a result, bandwidth capacity doubled every one to two years as bandwidth rose from 300 bits per second in the mid-1970s to speeds of one billion (one gigabyte) bits per second, the speed that Google announced in the spring of 2011 it would build into a system it would construct in Kansas City, Kansas.[11] Furthermore, international standards were adopted a few years earlier for an ethernet system that was ten times faster than that gigabyte-per-second Kansas City system, thereby giving engineers the room to begin developing it.[12] The cost of moving bits has fallen about 50 percent every year. Moreover, technologists were finding ways to compress computer files without losing much of the fidelity of the data.[13] This significantly lowered transmission time and costs for larger stores of data, enabling everything from experiences in virtual worlds to streamed feature movies with the click of a finger.

Fourth, similar, though less spectacular, improvements also took place in the radio spectrum, as engineers found more efficient ways to increase the flow of data in the spectrum and minimize interference in adjoining bands of the spectrum, thus allowing more data to be transferred wirelessly. This—plus lighter-weight computers and smartphones—broke the umbilical cord tethering users to their desktop connections. We'll discuss the Mobile Revolution in chapter 4.

Fifth, the internet remained an interconnected network. Like the worldwide telephone system, it did not balkanize into competing, mutually unconnectable bits and pieces, but remained a network of networks—interconnecting internet service providers (ISPs). It is harder to notice absences than presences, but consider what would have happened if different services had set up different networks with different domain names (one having .com and another having .biz) that couldn't find each other. Communication, information, and commerce would have fragmented and never reached the scale they did. Instead, even though censorious regimes try to ban or control the internet, they have not set up competing systems—so far.[14]

Sixth, most internet service providers in America (and increasingly elsewhere) provided all this bandwidth for a flat rate. It did not matter if a person was emailing "Hi Mom" or downloading a giant movie file.

This "all you can eat" environment was successful in bringing new waves of users online. So successful, in fact, that it might not hold up under the advances in tech innovation and continuing additions to the online population. The cost of providing bandwidth became great as movies and long videos began to move along the internet. Some firms began offering tiered services, charging different rates for different levels of consumption. Additionally, providers began to hint they would consider managing traffic flows and imposing "data-caps" in the mobile environment on the grounds that wireless spectrum is a limited resource that at some point could become saturated. This could change online culture, the commercial activity online, and the ecology of gadgets that are connected.

Seventh, storage vastly improved for personal computers, smartphones, and corporate installations. Like Moore's Law, the "law" articulated by Mark Kryder was that the density of information on digital storage devices has been doubling about every two to three years since 1956, the year the disk drive was introduced.[15] Since then, the density of information a drive can store has swelled from two thousand bytes to two trillion bytes (two terabytes), all packed onto a square-inch disk-drive storage space. That facilitates many things, including the economic viability of massive server farms that enable "cloud computing"—where personal and organization information are stored online rather than on personal computers: that is, things like Gmail accounts, Facebook profiles, and YouTube videos.

Eighth, all of these changes supported new and compelling applications ("apps") that led people to quickly embrace the internet. Email was the first—and still is—the most important app for internet users according to the number of people in America who use it every day. The earliest surveys of internet users showed that email was the most widely and eagerly embraced activity. A federal study in 1998 found that 78 percent of online Americans used email and over half (54 percent) of people with internet access outside the home used email. The numbers have been consistently high, regardless of income, race, gender, age, or any other characteristic.[16] Pew Internet (referring to the Pew Internet & American Life Project) surveys from their start in 2000 through 2011 have found that email is the single online activity that draws the most users on any given day. In mid-2011, 78 percent of American adults told Pew Internet that they were internet users. Of these users, 94 percent said they were email users—a proportion that has hardly wavered from the first survey in March 2000 to the Project's recent canvassings.[17]

The Personal and Connected Computer Supports Networked Individualism

We often take for granted the ways in which technologies work. But they didn't have to work that way. For example, some telephone pioneers thought the phone would be used to broadcast music, news, and sermons, while some radio pioneers thought it would be used for personal chats.[18] That's why sociologists have always cautioned against "technological determinism." People take technologies and use them in many ways— including some never dreamed of by their inventors.

Nevertheless, how technologies operate has important implications for how they are used. Not that their design determines behavior, but their design does create the *affordances*—"the possible actions a person can perform on an object." That, in turn, affects how people use them. What's happened with personal computers is a great example because design choices in its hardware, software, and connectivity have fostered networked individualism.[19]

The computer is personal. We call our computers "personal computers" or just "PCs." The implications of that are huge. For one thing, we normally each sit separately in front of our own computers, although as we show in chapter 6, there are times when household members show and share material side by side. For another thing, when we go on the internet, we login as individuals, with our own usernames and passwords. Although most of us take this for granted, this is a big change from earlier information and communication technologies (ICTs). Newspapers, magazines, books, and encyclopedias are often shared within a household; home phones can be answered by anyone in the household (although middle-class Americans often had a separate "teenagers' phone" to keep them out of adults' hair). The internet—followed by the mobile phone—became an information device for *individuals.*

The computer is connected. When personal computers first arrived, they were stand-alone devices: better calculators and typewriters. It was only later that the internet came along and personal computers became almost the only way to connect to it—so much so that personal computers and the internet have become just about synonymous. So individuals have become *networked* to each other and to all the companies, organizations, governments, and interest groups that have an online presence.

The computer and the internet have become humanized. Computers didn't always have user-friendly screens and keyboards. We communicated with the earliest ones via blinking lights, switches, and keypunched IBM cards.

Even when early personal computers came along, we needed to be programming wizard to use them and to access the internet before browsers and search engines. Now, we just need to know how to click on icons, buttons, and menus and, perhaps, how to type queries into search engines. Many children learn how to use the computer before they learn how to spell.

The computer has helped people more than intimidated them. Once DOS gave way to the friendly, graphical user interfaces (GUIs) of Windows and Macintosh, people began to see the internet as an aid to get their jobs done and to stay connected with family and friends. That's what much of part II of this book is about. To use reporter John Markoff's phrase, people regard the internet as "Intelligence Augmentation" rather than artificial intelligence. This augmentation can take thousands of forms, from helping people find their way with a GPS (global positioning system) to fixing their spelling as they write to helping them do their income tax returns.[20] To take an extreme case: nineteen-year-old Sheila Htoo arrived in Toronto in 2004 as a Karen refugee from Burma. She had grown up in a village without a computer, telephone, or even electricity. Yet she quickly and eagerly grasped the computer:

I was so addicted to chatting because I felt very lonely and desperate academically. I made so many new friends online, wasting time in chatting with them every night. I got to know many Karen friends online, formed weak ties and friendships. I communicated with my family through emails because my brother and sister were studying in Thailand towns. But the rest of my family had to travel about two hours on foot to a Thai-Karen village to phone me. Now I use Skype, MSN and Google chat to talk with my brother and sister as well as friends from different countries. I did not grow up in an internet-based society, but now I have been exposed and addicted to internet use. I do not think that I can live without the internet for a day now.[21]

Communication can be more customized and private. Before the internet, if we wanted to share some information with one person and not others, we would either tell that person face to face, call her, or send her a sealed letter. We had to monitor her communication to ensure that others would not know who we were talking with and what we were saying. By contrast, since the introduction of the internet, we can customize communication so that only the intended recipient/s get the information. This started with email, but it became true for all manner of communications—from password-protected YouTube videos to shared files in the cloud, and even to Facebook if people remember to fill out the privacy options.

The internet is decentralized and open to individual choice. Since the dawn of computing, there has been a fight between the centralizers—usually the information technology departments of large corporations—and the decentralizers—people who want to use their computers and the internet pretty much the way they want. The civil war will likely rage forever, but until now the decentralizers have more than held their own. Networked individuals' notions of personalization and customization have fit with the decentralized ethos of contemporary computing. Except when working for organizational giants, people are free to choose their own software and hardware, and to upgrade and modify them when they please.

The internet is asynchronous. Unlike in-person or telephone chats, people can go online when they want to access the information and communication that is waiting for them. Yet, when they do go online, the speed of information retrieval is both rapid and private, and the speed of communication responses is much quicker than by postal mail or exchanging voicemail. Asynchronicity means that networked individuals can personalize their choice of the times they want to connect without responding to others' schedules.

The Internet Expands: An Early Adopter's Tale

The growth of the internet's popularity is also the story of new features of the World Wide Web that have relentlessly renewed the online environment and enriched users' internet experiences. Entrepreneurs, technology activists, civic actors, educators, government agencies, nonprofit organizations, news operations, and millions of individuals have invested billions of hours and hundreds of billions of dollars in research and development to build the online world. There are countless stories about how users first discovered the pleasures of email and browsing and then deepened their online engagement as each new wave of innovation and tech improvement made the online environment more useful and compelling. For example, Leonard Witt began his networked life when, as the Sunday magazine editor at the *Minneapolis Star-Tribune*, he was introduced to the organization's internal messaging system in the mid-1980s: "It was almost magical that I didn't have to walk down the corridor or climb a flight of stairs or even dial the phone to have a conversation with colleagues. Once I learned the mechanics of messaging and the office 'norms' for when and how to use it, I could feel my collaborations and overall work improving."

When Witt became an email user in the Prodigy online service in the early 1990s and later in the open-web environment of the mid-1990s, he saw his outreach to a broader group of friends and work associates increase. "When I had an idea I wanted to share, and I have lots of those, wham! I could send it off to others instantly," he remembers. "Ideas are ephemeral so it is hard to tell how many would be lost forever without email. Now they were kept alive, not only with me, but with my friends and colleagues reacting to them and improving them."

By 2002, Witt had become a professor of communication at Kennesaw State University in Georgia and his enthusiasm for the internet marched along with the frequent innovations that took place as the internet became a much more commercial space with mass appeal. He recalls his early encounters with portals such as Yahoo! in the mid-1990s and his fascination that new kinds of content taxonomies were being assembled. There were links to news stories, health sites, software downloading sites, entertainment reviews of all kinds and fan sites, and religious material—all assembled by humans and entered in long link lists. Those portals were primitive, but their virtue to early internet users was that they provided a one-stop shop that imposed some order on the chaos of content that was flooding online. They also gave users the sense that there actually was a world wide web of information being created. Yet, Witt remembers that by 1997 or so he felt one-stop shops were too unwieldy to help those searching for particular information. Portals could not keep up with the explosive growth of web pages and other internet features.

Search engines such as AltaVista (1995) and later Google (1998) vastly improved users' abilities to navigate the rapidly expanding web. They allowed people to hunt down more specific material in the mass of web pages that were being built. Witt says: "Even back then, as clunky as the search engines were, it was easy to see that someday all the information we needed would be accessible on the computer, just as right now it is easy to see that all the world's information will be accessible on our smartphones."

The late 1990s into the early 2000s was the era of the early majority adopters. Many newcomers were attracted by a specific interest and they had learned from others in their circles that the internet was particularly good at meeting that interest. As they gained experience in the online environment, they found new ways to incorporate internet use into the rhythms of their lives. The excitement of that early discovery was enhanced by a sense that the internet provided many possibilities for enrichment, fun, and work-related benefits. In hundreds of interviews with internet users, we have found that these activities held a special place in attracting new-comers to the internet or enticing existing users to deeper engagement:

• Dot-com retailers such as Amazon (launched in 1994) and eBay (1995) quickly turned the online realm into a marketplace that offered new choices and convenience to consumers.

• Alternative news sites like the Drudge Report, which first started as a Hollywood and Washington gossip email list around 1995, were especially attractive to the early-adopter news junkies.

• Online games started during the 1980s as part of commercial timesharing services such as CompuServe and eventually went into more open online spaces in the early 1990s with such formats as massively multiplayer online role-playing games (MMORPGs). In the early days they were popular with geeks, especially boys and young men, but many more women became online gamers as the genres broadened to such things as online card games and board games in the mid- to late 1990s.

• The pornography industry was one of the earliest entrants into the online world, and the widespread availability of free adult websites was one of the factors that made the online world an appealing place for young men.

• Health sites like the National Institutes of Health's Medline and WebMD (1995) were particularly compelling to women, who notably trailed men in the early adoption of the internet but by 2000 had reached parity.[22]

• Online radio stations started popping up as early as 1994 and eventually allowed millions of online listeners to sample new genres, control their playlists, return to their favorite childhood radio station, or listen to out-of-town sporting events.

• Specialty sites of all kinds found new audiences and allowed the like-minded to commune. Financial news and tips sites drew legions who were anxious to hear about every turn of the market and insider gossip about companies. Hobbyist sites of staggering diversity, especially for genealogy, crafts, outdoor activities, clubs, do-it-yourselfers, and nostalgia themes grew communities of participants. Sports sites gave fans new outlets for trivia chatter, dissection of statistics, fantasy leagues, and trash talk. Celebrity and gossip sites mushroomed in popularity. No niche was too small or too exotic to be left out of this long tail of material.

• Instant messaging clients such as ICQ (created in 1996) and AOL's AIM (1997) became addicting to teenagers and young adult users.

• Napster's peer-to-peer file-sharing program (launched in 1999) prompted a new wave of internet users who flocked online to download and share free music. Even when legal problems closed Napster, millions of users had become accustomed to getting music, movies, and videos from the web.

Witt remembers all of these new developments and says that each "was a 'wow' for me," echoing the views of many veteran internet users who came to believe that the sheer growth of the material on the World Wide Web made it a more compelling place to visit and to interact with others. Google's first run at measuring the size of the web in 1998 counted twenty-six million pages. In two years, the number of pages on the internet grew to over a billion. It hit more than one trillion pages in the first half of 2008—although some carping critics pointed out at that time that "only" forty billion pages were actually indexed by Google.[23] The unsearchable remainder has become known as the "dark internet," named after the unseeable "dark matter" that pervades the universe.

A virtuous circle between supply-side creation and demand-side partici-pation has emerged. As networks have grown, the value of being connected to the network not only has grown, but also has grown exponentially, constantly producing fuel for further expansion. The growth of the inter-net, in particular, shows how a powerful set of self-reinforcing conditions cascade. As more people get email addresses, the value of email to personal communication grows. As more and more commercial, civic, educational, governmental, nonprofit, and individual websites are created, the value of going online to seek information and perform transactions mushrooms. As more tools develop to encourage user participation and customization, more people find reason to engage with others online, and to share and to mash up media. In the early days, people often had to be coaxed or compelled to go online because their employers or their schools encour-aged or even required internet use. In the early 2000s about half of internet users said they first went online either for work or for school. Yet, the majority increased their online activity on their own as they became veteran users because they found the internet served a variety of personal and professional needs.

The growing usefulness of the internet became a boon to both parts of the phrase "networked individuals." The internet allowed users to be both more *networked* and be more assertive as *individuals*. The vast majority used email to deepen and improve their bonds with their strong ties—partly because they felt email allowed them to communicate more often and more easily with close family and friends. In addition, many used email to stay connected to their weaker ties or to develop new relationships. At the same time, as the web developed, people reported that their internet use helped them become more individually empowered. Notably, in Pew Internet's first survey in March 2000, 78 percent of online Americans said their internet use had improved their ability to learn new things, 76 percent said it had improved their connection with their friends, and 55 percent said it had improved their connections with members of their family. Roughly a third of the internet users noted improvements in their capacity to manage their health, their ability to shop and their ability to manage their finances.[24] Those figures rose in a follow-up survey a year later and rose yet again in subsequent work.[25]

A Pilgrim's Progress Continues as Broadband Brings in the Early and Late Majorities

In the decade-long span from the mid-1990s to the mid-2000s, the internet's role in most users' lives changed from being a dazzling novelty to being a commonplace, easy-to-use utility. It became embedded in people's everyday lives—pervasive and domesticated.

For journalist Leonard Witt, there were several signposts of this change: The amount of time he spent online grew and the array of things he did online expanded dramatically. He transferred to his online repertoire activities that he had previously pursued by other means. By now he and his wife were doing everything from buying books on the internet to tracking down a cozy little vacation apartment in the Montmartre section of Paris by contacting the owner directly through a website and email.

The biggest change was about to come. From 1996 to 2002, while Witt was the executive director of the Minnesota Public Radio Civic Journalism Initiative, he was producing content that would be used on the internet. At first, he always needed a wizard-like webmaster to be his intermediary. Yet, Witt soon became an active content creator with no intermediary needed. He started blogging in 2003 about "public journalism,"[26] a journalism reform movement that had started in the early 1990s. In his blog, he explored the predicament of mainstream news organizations and began promoting an alternative vision of journalism that focused on the community role of news organizations and individual reporters. He exults: "Once I started blogging, the world changed for me. I had a new voice. I had a new—and much bigger—audience for my thoughts. I had readers and comments from people on every continent. My universe got a lot bigger. I used to connect with many dozens of people in my early email days. After I started blogging, that number grew to many thousands." He could literally see his influence spreading as conference invitations grew, as he was invited into more newsrooms to address the staff, as his work was linked to by prominent media analysts, and as his ideas started popping up in articles, blog posts, and presentations by others.

The arrival of higher-speed broadband connections at home and at work was a major catalyst to Witt's intensified use of the internet. "Broadband enabled everything that mattered to me," he says. "It cut the friction of my online work and experiences. It also allowed me jump around a lot more. I don't think in a linear fashion. So, the faster the connection, the more I can jump around and the quicker my work gets done."

Witt was talking about both his browsing and his communicating. His daily routines were shaped by his browsing through networked information ("I enjoy traveling down link holes") and his social and professional networking through a variety of communications strategies. Sometimes he called colleagues for consultations; sometimes he emailed listservs seeking advice or offering it to others; and sometimes he broadcasted his thoughts to the growing readership of his blog.

Although Witt is an especially active networked individual, his growing reliance on the internet is quite typical. A crucial factor in the change from internet-as-novelty to internet-as-utility was the rapid switchover from dial-up access to high-speed broadband connections. Less than 5 percent of the adult population had broadband connections at home when the Pew Internet & American Life Project began its studies in March 2000. By the summer of 2011, 61 percent of all American adults had broadband connections at home (figure 3.1). In the process of converting to better connections, many Americans were like Witt: They became different kinds of internet users—more purposeful in their use of the internet, more

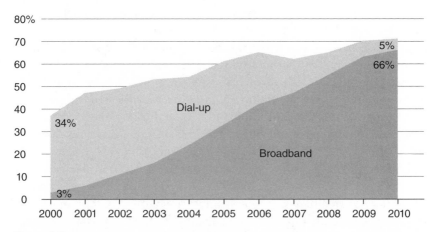

Figure 3.1
Percentage of U.S. adults with home broadband or dial-up access.
Source: Pew Internet & American Life Project surveys 2000–2011.

serious in their pursuits, and at the same time, more playful, more committed to using online tools to accomplish tasks big and small, and more likely to participate in the online commons.[27]

Higher-speed networking has also allowed users to add abundant use of video to their cyber activities. As larger files were able to be transferred more efficiently, peer-to-peer exchanges flourished, allowing people to send each other videos and book-length texts instead of buying them in stores. By mid-2011, 71% of internet users regularly watched video online.[28] More bandwidth and abundant, cheaper storage has allowed cloud applications to multiply: People now can use their computers to access software (such as Word) and their own files on vast server "farms." This represents a shift from the personal computing environment where most personal files are stored on an individual's computer (see chapter 4).

In every dimension probed by Pew Internet, broadband users have become heavier consumers and creators of online material. Their computers *are* the internet now, in the sense that the computer no longer exists as a stand-alone piece of technology. This new reality showed up in survey findings that separately tracked computer users and internet users. That early-1980s phone company survey we discussed earlier in the chapter showed that about a tenth of computer users also exploited networking applications. By 1994, about a quarter of computer users went online. At the time Pew Internet began its surveys in 2000, 76 percent of computer users went online. And by 2011, more than 98 percent of computer users were internet users. Broadband connections have made the network the dominant realm of computing, of pursuing data and media, and of using tools for collaboration and creativity.

Once users converted from dial-up connections and got some experience in this more attractive networked environment, broadband users reported greater satisfaction with the way the internet was serving their needs. In 2002, 31 percent of internet users said it would be "very hard" for them to give up the internet. By 2007 that number had grown to 45 percent. In addition to saying that the internet was harder to give up once they adopted broadband, users acted as if the internet were more important in their daily lives: The frequency of internet use increased dramatically. The percentage of Americans who used the internet on a typical day doubled from 29 percent in 2000 to 60 percent in 2011.[29] The percentage who said they logged on multiple times a day from home grew from 18 percent in 2004 to 38 percent in 2011. The percentage who said they logged on multiple times a day from work doubled from 17 percent in 2004 to 34 percent in 2011. For nearly a third of the population, the internet has become a place simply to hang out—a destination for diversion and entertainment.

When it comes to particular online activities, the story of increasing use of the internet is the same. As part of its regular survey work, Pew Internet asks respondents about their online activity "yesterday"—or the day before they took the survey. Using those "yesterday" findings as an estimate of a "typical day" online shows substantial growth in the people doing many internet activities on any given day:

• The 21 percent of American adults exchanging emails on a typical day in 2000 more than doubled to 53 percent by 2011.
• The 18 percent using a search engine on a typical day in 2002 more than doubled to 43 percent by 2011.
• The 11 percent who got news online in 2000 more than tripled to 41 percent in 2011.
• The 8 percent who checked weather reports online in 2000 more than tripled to 32 percent by 2011.
• The 2 percent who did online banking in 2000 grew more than ten times to 26 percent by 2011.

With the expanding activities and hype surrounding the internet, public expectations about its capacities soared. Asked in a Pew Internet survey in September 2007 what sources of information they used to solve any of several major challenges or problems in life, the internet was the first source on the list, followed by professionals, family and friends, newspapers and magazines, government agencies, television and radio, and libraries.[30] In another research project, Pew Internet and the Pew Research Center's Project for Excellence in Journalism asked Americans in early 2011 about their major sources of local information on a variety of topics and among internet users, the internet was cited as a leading source for material about schools and education, business news, jobs, housing, weather, local politics, arts activities, social services, and community events.[31]

The power of all this newfound connectivity came to a climax for Leonard Witt in 2008, when he was blogging about a new idea he was then calling "Representative Journalism." Ruth Ann Harnisch, president of The Harnisch Foundation, found her way to Witt's ideas through her own internet browsing. Within a year, she had made a $1.5 million pledge to establish the Center for Sustainable Journalism to help advance Witt's ideas. "Without the internet, this simply would never have happened and yet on the internet connections like this both big and small happen every day. Every day I say to myself this can't really be happening, but, of course, it is," Witt says.

The State of Digital Divides

By mid-2011, the Internet and Mobile Revolutions had become main-stream. Pew Internet surveys show that 78 percent of American adults and 95 percent of teenagers use the internet; some 61 percent of Americans are home broadband users and 59 percent are connecting via wireless connections.

Still, there is another, less mainstream side to the stories: Pew Internet's May 2011 survey found that 22 percent of adults *do not* use the internet; 39 percent *do not* use high-speed connections at home; and 41 percent *do not* use mobile connections. Indeed, 17 percent of adults and 25 percent of teens *do not* have mobile phones. Several factors stand out as the significant predictors of *non-use* of the internet: age, socioeconomic factors such as educational attainment and household income, English proficiency, and disability. Rural residents are also statistically significantly less likely than urban or suburban residents to have home broadband.[32]

Over the years, Pew Internet surveys have probed why people did not use the internet or broadband, and the answers have been remarkably consistent even as the non-user population was changing. Roughly half of the non-users cite irrelevance for their nonadoption. They say they did not want or need the internet or see that it would add significant information resources or useful communications channels to their lives. About a fifth say the price of computers and internet connections is a problem for them, and a roughly equal number talk about the hassles they have had or expect to have using computers. And 5 percent to 10 percent say they do not think it is easy to connect where they lived. One surprise in the surveys is that 13 to 21 percent of the non-internet users consistently say that there is an internet connection *in their home* and that other household members go online at home. For one reason or another, these non-users have opted out of online life—sometimes because other family members are more competent and aggressive computer users; sometimes because of a more deliberate choice to avoid the online world.[33]

Even many internet users engage only in a small list of activities: 24 percent of internet users say they have never looked for news online; 42 percent have never done online banking; 35 percent have never checked out social networking sites such as Facebook or MySpace; 34 percent have never made a travel reservation via the internet and a similar proportion have never bought a product online; 46 percent have not hunted for a job via the internet; 47 percent have never looked at a Wikipedia page; 13 percent have never used a search engine; and even 6 percent do not use email—a figure that has not varied much since the earliest surveys of the Internet Project in mid-2000.

Digital differences are evident everywhere among demographic sub-groups. The differences often echo media preferences that were measured in the pre-internet age. Women often are more likely than men to use some of the communications and highly social online activities such as email or social networking sites. Similarly, women are more likely to seek health information online and get spiritual material. Men are more likely than women to get news online, do research related to their jobs, get sports and hobby information, gather material about politics, and perform civic activities. Users under thirty, both male and female, are more likely than their elders to do these things online: use social networking sites, participate on dating sites, play games, seek jobs, create videos to share, and participate in virtual worlds. Older users are more likely than younger users to get news, health information, and government material; buy and sell goods; and do online banking. Some of the most striking differences among racial and ethnic groups relate to communications preferences. Whites are more likely than minorities to use email. African Americans and Latinos are more likely than whites to use all the nonvoice data activities on cell phones such as texting, shooting and sharing pictures and videos, listening to music, and playing games.

As more and more things become networked and so many everyday activities are tied together via the internet, people are finding it harder to be a non-internet user. Even if they never sit down at a computer, their phones, cars, and gadgets have become more connected. Moreover, it is likely that many of the existing access gaps will shrink more. The age gap is shriveling as yesterday's sixty-year-old active internet user becomes tomorrow's seventy-year-old active user. Rural access is increasingly available with advances in wireless and satellite connectivity. At the same time, the difficulties of poor people in gaining internet access could decrease as internet service becomes less expensive and as government and nonprofit groups provide assistance.

While (almost) everyone can use the internet, many cannot use it effectively, lacking what sociologist Eszter Hargittai calls "digital skills" and analyst Howard Rheingold calls "internet literacy" and "net smarts." It's the new digital divide, where differences in skilled use of the internet can worsen social inequalities. At present, income and education levels are associated with internet skills. Yet, those who are more skilled online are more apt to be hired for good jobs. They are also more likely to use the internet to look for political, financial, corporate, and government information. As a result, they are better able to know what sort of information they can find online and where, how to evaluate that information, and how to employ it effectively.[34]

The Culture of the Internet

While the nature of the online world is as varied as people are varied in their moral views, their economic circumstances, and their social structures, there are certain broad elements that have shaped the internet into something that is especially hospitable to networked individuals. Sociologist Manuel Castells has suggested that at least four distinct cultures have shaped the nature of the internet. Although the participants in these four cultures are only a small minority of internet users, they have largely shaped the ethos of the online world and how the internet has afforded networked individualism.

Techno-elites, the first culture, have baked the ethic of open scientific and technological development into the internet's protocols and constantly affirmed a value system that rewards improvements in the technology. They are often innovators or early adopters. Their culture awards privilege to the best programmers who further the goal of building global, interactive communication centered on the open communication of software. Credentials matter less and reputation matters more in this world of incessant peer review. Financial reward is not a marker of privilege; rather it is the authority awarded by the community of computer geeks. "The culture of the internet is rooted in the scholarly tradition of the shared pursuit of science, of reputation by academic excellence, of peer review, and of openness in all research findings, with due credit to the authors of each discovery," Castells writes.[35] He could also have argued that the open architecture the techno-elites have designed enables networked individuals to have an extensive exchange of ideas.

Hackers, the second group, are not the bad actors who try to steal, wreak havoc, or bring down the network. (In common parlance, "hacker" is often used to mean "bad guys," but those are more properly called "crackers.") Hackers, Castells argues, are the programmers who contribute to upgrading the internet through work not tied to corporate or institutional assignments. As innovators, the hacker community is devoted to expanding software that is able to run on all kinds of machines and internet servers. They aspire to reinvent ways for people to communicate using computers and believe that convergence between humans and machines is a good thing that is fostered by unfettered interaction. They believe in innovation without prior permission. The hacker culture's central value has been articulated by Castells as free speech in the computer age, and later had its meaning expanded to become "freedom to create, freedom to appropriate whatever knowledge is available, and freedom to redistribute this knowledge under any form and channel chosen by the hacker." This could serve as the credo of some utopian networked individuals.[36]

Virtual communitarians: If techno-elite and hacker cultures have provided the technical and political foundation of the internet, the virtual communitarians have shaped its social forms, processes, and uses. This cohort, the third of the four cultures, has its roots in the San Francisco Bay area's counterculture. Although there was great variety to the types of communities that began to form online around the world in the 1980s and early 1990s, the early communities shared a commitment to two values. First, they cherished "horizontal, free communication," often standing opposed to culture defined by corporate mass media and large government bureaucracies. Second, they upheld "self-directed networking. That is, the capacity for anyone to find his or her own destination on the Net, and, if not found, to create and post his or her own information, thus inducing a network." In short, their implicit goal was to foster networked individualism.[37] Indeed, "inducing a network" could easily be the T-shirt motto of networked individuals. This cultural endowment from early internet users helps explain why the new technologies found such a ready audience once they began to break into the mainstream. The affordances of the technologies helped them fit beautifully into the lives that users were already leading. As Castells wrote, self-directed networking is a tool for organization, collective action, and the construction of meaning.[38] He could have added that it is also a tool for getting problems solved and emotional needs addressed.

Entrepreneurs, the fourth group, have been the ones who moved the diffusion of the internet into society at large, mostly from their Silicon Valley lairs south of San Francisco. Their chief cultural legacy to the online world has been their reverence for making money—preferably outlandish sums of it—out of ideas about the future. "The foundation of this entrepreneurial culture is the ability to transform technological know-how and business vision into financial value, then to cash some of this value to make the vision a reality somehow," said Castells. The increasing public awareness of what the internet can do for them whets their appetite for more. Tech entrepreneurs have made the commercial build-out of the internet possible. Their approach infuses online culture with the spirit of individualism exercised through looser networks.[39]

The participators are another, distinct culture that is not part of Castells's typology of four. They are the internet users who create and share material online. This widespread category of influential users is evident in Pew Internet's research. While many individuals post substantial material online, roughly one-third of internet users are participators who actively post material that is meant to influence others or be helpful to them. These engaged users include users who compose blogs, upload pictures and videos online, create avatars, and contribute substantial content to social network sites such as Facebook. They belong to online support groups. They critique, rank, and rate everything from books to movies to news personalities. They advocate for political and social causes through their social network profiles and group affiliations. They explain their work or worldly insights in their blogs. They mash up existing media into video parodies, and they chronicle their travels through picture albums on photo-sharing sites. They provide tips and news nuggets about their hobbies or their passions. And they do much more. This distinct sensibility of networked creators and their publication practices are covered in Chapter 8.

The active participators are the multi-ideological offspring of Castells's virtual communitarians. The most active in the participatory class represent the vanguard of networked individuals online. They are creating what William Dutton of the Oxford Internet Institute calls a Fifth Estate in civic life. Recall that the French thought that society was divided into three estates: the clergy, the nobility, and the commoners. A nineteenth-century conceit arose that newspapers and reporters had a distinct civic role and set of sensibilities, constituting a Fourth Estate. Dutton expands on those historic classifications by arguing that the internet is enabling people to network in new ways with other users and with a vast range of information, services, and technical resources. This is creating a new class of civic actors—a Fifth Estate—with distinct sensibilities and interests in pursuing accountability in government, other institutions, and other people.[40] Dutton notes that the internet reconfigures these networked individuals' access to information, people, and other resources, which allows them to "move across, undermine and go beyond the boundaries of existing institutions" to seek and enforce new levels of institutional and personal transparency.

Dutton's British data closely parallel Pew Internet findings in suggesting that a modestly sized core of online activists dominate the participatory class in imprinting the character of the online marketplaces of commerce and ideas. Much of the content they create is inspired by the public disclosures of institutions and news stories about people and organizations. At the same time, much content creation and personal online commentary emerges from ongoing "soft surveillance" of others: watching who has posted what on blogs, Twitter, Quora, and other online forums.

Another new reality is that many internet users are leaving considerable digital footprints, advertently and inadvertently, for others to follow.[41] And follow they do. As we elaborate in chapter 9, they search for information about people they are going to meet, about institutions they are going to engage, and especially about themselves. Not only are other users "creeping" and "stalking" each other, but also governments and large organizations have the capacity to surveil individuals.

The Evolution of Networking

The internet's evolution has been shaped by user innovations, many of which have been pushed along the needs of networked individualism: that is, for engaging their social networks and for gathering information that is personally important. In little more than a decade, the internet moved from being a plaything of computer scientists to becoming an important force in ordinary people's lives. As a hub of communication for networked individuals, the internet has served the needs of people who want to expand their networks and more deeply embed themselves in their existing networks. As a pathway to information and participation, the internet has provided networked individuals with new power to pursue the things that interest them.

The societal trend toward the networked operating system was already underway before the internet became popular, but there is no doubt that that arrival of the internet pushed this trend to new heights and in new directions. The merger of the Internet Revolution with the Mobile Revolution has pushed it even further, and that is the tale we explore next.

4 The Mobile Revolution

Traditional research has not fully captured the changes that mobile phones and wireless computers have introduced to the network operating system.[1] One way to grasp the magnitude of these changes is to remember scenes from the "old days" and how people functioned in the pre-mobile age.

Almost any movie or TV show from before the Mobile Revolution will illustrate what we mean. The 1970 Neil Simon movie, *The Out of Towners*, is a great example of how much the world has changed because its comedy depends on the lack of mobile connectivity. Unable to get to a landline phone, Jack Lemmon and Sandy Dennis *cannot:* hold their hotel room reservation and the room is given to someone else; call ahead and reschedule Lemmon's crucial job interview when they run into trouble; check on a con artist's false story; summon help or seek follow-up assistance when they are mugged twice, kidnapped, and abandoned in Central Park; or let their children know where they are. One telling moment comes when a police sergeant asks, "Where can we be in touch with you?" if the police recover Lemmon's stolen wallet. Lemmon snorts and tells him they cannot be reached because the couple cannot locate a place where a phone is.

"Where can we be in touch with you?" The question seems quaint now. Yet, before the mid-1990s, almost all phones were place-bound. When the Mobile Revolution took hold, that relationship between place and phone became unhooked, and this has changed the way people connect with each other and with information. Think back to some other cinematic plots— and real-world activities—that rarely occur in our mobile world, such as the examples that follow.

Being attacked while alone: most suspense movies hang on the inability of isolated victims to reach out for help. Think of Alfred Hitchcock films, such as *Rear Window* (1954), when wheelchair-bound Jimmy Stewart impotently cannot warn Grace Kelly as he watches a murderer stalk her in the building across the way; or *The Rocky Horror Picture Show* (1975), when the innocents Brad and Janet go to a decrepit house to get help with their flat tire—but are then trapped by aliens. Such movies are no longer credible when mobile phones are like having an extra appendage.

Coping with and documenting disasters and traumas: For instance, hotel guests trapped by terrorists in Mumbai in November 2008 used their mobile phones to access hotel floor plans, find escape routes, and alert friends.[2]

Running frantically to the scene of the action to convey important information: Think of Dustin Hoffman in *The Graduate* (1967) sprinting down the street to stop Kathleen Ross at the altar before she marries the wrong guy. A text message would do the same job these days.

Depending on others being uninformed and out of touch: *Ferris Bueller's Day Off* (1986) is a teenage slacker's dream come to life. In the movie, Matthew Broderick successfully skips school because his parents, priggish sister, and the dean of students cannot contact him—and each other.

Getting away with capers because the good guys cannot coordinate: Pew Internet correspondent Betsy notes: "I can't read the alphabet mysteries anymore, such as *A is for Alibi* [by Sue Grafton, 1982]. I keep on saying, 'Where is your cell phone?'"

Failing to communicate in a timely fashion: Romeo and Juliet each committed suicide because of the lack of timely communication. Romeo killed himself because he thought that Juliet was dead. The letter alerting him to Juliet's special sleeping potion never got to him. If only they had texted. The play would have to be rewritten for the mobile era—but would it be more comedy than tragedy if they died because their batteries ran out of power?

The First Mobile Phones Were Heavy Loads

The earliest public mobile communication in the United States was a comedy—at least to one of the participants. In the telling of Motorola engineer Martin Cooper, the first mobile call took place on April 3, 1973, using a two-pound instrument that had a maximum talk-time of thirty minutes and took a year for the battery to recharge. Cooper's version of events is not fully corroborated by others, though it has not been fully refuted either. He says he was accompanied by reporters on a walk in Manhattan and placed the call in front of reporters as a publicity stunt to a longtime rival at Bell Labs, Joel Engel.

Cooper began: "Guess who this is, you sorry sonofabitch?" Cooper says he could hear Engel whisper to a colleague, "It's him again" and the Bell official then hung up. Cooper continued to roam around mid-town Manhattan with reporters in tow, dialing in to Engel's office every once in a while and asking, "Can you hear me now?"

The calls were especially sweet to Cooper for two reasons. The first is that Bell had developed mobile phone technology, but had little idea how to exploit it. Motorola did. "We desperately wanted to avoid having a Bell monopoly of this new technology," Cooper said in an interview. The second is that Engel had been a longtime tormentor of Cooper going back to their high school days. So Cooper made sure to initiate one of the calls from a men's bathroom. "I wanted a way to get back at him; show him that I wasn't just 'Farty McCooper' as he used to call me. I thought I could teach Engel a lesson or two with a real cellular phone." By 1983, Motorola had created a one-pound phone that sold for $3,500.[3]

It is fitting that these first mobile calls were annoying interruptions, given the ambivalent feelings many people have now about the intrusiveness of mobile phones. These early calls came after nearly a century of breakthroughs in radio communication that started with wireless links among ships. They advanced sharply when transistors became a part of mobile telephony in the 1950s and global standards for wireless digital transmissions were established in the 1980s. Citizens band (CB) radios proliferated in the 1970s, affording short-distance broadcast chats for lonely drivers and speed-trap avoiders.[4] Car phones came into being in 1946. In 1955, the American TV show *Highway Patrol* made these phones famous when Broderick Crawford repeatedly barked out "10-4: over and out" to end a mobile conversation. Yet, these phones used eighty pounds of equipment in the early years and needed operator assistance.

Things took off in less than a decade as better technology emerged: Transistor and battery improvements reduced the size of the phones. Signaling capacity also improved and vastly sped up telephone networks' capacity to transmit calls. Cell towers sprouted up quickly in cities and then suburbs as demand grew. Technology switched in the 1960s and 1970s from rotary dial phones to pushbutton touchtone phones. Eventually phones could become "smart"—with pushbuttons used as inputs to computer applications and the internet. Low-cost text messaging, using the phones' pushbuttons (and eventually keyboards), complemented voice calls for many users by the early 2000s. Digital cameras, using charged-coupled devices (CCDs), became standard phone features, allowing users to take pictures they could share with friends or put on the internet. Increasingly powerful computer chips allowed mobile phones to become smartphones: connecting to the web and hosting a variety of applications such as GPS routing systems.[5]

The World Goes Mobile

Mobile phones have become key affordances for networked individuals as they have become easier to carry, cheaper to use, and able to function in more places. With the proliferation of smartphone applications ("apps"), they have become more than just a phone or a sidekick to computers. Indeed, apps have developed a life of their own and serve users in different ways than personal computers. At the same time, wireless computers have become lighter in weight and easier to use.

The explosive growth of mobile devices—phones and wireless laptops—reflects their routine use. The number of mobile users—and supporting infrastructure (towers, switches)—picked up in the 1990s and accelerated into the early 2000s as prices fell (figures 4.1, 4.2, and 4.3). The number of American subscribers grew from 340,000 in 1985 to more than 302 *million* in 2011, comprising 83 percent of the adult population and 75 percent of teenagers. At the same time, iPhone and BlackBerry became household words. These smartphones were used by 35 percent of adults by mid-2011.[6]

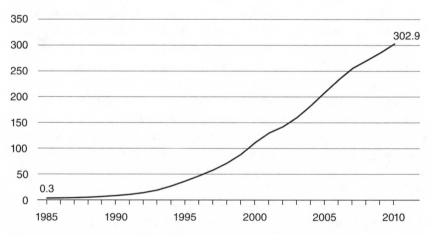

Figure 4.1
U.S. mobile subscriber connections (estimated, in millions).
Source: CTIA.

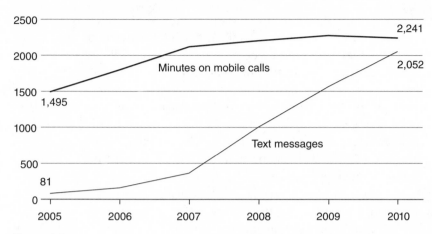

Figure 4.2
U.S. wireless usage: number of minutes and text messages (in billions).
Source: CTIA.

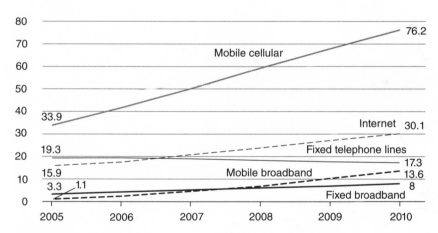

Figure 4.3
Global ICT ownership growth (per 100 world inhabitants).
Source: International Telecommunication Union.

The evidence shows that the value of mobile phones has grown in several ways. By the end of 2010, the number of American households that had no landline phone and whose occupants were "cell only" rose to 30 percent. Another 16 percent of households with both mobile and landline phones receive almost all of their calls on their mobile phones.[7] Most other Americans use both landline and mobile. At the same time, devotion to mobile phone use has grown. In a 2009 survey, nearly half (45 percent) of Canadians said they "can't leave home" without their mobile phone, with 10 percent of Canadians saying they "can't live" without it.[8]

Most demographic groups show heavy adoption of mobile phones, according to Pew Internet data. Yet, there are still statistically significant gaps when it comes to the poor (those living in households earning less than $30,000), those who are age sixty-five and older, and those living in rural areas (table 4.1). By spring of 2011, the great majority of American adults owned mobile phones, but the same demographic groups that had lagged in 2004 still lagged in 2011, although with higher percentages

Table 4.1

Percentage of U.S. Adults Who Own a Mobile Phone

	March 2004	May 2011
All	74%	83%
Men	74	85
Women	73	81
Whites	74	80
Blacks	73	89
Latino*	76	86
Ages 18–29	79	94
Ages 30–49	82	90
Ages 50–64	75	82
Ages 65+	46	55
<$30,000	56	77
$30K–$50K	76	87
$50K–$75K	84	88
$75,000+	94	96
Urban	75	84
Suburban	77	83
Rural	63	75

*2004 figure is for English-speaking Latinos; 2011 figure is for English and Spanish speaking.

Source: Pew Internet & American Life Project surveys.

of users. But the mobile digital divide is decreasing even more quickly than the earlier digital divide in internet use, and users have become adept at typing on tiny keyboards. Teens are showing the way. Since 2005, mobile phone ownership has become mainstream among even young teens. Three-quarters (75 percent) of teens and 94 percent of young adults aged eighteen to twenty-nine own a mobile phone by mid-2011.[9]

The turn to mobile phones is muting the racial and ethnic digital divides that have worried American policymakers since the mid-1990s. While African Americans have consistently been less likely than whites to be wired internet users, they are more likely to use a mobile phone to access the internet. A May 2011 Pew Internet survey found that while 41 percent of white cell owners go online via their phone, some 53 percent of cell-owning blacks do. Mobile internet connectivity reduces the overall internet-use gap between blacks and whites. Latinos are also heavy users of mobile phones to access the internet, similarly reducing their digital divide with white Americans. Moreover, as teens and young adults grow up and displace non-using seniors, mobile phone use is becoming almost universal in North America, ending at least one digital divide.

The numbers of mobile phone users are way up, but numbers themselves don't make a revolution. As communication scholars James Katz and Marc Aakhus point out, the popularity of this technology changes "*apparatgeist*"— the relationship of people to digital technologies and how that changes the way people relate to each other and to larger social institutions.[10] Pew Internet surveys in 2006 and 2007 confirmed this when they found two broad types of mobile technology users.

The first category, those "Motivated by Mobility," have positive and improving attitudes about how mobile access makes them more available to others. Two-fifths (39 percent) of the survey respondents are heavy participants in the Mobile Revolution. They—and their number has grown rapidly since 2007—are the leading edge of the Mobile Revolution. The data show they are a mixture of young adults, road warriors, and teleworkers—people who tend to desire instant information and those in the survey, disproportionately women, who cherish quick communication. Further, they are more likely to be minorities than whites.

The second category, with 61 percent of the adult population, are the "Stationary Media Majority," who do not feel the pull of mobility—or anything else—drawing them further into the digital world. Landline connections are the norm for them, on their phones and computers. Within this category, one-quarter of the survey population (27 percent) are actively involved with the internet, their mobile use mostly limited to basic talking and texting, while one-third (34 percent) hardly ever use mobile devices. They tend to be older, poorer, have less than a university education, and are more likely to live in a rural area. Many are "ambivalent networkers," wrote study author John Horrigan. They "bristle at all their gadget-facilitated connectivity, but don't give it up."[11]

Canadians have also widely adopted mobile phones. The Telus Canadians and Technology survey found in July 2009 that about one-quarter of Canadians, aged thirteen-plus report that their mobile phones are the primary way to keep in touch with friends and family (28 percent) and to organize their social life (22 percent). But the percentages double with the young adult generation, aged eighteen to twenty-four. Nearly half look to mobile phones for contact with friends and family (49 percent) and organizing their social life (44 percent).

Although we focus in this book on North America, the Mobile Revolution is a global phenomenon; mobile connectivity around the world has grown even more explosively than in North America. By 2009, there were more than three billion mobile phones in use and cell towers were probably within reach of 80 percent of the world's population (see figure 4.3).[12]

Three economic factors have made global use expand more rapidly than in North America. First, the cost of fixed landlines has always been higher outside of North America—even in the developed countries of Western Europe, Japan, Australia, and New Zealand.

Second, few countries had the extensive—and expensive—copper wire/fiber optic infrastructures that landlines need. Because cell towers are cheaper to build, especially when the density of mobile users is low, many places in the developing world have leapfrogged their shortage of landlines and plunged right into the Mobile Revolution. People prepay for their calls and share their phones. Some use schemes to put cash into their phones for money transfers and purchases.[13]

Third, mobile phones are crucial in less-developed countries because they are often the first means of telecommunications that people have ever had. While mobile phones increase connectivity to people in the developed world, they provide even greater improvements in connectivity and social capital in the less-developed world. They intensify contact with dispersed family members, expand networks, and enhance sociability and support. They substitute for often-difficult travel, provide price information to marketers, and extend business and family relations.[14]

By 2011, more than three-quarters of the world's mobile phones were in less-developed countries, with China alone having some 879 million subscribers (and even more users of shared phones).[15] "For you, it was incremental—here it is revolutionary," asserts Isaac Nsereko of Africa's largest mobile operator, MTN.[16]

Texting Joins Talking

There was comparatively little public and media fanfare in North America about the increasing adoption of mobile phones until users started doing un-phonelike things on their handheld devices in the late 2000s. The most prominent of those has been texting, also called SMS for "short message service." The first texting schemes were created in the late 1980s as data additions to the emerging mobile phone market.[17] Texting took off when pricing plans in Europe and a decade later in the United States started applying relatively cheap rates to 160-character text messages.

Pew Internet surveys show how texting became a mainstream activity for all types of Americans between the spring of 2006 and the spring of 2011, nearly doubling from 31 percent of the population age eighteen and over to 59 percent. As is the case for mobile phone ownership itself, older, poorer, and rural people text the least. Yet, most demographic groups doubled or tripled their texting in this short period.

Teens are especially networked via texting. A 2011 Pew Internet survey of those ages twelve to seventeen shows that the average teen texter sends and receives fifty texts a day (1,500 per month) and one third handle double that volume—over three thousand per month. About two-thirds of all teens use text messaging, "mostly due to its simplicity as well as the privacy of being able to communicate without being heard," says Amanda Lenhart, the principal author of the Pew Internet report. "If teens are a leader for America, then we are moving to a text-based communication system. For them, there is less interest in talking."[18]

With increased texting, the sheer volume of communication greatly increases, and individuals become more networked. Each new communications medium adds onto people's connectivity. It doesn't fully replace the old media—so that the total amount of communication goes up using a greater variety. Pew Internet surveys show that all forms of mobile communication have overtaken the frequency of other kinds of ICTs (information and communication technologies) and even in-person contact—and on mobile phones texting has overtaken talking as the most frequently used teen communication.

Table 4.2
Percentage of U.S. Adults Who Send or Receive Text Messages

	March 2006	May 2011
All	31%	61%
Men	31	61
Women	29	60
Whites	26	56
Blacks	39	68
Latino*	47	71
Ages 18–29	56	89
Ages 30–49	37	77
Ages 50–64	18	48
Ages 65+	3	13
<$30,000	23	52
$30K–$50K	34	64
$50K–$75K	35	67
$75,000+	42	80
Urban	33	61
Suburban	31	58
Rural	22	47

*2006 figure is for English-speaking Latinos; 2011 figure is for English and Spanish speaking.
Source: Pew Internet & American Life Project surveys.

The data about teens are the most compelling, as table 4.2 shows. When asked about the ways in which they communicate with friends outside school on a daily basis: 54 percent of all those ages twelve to seventeen say they used texting on a daily basis; 38 percent use mobile voice contact daily; 33 percent say face-to-face meetings outside school daily; 30 percent report talking on a landline telephone daily; 25 percent use daily contact through social network sites like Facebook; 24 percent use instant messaging. By contrast to longstanding patterns, email is the least used communication activity, with only 11 percent reporting that they use it on a daily basis.

Teens prefer mobile texting and talking because they can do it privately from their personal phones, and because texting is unobtrusive—it can be done silently while in a class, out with friends, or even at home with parents. Unlike phone chats, texting can be asynchronous: Busy teens can leave messages for each other. More than any other age group, teens need to be both individualists and networked. They want to forge their own identities independently from their parents. Yet, they have real social, instrumental, and nurturance needs for connections to their peers—and also to their parents.

Teens and young adults also use their mobile phones to micro-coordinate their lives. Information scientist Rhonda McEwen found that Toronto undergraduates do not use landlines to call close friends even if it is available. Although three-quarters of those surveyed had access to a landline, four-fifths of them would still call a mobile number even if they knew the recipient was within range of a landline. Yet Toronto teens perform an ambivalent approach-avoidance duet when they meet others. They immediately exchange mobile phone numbers, but they implicitly understand that neither will call the other until the relationship becomes more serious. In general, teens see the mobile phone as an instrument of intimacy. They use Facebook and instant messaging for more distant or newer relationships.[19]

Beyond Talking and Texting: The Smartphone

The evolution of the mobile phone hasn't stopped with texting. In the mid- to late 2000s, there was a convergence of improvements in computing, storage, and radio-spectrum management that made mobile connectivity easier and cheaper. Phones themselves became more versatile as cameras were added and apps were developed. These turned the former two-pound "mobile" calling device into a light, compact multifunctional Swiss Army–style tool, able to communicate, browse, create, and amuse—and to be in touch with social networks in an instant (table 4.3). The social-sharing functions were becoming particularly

Table 4.3
Percentage of Mobile Users Who Use Their Phones for These Activities

	2007	2009	2010	2011
Take picture			76	85
Texting	58	68	72	85
Access internet	19	32	38	51
Record video	18	19	34	40
Play music	17	29	34	39
Email	19	29	34	44
Play game	27	27	34	41

Source: Pew Internet & American Life Project surveys.

important to mobile phone users by mid-2011 as they sent photos and videos to others (74 percent) or posted them online (31 percent), accessed social networking sites (48 percent) and Twitter or other status-updating sites (20 percent), and even made charitable contributions via text (10 percent).

This expanding functionality makes mobile phones useful in new ways. In Toronto, disgruntled transit riders have made a habit of photographing sleeping station agents and bus drivers taking coffee breaks. They actively share their photos with newspapers to force more customer-oriented service. In response, the transit operators set up a counter-Facebook site, "Toronto Transit Operators against Public Harassment," where they post pictures of obnoxious riders.[20]

There is another story to tell in the emergence of mobile apps, first widely introduced by the iPhone but now being built by the many thousands to serve a growing number of smartphones with customized information, games, and other activities. "If the cell phone kept us connected to each other, then the smartphone kept us connected to the world," muses *eWeek* editor Debra Donston. Easy-to-use apps are leading to vastly increased and diversified mobile phone use.[21] The first Pew Internet survey on the subject in the spring of 2010 showed that 35 percent of all U.S. adults—or 43 percent of mobile phone owners—have apps on their phones. Supporting data from the Nielsen Mobile Insights group looked at the subpopulation of those who had downloaded an app in a month. The most popular apps are games (especially puzzle/strategy games, card games, and arcade games), social media websites, maps and directions, and weather reports.[22]

Yet, for all the developers' media excitement about the apps, many mobile users are not fully plugged into this world: The Pew Internet survey found that 11 percent of mobile owners do not know if they have apps on their phone; only 24 percent of Americans actually use the apps on their phones (even though 35 percent say they have apps on their phones); and 18 percent of those who have apps do not know how many apps they have on their phones.

That situation will, of course, change as people become familiar with all the capabilities that are being built into smartphones. Indeed, it was not farfetched for *PC World* writer Jeff Bergolucci to write that multifunctional smartphones will likely eventually kill off several major stand-alone consumer technologies: MP3 music players like iPods, portable game consoles; point-and-shoot cameras; personal video players; voice recorders; portable GPS navigation devices; personal digital assistants; wristwatches;

paper maps; and 411 directory assistance services.[23] The rise of smart-phones and the surrounding apps ecology has prompted spirited debate about whether non-web exchanges that run on the internet but not on the web—such as mobile apps, peer-to-peer services, video exchanges, and downloads—would supplant web applications as the dominant form of media and communication exchanges. *Wired* magazine kicked off the debate with a provocative cover story, "The Web is Dead. Long Live the Internet," laying out a credible scenario where people turn away from the sprawling, browser-based, search-oriented web in their search and content-creation activities toward the more customized world of non-web, mobile apps. The argument about the validity of its thesis rages on through this writing.[24]

Computers Have Become Mobile and Wireless

Do a stimulus-response test: Ask people what their personal computers are for, and they will usually say "the internet." That's not always been so. It's only in the past fifteen years that computer use has become synonymous with internet use. When the Pew Research Center for The People & The Press did its first internet-related survey in 1996, only 19 percent of computer users were also internet users. Most used their computers for stand-alone programs: word processing and spreadsheets. Yet, by the spring of 2011, 98 percent of computer users were internet users. In effect, the internet had become the computer.[25]

Even now, many personal computers are tethered to the internet via cables: reliable and secure. Yet wireless connectivity is now something a majority of Americans enjoy. When the Pew Internet project adds up the number of laptop owners who connect through a wireless card—88 percent of laptop owners—and the number of smartphone owners who connect with their mobile handhelds, the project finds that 63 percent of all Americans are wireless connectors, as of mid-2011. Wireless access has allowed the internet to travel with users, so much so that many Americans use multiple devices to connect to the internet. For instance, 32 percent of Americans said in a mid-2011 Pew Internet survey that they have gone online wirelessly using both their mobile phones and their laptops. Some of coauthor Wellman's students sit with both a laptop and a smartphone at their seats: one to take notes and the other to chat with their friends.

This greater use of mobile connectivity has also encouraged greater internet use. Wireless users are substantially more likely than those who only have broadband landlines to do more internet activities. Among other things, Pew Internet surveys have shown that mobile connectors are 41 percent more likely to be online news consumers than those who only have fixed, wired broadband connections; 64 percent more likely to have done online banking; and 92 percent more likely to have made a charitable donation online.[26]

Living in the Cloud

People can do some things by themselves with a portable computer. Just as in the old days, they can write a document or analyze spreadsheets. And with minimal internet access, they can send emails and instant messages or browse the web. But to do anything more heavy duty, they need access to software and materials that are stored online, in the "cloud." Cloud computing applications became popular in the early 2000s. Still, it wasn't until MySpace and Facebook took off that people started living in the cloud—often without realizing it. The general technology to "push" to people the digital material they might like had existed for many years, but the "killer app" for that function did not arise until social networking sites made this push compelling by allowing it to help users answer the question: "What are my friends doing now?" At the same time, cloud functions have become more compelling with the rise of mobile connectivity because they enable people to have access to their files and business applications wherever they can grab a connected device—or pull out one from their bag. And they can work together, using a shared password to coedit a document or edit an online calendar showing when they are available.

Using the cloud has its risks: Cloud-service companies may disappear; the internet connection can go down; surveillance is easier; cracking (the odious form of hacking) and identity- and data-theft can be more devastating. For example, Gmail has gone down at critical moments for some users. In October 2009, the wireless phone company T-Mobile wrote to its customers that "personal information stored on your [mobile] device—such as contacts, calendar entries, to-do lists or photos—that is no longer on your Sidekick almost certainly has been lost as a result of a server failure at Microsoft/Danger." Microsoft Danger's servers had crashed days before without backup copies of users' items. Another privacy invasion occurred when hackers took over Twitter on December 17, 2009, replacing its content with: "THIS SITE HAS BEEN HACKED BY IRANIAN CYBER ARMY iRANiAN.CYBER.ARMY@GMAIL.COM."[27]

Continuous Access and Hyperconnectivity

The Mobile Revolution has extended the cultural changes that were already underway as the Social Network and Internet Revolutions took hold. A large number of people have emerged who are almost always online or on their mobile phones: available to others, capable of searching for information, and usually able to create online material if they wish. They have built continuous access into their lifestyles and expectations. Additionally, their access nudges them into an internet-first frame of mind, encouraging them to use their smartphones, laptops, or desktops to access the internet when they have a question to research or something to publish—a status update, a picture, a video. This level of connectedness also leads them to prefer to text and chat on mobile phones as they share their stories.

The small size of mobile phones also gives users a sense that their social networks are easily accessible wherever they are: The diminutive device potently symbolizes a network in their pocket. Some 84 percent of cell-owning teens say in a Pew Internet survey that they take their phones to bed with them to make sure they are aware of messages and status updates throughout the night. Others confess that their phone is part of their body. As sociologist Manuel Castells argues: "We now have a wireless skin overlaid on the practices of our lives, so that we are in ourselves and in our networks at the same time. We never quit the networks, and the networks never quit us; this is the real coming of age of the networked society. . . . People can now build their own information systems."[28]

This easy and constant accessibility changes how people relate. For networked individuals, this switch to perpetual access that is untethered from places gives them more control of their outreach to others and their availability to others. This also affects people's sense of time, place, presence, and social connectedness. This, in turn, leads to new notions about when it is possible—and permissible—to be in touch with others. People's expectations about the availability and findability of others have sharply expanded since the Mobile Revolution began. In one poignant example, researchers Scott Campbell and Michael Kelley have shown how alcoholics and their mentors are always on call to each other for moral support and expertise.[29]

For better or worse, mobile hyperconnectivity means that people do not have to walk—or sit—alone. They are *networked* individuals. At times, people use their mobile phones to communicate to onlookers that they have friends and that they are not lonely losers. They may physically be alone, but they are not socially lonely. At times, they even fake it. Some 13 percent of U.S. adult cell owners say they have pretended to be using their phone in order to avoid interacting with other people around them.[30] Others have pretended to be on their phone when they feel endangered and want to ward off trouble. Moreover, as people use their mobile phones to reduce loneliness or kill boredom (as 42 percent of U.S. adults have), they reinforce their existing relationships. This intensification creates a cocoon-like zone of intimacy in which people can continuously maintain their relationships with others who they have already encountered. Thus, mobile phones both liberate and reassure.

Controlling the Volume and Social Interactions

The reality of perpetual connectivity is well suited to networked individuals because it greatly increases their opportunities to network. But what about what sociologists call "work-life balance"? Language scholar Naomi Barron notes that mobile communication—combined with caller ID, voicemail, away messages, and other technologies—allows people to "control the volume" in their social lives. They can turn their phone on and off, screen their calls, or manage others' expectations about their availability. Yet, the same power that they have to regulate the access others have to them means that they need to work harder to gain access to others.

The expectation and reality of perpetual access also creates stresses. Jeremiah, a tech-sector worker interviewed by Pew Internet (who only wanted us to use his first name) described his evolution as a manager of his social relationships. When he first bought a mobile phone in 1997, he was "on on on all the time, and it didn't matter who I bothered or who called me at any hour." He says it was intoxicating to be plugged into his social and work environments "365/52/7/24/1440"—every minute of every day of the year. Then, as more of his colleagues got mobile phones, the number of calls began to rise, and the number of overlapping or back-to-back-to-back interruptions started growing. He recalls: "I finally lost it sometime in 2000 when I got a middle-of-the-night call [in San Francisco] from someone in Asia and I started screaming, 'Don't you know what time it is here?' and the guy replied, 'I thought you were close to my time zone in Singapore.' I realized he didn't know where I was and didn't particularly care."

From that point onward, Jeremiah began to regulate his accessibility. First, he started using email "away" messages to inform others when he was focusing on particular tasks and to reduce the pressure he felt to respond quickly to all emails—both work and personal. After that, he likewise used away messages on his mobile phone to let callers know what he was doing and when he would be able to receive and respond to voice-mail messages. He also created several email accounts to share with close friends and colleagues to allow them different pathways to him that he monitored more frequently. "In the beginning, some of my friends were insulted that I was actually daring to limit their access to me," he explains. "Over time, though, I think they began to face the same time-management hassles and adopted my self-defense techniques. They definitely stopped bitching to me about my strategies to get a little more control over my time."

When Facebook opened up to the general public in 2006, Jeremiah created a profile that was designed to be a "public address system" about his whereabouts and availability. He says that by posting status updates regularly on Facebook, he could announce his "office hours and office-closed" notices to a wide range of friends. Most honored his wishes, especially since he made it clear to his closest pals and most important clients that he was available at all hours in urgent situations.

Jeremiah has seen those around him take even more dramatic steps to try to manage outside contacts. More than a dozen friends have declared "email bankruptcy" by saying they have given up any hope of responding to the hundreds of unanswered emails in their inboxes. Some have started over. One has said he will read select emails in the future, but hardly ever respond. Several have quietly let select acquaintances know that the way to start a conversation is via texting or IM-ing (instant messaging). Jeremiah writes: "Challenges over personal access are universal in my business. We've really gone from the anytime, anywhere ethic to one where you have go through protocols and permissions to get to deal with someone. The access gates have slammed shut." A *New Yorker* cartoon nicely sums up the situation, with one man saying to another, "I used to call people, then I got into emailing, then texting, and now I just ignore everyone."[31]

At the same time, others are organizing their communications based on the context of their contact. People use multiple media to communicate and can choose the one that is most suitable for the moment. If they don't know where the other person is, their first questions usually are: "Where are you? Are you OK to talk? Is there anyone with you?" Discreet text messages are handy: Torontonian Julia Madej exchanges romantic texts with her husband Luke "about 50 times a day."

Ad Hoc Communities Using Mobile Communication

Ad hoc communities are created in an instant, thanks to mobile communications. Technology analyst Howard Rheingold gave birth to the idea that "smart mobs" are a hallmark of this new age.[32] Groups no longer require centralized decision making and top-down information flows to gather information that allows group members to act in a coordinated fashion. This information is now distributed and conveyed by group members contacting each other when they have the urge. The nature of such ad hoc community is well illustrated by an episode communications scholar Rich Ling recounted, about when he came to the aid of a woman who had just fallen and hurt herself:

A woman fell on a stairway and hurt her leg when she was rushing to get her groceries into her apartment while her two-year-old son was asleep in the back seat of the car. Aside from a banged-up leg and the resulting shock, the woman was not otherwise hurt. To confirm this, however, she needed to go to the emergency room. In addition, her son needed to be cared for. Thus, there were a whole series of communications to be made.

When lying on the stairs, before other[s] had even recognized that there was a minor emergency afoot, the woman had incidentally received a call from a friend who had rung for a chat. After being alerted to the situation, this friend was on her way to the apartment but was still a half hour away. After this call I chanced by and was drawn into the situation. I helped her to a more comfortable position, she was able to call another family member to come and help with taking care of her son. This family member was en route to another location at the time, but it was arranged that he could come and get the child. Although it would take approximately an hour. A short-term babysitter was found—my daughter. In addition, another friend was alerted and he was able to meet the woman at the emergency room after he had retrieved his own child from day-care. Finally, a call was made to my wife in order to postpone my picking her up from a shopping trip.[33]

Ling was struck by the efficiency of all the interaction. People were alerted and activities were rearranged on the fly in real time. "Underlying all of this was the assumption that each relevant person had a mobile phone and was accessible via that form of mediation," he wrote. "This assumption has become a part of the logic of a real-time form of coordination." Chapter 8, "Networked Creators," goes into more detail about the power and impact of such networked interaction in more global and civic activities.[34]

The New Choreography of Physical Gatherings

Before the mobile-ization of the world, time and space were critical factors for in-person contact. People needed to specify when and where they would meet. Coordinating a rendezvous, a party or a business meeting was a formal negotiation yielding firm coordinates. Early in the twentieth century, sociologist Georg Simmel pointed out that a similar, large-scale change occurred with the nineteenth century's Industrial Revolution. With the coming of big machines, cities, bureaucracies, stores, and railroad lines running on strict timetables, people had to be at precise places at precise times—or else the machines wouldn't be operated, papers wouldn't be pushed, customers wouldn't be served, and trains wouldn't be boarded. Public clocks—and private wristwatches—regulated the industrialized world. This was a profound change from preindustrial village life, where people went to their farms, shops, or pubs according to their needs—not their clocks.[35]

To some extent, mobile phones allow us a slight return to this more casual negotiation of time. In the age of mobile connectivity, time is more fluid and people's expectations have changed. In the felicitous phrase Ling uses, "hyper-coordination" is now possible and preferred, especially by younger mobile users.

Within a decade, we have come to take mobile connectivity for granted. When you read "Interlude: A Day in a Connected Life," following this chapter, notice how much back-and-forth goes into Maya's getting together with her friend Geri. Rather than people stating precisely where they will be and when, people use their mobile phones as they draw near a gathering, repeatedly reporting their whereabouts and approximate arrival time, and often pointing out landmarks so that those meeting them will be able to place them and even see them as they approach. They understand from the beginning that the initial time and place for the meeting are approximate and changeable. They are more careless about arriving at the proper time and they fuss less about knowing the proper place ahead of time. Sociologist Bernie Hogan calls this "soft time" and "soft location." It is part of networked individuals' shift from place-based connections to person-based connections, with "a flexible lifestyle of instant exchange and constant updates."[36]

Longer Encounters

In the era of perpetual connectivity enabled by mobile communication, social encounters can be prolonged and elaborated. Pew Internet respondent Maxine Clarke gave a good example of this:

Every time I leave someone, I remember things I wanted to say or I have reactions to our discussion that I want to make sure I register. Before I had a cell phone, I would have just let them pass or I would have brought them up the next time I saw the person or had a [landline] call with her. After I got a cell phone three years ago, I realized I could just ring the person right back and we could pick up just where we left off. Sometimes I'll just call someone I've just seen just to say, "That was fun, let's do it again." This is a more spontaneous and human way to be in touch with others. I don't have to make an effort to reach out. I can do it on the spur of the moment.

Information scientist Rhonda McEwen noticed a ritual when Toronto students get together. A mobile call or text message preceded the in-person meeting by a few minutes, and a second call followed after the friends departed. "McEwen reports that first it's: 'I'm just calling to make sure we're on for today' or 'remind me to tell you about whatever.'" The students then recount: "We meet, then afterwards, 'Hey thanks very much for today; that was great.' It's more of a cell phone thing. I'd say I'm more prone to call before, meet and text after, that's sort of my habit."

This ritual sandwiching of mobile chat with the meeting stretches the interaction beyond the physical meeting. The pre-meeting call lowers interaction barriers before the in-person meeting because the participants have something very recent to reference. The call afterward politely ensures that the interaction lingers on via the mobile phone. One student described the pre-meeting call "an appetizer before the main course" and the post-meeting call "the dessert."

The Weakening—But Not the Death—of Distance

Ages ago in internet time—1997—*Economist* writer Frances Cairncross published a book provocatively titled: *The Death of Distance: How the Communications Revolution Is Changing Our Lives*. Her thesis: "New communications technologies are rapidly obliterating distance as a relevant factor in how people conduct their business and personal lives."[37]

More than a decade later, we can see that Cairncross was both right and wrong. Our book presents many examples of people connecting over great distances: at work, in friendship, and even in families. Distance no longer means that communication has (almost) died. For some things, such as online games, distance does not even matter—except when collaborating players get out of sync because they are sleeping in different time zones. For some things, time-zone differences are even beneficial, as when medical secretaries in India enter American doctors' notes during the American nighttime and the Indian daytime.

With communication being personal and mobile, location often is not apparent. Mobile connections can become "places." In some circumstances, people can become more defined by their mobile phone numbers and internet aliases than by where they physically live and work. When graduate student Kris Thomas went to Addis Ababa in 2008 to deliver food to an orphanage, many people gave their mobile phone number as their "address":

We asked our driver for his address, so we could hire him again. He said, "sure, sure," and he took our pen and paper, wrote something and handed it back. It was a cell number, complete with the international calling code prefix. His address was a phone number. That is why so many residents had cell phones. Their place is not tied to a home, an address, a permanent place on earth, but to their phone number. "I am here," but "here" is where you can reach me, via my phone. Our driver couldn't give us any addresses where he could receive mail—neither his home address nor the orphanage's. He seemed surprised that we would want such a thing—why would we need it, when we had his phone number? I got the sense in Addis that maybe the socio-economic divide is not "have a home versus homeless/ shanty living" but instead perhaps "have a phone number and therefore a place, versus do not have a cell phone, truly without a place."[38]

Nor is this only a phenomenon of the developing world. Oxford sociologist Bernie Hogan tweeted on February 4, 2010: "A friend asks for my address & phone. I give him email & cell. It never dawned on me that he meant 'home' address."

Yet, distance still matters in many situations. We show several times in this book that the closer that people live and work to one another, the more contact they have.[39] Moreover, the emergence of location-aware software means that place remains important as long as we think of place in the way that networked individuals do: as the locations where they are at that moment, and where they are heading.

Connected Presence, Absent Presence, and Present Absence

People can initiate multiple social contacts and information searches so rapidly that time is basically "timeless"—what communications scholar

Manuel Castells calls "the space of flows."[40] This is a realm where multiple near-simultaneous communications are possible and can be consummated at any moment—including times when people are standing in line, walking down the street, or driving in their cars. Time sequences need no longer be as distinct as they were when parts of the day had different characteristics: Waking up was followed by breakfast, traveling to a job, work time, lunch break, traveling home, dinner, and evening leisure time. Unplanned traffic jams and waits in doctors' offices are especially empty. Mobile devices can now fill these heretofore useless waiting times with all manner of activity enabled by mobile devices—and the sanctity and separateness of different times of day can easily be interrupted. Mobile networked individuals have more room to maneuver and more opportunities for interaction. Even when not physically together, they have what communication scientists Scott Campbell and Yong Jin Park call a "connected presence."[41] For instance, people can update their friends on aspects of their lives without having to wait for the next time they see each other in person. There is less backlog of information.

One caution is that intensive ICT use means that people can be physically in one place while their social attention and communication focus is elsewhere—a state that social psychologist Kenneth Gergen calls "absent presence."[42] This can create awkward, annoying social discontinuities as people "leave" the group they are physically a part of to take a call or respond to a text message from someone afar. "Distracted driving" has become a policy concern, with states and provinces outlawing holding a mobile phone while driving. Pew Internet surveys have found that 47 percent of U.S. adult texters and 34 percent of texters aged sixteen to seventeen (of driving age) have sent or received texts while driving. Some 49 percent of all adults and 48 percent of all teenagers have been passengers when the driver was sending or reading texts. Finally, 44 percent of adults and 40 percent of teens said they were passengers in cars when the driver put them in danger because of the driver's use of a mobile phone. The plaintive cry of those ignored or abandoned by their "absent present" companions was sounded by Pew respondent Michael Jamison:

I've had a number of arguments with my family about how much I feel disrespected when they check their text messages or crackberries [BlackBerrys] when they are with me. They understand how I feel now and have (mostly) stopped doing it. I worry about my son in college because he and his friends don't seem to ever be totally present in their live interactions with each other. They are all constantly texting others instead of fully engaging in conversation with the people they are with. A key aspect of true friendship is that friends will really listen to each other. I don't know how you can do that if you are texting others at the same time.[43]

Scattered studies suggest problems. For example, distracted drivers do not need to be holding mobile phones to have higher accident rates—it is the act of talking on the phone that is the issue. While walking, mobile phone users are more likely to ignore key things happening around them. Pew Internet found that 17 percent of mobile owners had bumped into another person or object when they were distracted by talking or texting on their phones.[44] In one experiment, more than half of participants did not notice a clown unicycling nearby.[45] Two-thirds of employees want to ban smartphone use at meetings as distracting and impolite. And one women wrote to advice columnist "Dear Abby" complaining about others chatting on mobiles while using public toilets.[46]

The array of both positive and negative feelings that people have about the role of mobile phones was nicely captured in the words of an anonymous Pew Internet respondent in April 2009:

My husband has a heart condition. Last fall he had an episode and ended up in an ER about 45 miles away. I had my phone turned to privacy because I was in a meeting, but after I heard it vibrate for the third time in about ten minutes I knew I should answer it, so I was able to get to my husband on a timely basis instead of finding out after I got home. Yet, my husband's cell phone bugs the hell out of me! He always raises his voice, never wants to let it go to voice mail, and always has it in his pocket. He stops everything to answer his phone, while I do not.

Norms, expectations, and habituation are part of the issue, for people have multitasked while driving for generations. Moreover, Europeans often sit among friends at cafés while simultaneously using their mobile phones to incorporate absent friends into their group conversations. To complement Gergen's notion of "absent presence," we call this "present absence."

The Blurring Boundaries of Public and Private Spaces

The boundaries that used to exist between public realms and private havens are no longer as rigid. People now engage in intimate mobile phone conversations as they stand on sidewalks. Work supervisors now have more ability to interrupt family gatherings. The private is more likely to become public. Several Pew Internet respondents discussed situations where they had confronted individuals who were inflicting their private lives on others in public places. Nikki Waters described how she and several other passengers listened to a woman curse out her boyfriend for several minutes on a BART train and then confronted her with the blunt request: "Lower your voice and move on." Even more revealing about how private matters have colonized public spaces in the age of mobile phones is this story a lawyer recounted about an overheard conversation on a train from Washington, DC to New York City:

> I, along with all of the other passengers, were sitting quietly when the man directly behind me decided to make a phone call using his Bluetooth [wireless ear piece link to his mobile phone]. He was talking so loudly that I think most people in the car were able to hear him. His conversation, though he stressed how necessary it was to be kept secret (ah, the irony), detailed the current plans of Pillsbury [Pillsbury Winthrop Shaw Pittman law firm] to lay off somewhere in the range of 15–20 attorneys from four offices by the end of March, including a few senior associates with low billable hours and two or three first-year associates. I wouldn't have believed it except for the fact that he identified himself to the caller as Bob Robbins, who I learned is the leader of the firm's Corporate & Securities practice section, and that he was talking to Rick Donaldson, who I learned was COO (chief operating office). What's more, he was NAMING NAMES over the phone![47]

While some people do not notice—or care—that they are in public, others are taking steps to preserve some privacy in public spaces. Sociologist Keith Hampton and associates watched people using wireless laptops and mobile phones in public places, such as parks, and semipublic places, such as coffee shops. Some users maintained open glances while they looked at their laptops and mobile phones, inviting conversations; others surrounded themselves with laptops, books, and outerwear as visible barriers to interaction. In figure 4.4, Nelu Handa smiles happily at Ezra's Pound café in Toronto while surrounded by her wireless laptop, two mobile phones ("one is for friends, and the BlackBerry is for business"), an iPod music player (attached to her earbuds), a cup of coffee, a camera, eyeglasses as an aid to visual communication, and anachronistically, a large notebook for writing ideas. Although surrounded by tech gear, she was quite happy to chat with other diners.

Mobile hyperconnectivity in fuzzily bounded public-private space changes individuals' expectations about the availability of other people and the accessibility of information. As personal autonomy grows with new tools, there is a counterpressure for people to stay connected. This is partly driven by *social striving*: Who wants to be out of the loop? A new formulation of that concern in tech circles is "FOMO"—Fear of Missing Out. In addition, the imperative to connect is partly driven by *social needs*: Who wants to miss a call from someone who might offer something useful? It is also partly driven by *social obligation*: Who wants to get a reputation as being a wallflower?

Figure 4.4
Nelu Handa at work at Ezra's Pound Café, Toronto, May 2009.
Source: Barry Wellman. © 2009, used with permission.

The old rules of etiquette and courtesy are reconfiguring in this new environment that enables users to conduct their private business in public places. Evolving mobile etiquette—"metiquette"—injects new realities into social events. When, if ever, is it permissible to interrupt a conversation to accept a mobile call or a text message? When, if ever, is it okay to check email on a mobile device while a meeting is taking place? When, if ever, is it permissible to browse a social network site when a teacher is giving a lecture? When, if ever, can you scream your dismay into the phone while you are waiting in line for the bus? The norms of networked individualism have not caught up to the practice of networked individualism.

The rebalancing of public and private means renegotiating the norms of absent presence. Many people expect to get undivided attention when talking in person. Yet, some of coauthor Wellman's students think it is okay to text while meeting friends in person. They think it rude to have extended phone chats, but a murmured quick call to arrange something is okay. Sociologist Erving Goffman has pointed out that people must practice "civil inattention" in order to get through life in public spaces.[48] As Crocodile Dundee did not realize when he moved from the Australian

Figure 4.5
Laptop users in Bryant Park, Midtown Manhattan, June 21, 2010.
Source: Oren Livio; © Keith N. Hampton 2010, used with permission.

Outback to New York City, big-city people would get overloaded if they paid attention to everything. But what if people are thrusting heretofore private matters into our faces and ears: talking loudly on phones in public or texting incessantly? Look in the second picture (figure 4.5) at how a number of networked individuals mark off private space in the midst of a crowded New York park. Each is alone, but each is connected to an outside world.

As with earlier technologies, societies are still adjusting to what is acceptable behavior while using mobile devices. When someone reaches for her mobile phone during a lunchtime meal to check in with her spouse or colleague, people may not make the snide comment that they would have a few years ago nor may they feel as offended by the action as they would have not long ago. When a mobile phone rings during the toasts at a wedding, some people chuckle and shake their heads, while others glare. Yet, a YouTube video records the gasps at the church when "Louchester" pulled out his BlackBerry at the altar to change his Facebook status to "married."[49]

The Triple Revolution Pushes on: Mobile + Internet + Social Networks

Most North Americans use mobile phones. But the extent to which they are in perpetual contact varies. Many are motivated by mobility to deepen their relationship with digital resources; many are in a holding pattern. Some in the holding pattern are becoming more involved, some will remain steady users, and some are not likely to become more active.

One implication is that there is an *inflection point* that comes when North Americans go beyond using their mobile phones only for talking, chatting, and snapping—and start using their mobile devices to access the internet. This is when the value and impact of mobile connectivity will become most pronounced. Indeed, people now report a growing reliance on mobile devices. For example, Pew Internet found that the mobile phone went from the device that was the fourth "hardest to do without" in 2002 to the number one slot by 2007.[50]

Through developments such as these, mobile connectivity has increased the ability of people to act as networked individuals by giving them more control over how they can reach out to others for information and support, share ideas, create personal networks around similar interests even if the network members live far apart, and switch between portions of their networks. In the process, mobile connectivity has lessened individuals' perceptions of themselves as embedded members of fixed groups.

Mobile connectivity is a social lubricant. The global uptake of the mobile phone is probably the most rapid embrace of a consumer technology in history. Important in itself, it has come together with four other developments to enable widespread mobile connectivity that have profoundly affected behavior: (1) the emergence of lightweight portable computers: laptops, even smaller netbooks and tablets, and smartphones; (2) the rise of wireless connections so that people can connect to the internet wherever they can get a broadband phone or computer signal; (3) the emergence of cloud computing that enables people to store email, documents, and media, and to use social media on remote servers that are accessible from any connected device; and (4) the boom in apps that have turned smartphones into diversified personal and portable computing devices that can access the internet.

We close part I of our book with the hope that we have made clear that the three revolutions intertwine and affect each other in the network operating system. The Mobile and Internet Revolutions are not either/or: They reinforce each another. The always-connected layer of mobile access has enhanced the ascent of broadband and the always-on internet. The way those motivated by mobility use both their wired and wireless access suggests a new era for many users, where the norm is continual access to information and communication. Indeed, as the internet and mobile access converge, we are finding some networked individuals whose smartphones and highly portable computers satisfy all of their needs.

The implications of a significant portion of the population being involved in continual access are only partially understood. Space and time are becoming softer, with people finding their way to each other in due course—to socialize, work, or organize. Location is becoming important—but now mobile apps find people wherever they are. As always, distance is not dead, it is just being renegotiated. Physical presence and absent presence are becoming integrated as the character of public and private spaces changes. We are seeing people spending more time away from their home and office desktops and more time with their mobile appliances. The internet is becoming the mobile internet. Your place is where your connectivity is.

Networked individuals are using both the internet and mobile access to orient their "continuous partial attention" to a variety of social networks and information sources.[51] This mobile-ization strengthens the three pillars of online engagement: connecting with others, satisfying information queries, and sharing content with others. In part II, we describe in more detail how important realms of human activity are being shaped by the Triple Revolution and how networked individualism is playing out in relationships, families, workplaces, and creative and knowledge spaces.

Interlude: A Day in a Connected Life

What are the everyday realities of hyperconnectedness? How do information and communication technologies affect home life and work life? Coauthor Wellman asked his Toronto students to complete time diaries and reflect on these questions. Maya Collum's submission was exceptionally detailed and thoughtful. It also brought together trends that were common in all the papers.

Wednesday, September 19, 2007

Today was an absolutely ordinary day, and, in being an absolutely ordinary day, it was inundated by technology, although I didn't realize to what extent until I reviewed my notes.

8:00 am The day began as any other weekday, with a brief check on what is happening in the world. The process is automated at this point: Check Gmail, check Facebook, check news on the CBC [Canadian Broadcast Corporation] website. Gmail proved promising: no forward or junk, just two emails from political and environmental lists I subscribe to which I promptly delete and never bother to read, and one email from my good friend, Gratia. Gratia lives nearby and we see each other a fair amount. As such, no catch-up is required and this email is right to the point. She wants to know what the plan is for tonight's weekly get together. I respond with an equally to-the-point email. Other discussions can wait until we see each other tonight. . . . I don't have the time or need to post on anyone's Facebook wall.

8:30 am My boyfriend (now husband) Matt comes downstairs and we have our brief morning chitchat before heading to work. The conversation is quick—what are you doing tonight, what is the dinner plan, etc. We leave more in-depth conversations for tonight and we both know we'll discuss any details or questions that arise during the day, on IM (instant messaging).

9:00 am   Log onto IM.

10:30 am   I receive a mass group email from Chris, an acquaintance, regarding his upcoming show. I do not respond as it's almost etiquette not to respond to emails like this.

11:30 am   Cyril, a colleague and friend, pops into my cubicle for a visit. These pop-ins are sometimes too frequent but with Cyril, I never mind. He is my mentor and after our work matter has been settled, we chat about the future of my career at our organization. Cyril is professional and not one to waste too much time; he leaves about fifteen minutes later . . . in perfect time for Kate, a good friend and colleague, to pop in. We have our morning catch-up session featuring a lot of gossip but that's how our interactions seem to flow. She leaves about ten minutes later and our IM conversation begins from her desk.

Noon   Since I don't usually take an hour for my lunch break, I feel justified in taking a short break to putter around the internet. Again, I check Gmail, Facebook, and CBC. Gratia has responded to my morning email on Gmail. I respond. Facebook is a tad more eventful: Kate has posted a group chat about our plans tonight and people are responding. There are a few comments to read; I respond directly to Kate's. The IM conversation with Kate continues and one with Matt begins. IM conversations with these two interweave sporadically throughout the entire day. . . . The conversations are short, to the point, sometimes just goofy remarks or emoticons, but never anything too in depth or that requires too much attention. I also receive an email from my mom but she's asking too many questions to warrant a response right now.

12:30 pm   An IM ping from Alex comes in. Alex is also wondering about plans tonight and has a few questions. I answer them but wonder why he doesn't just read the Facebook thread; perhaps it's blocked by his firewall at work. . . . The conversation with Alex continues for another fifteen minutes, though it's not continuous. A response takes at least three minutes.

1:00 pm   Connor IMs me. He's obviously bored at work because the questions are a bit too demanding. I quickly tell him I'm busy but will ping him later. I must admit, at this point I'm irritated by all of the interaction. I set my IM status to BUSY.

2:30 pm   Another influx of interaction begins but I am aware that I inadvertently ask for it; if I didn't insist on checking my email and Facebook obsessively, and if I didn't leave IM running, this wouldn't happen.

I receive an email from Jen wondering if I'm attending the group event tonight; I respond that I am. Alex pings me again about plans tonight. Kate has responded to my Facebook thread post.

3:00 pm   Kate pings me on IM and we agree to meet for our daily coffee break. It lasts only about fifteen minutes. A few more emails filter in from Jen and Kate about plans tonight and other miscellaneous matters.

3:30 pm   I send Kate an email with a Craigslist posting for our friend Amanda who is apartment hunting. I found this ad yesterday and set a reminder on my eCalendar to email it today.

4:00 pm   A few more emails are exchanged regarding plans tonight. This is getting a bit ridiculous!

5:00 pm   The work day is done and I'm off to school.

5:30 pm   Waiting for class to start, I check my cell phone voicemail. There is one message from Jamie asking to borrow my bike tonight. I respond via text. Why bother spending several minutes returning the call when the entire conversation can be summed up in one text?!

6:30 pm   I exchange four text messages back and forth with Gratia. She's in class as well, and I suspect equally as bored as I am.

8:00 pm   I call Gratia as soon as class lets out because I have changed my mind about how I want to spend the evening. Ironically, after so much time spent planning my evening, all I want to do is go to the local pub for a pint. She agrees.

8:15 pm   After biking home, I find my roommate's entire band, Fjord Rowboat, sitting on our porch. I socialize for a bit and head to meet Gratia at the pub around the corner. A quick text is sent to Kate informing her that neither Gratia nor I will be attending tonight's festivities. I'll admit, I text instead of calling because I won't have to deal with her response if she's not pleased.

10:00 pm   Gratia and I finally chat in person. One beer becomes two and our bartender/friend, Andre, joins us for a quick conversation. I text Matt when Gratia is in the washroom, to ask if he wants to join us. No response.

11:45 pm   Home. I find the Fjord boys still on the porch as though their rehearsal never started. We all have a final beer and I commit to calling it a night around 12:30.

1:00 am Matt and I have our final chat about the day, although it dies quickly as we're both tired. Some ten minutes later I collapse in a heap of exhaustion—good night!

Looking at her day, Maya says that three things strike her. First, she is surprised by "just how much personal communication interweaves into my day." Moreover, many of those communications are covert: "I technically shouldn't be having so much personal communication during the work day, and I shouldn't be having any during class." Still, technology has made all this easy to conceal, and in so doing, her personal communications are not very disruptive to anyone other than Maya. She is carrying her personal network with her and maintaining ties efficiently.

Second, Maya does not feel much guilt about this. "Despite what seems like a massive amount of personal interaction throughout the day, I manage to be extremely productive and get everything accomplished. None of this technology distracts me from the day completely," she wrote. "These personal interactions never receive my full, undivided attention. Face-to-face conversations do, but they are scarce and it is likely for this reason that I prefer non face-to-face mediums like IM, email, Facebook, etc., because I can control the timing and level of attention they receive. Thus, it is not an issue of distraction but rather, the art of multitasking and control."

Third, Maya uses multiple communication media. For example, she interacted with Kate all day: face to face, on IM, through email and Facebook, and when a work matter came up, they used a landline phone. Maya concluded:

It is somewhat striking that I exhaust every possible medium when communicating with individuals. There are certain people whom I contact by using just one means of communication such as Facebook. But, for all of my regular and close contacts however, we communicate via whatever medium is easiest or more appropriate. To a large extent, I can decide who I talk to, when we interact, and how. I can decide to what extent these interactions affect or interrupt my day. There's an immense amount of freedom in this.

Maya—like many young adults—is *hyperconnected*, using many media for (almost) continuous communication presence. They—and their friends —choose the medium that is most suitable for their relationship with the other person, their social context (work, at home, the street), and the nature of the message—it's less confrontational but tacky to break up by email or texting, for example. Often, a conversation started with one medium will switch to another, as when Maya's conversation with Kate moves among email, Facebook posts, and texting on IM. For better or worse, mobile hyperconnectivity means that people never walk—or sit —alone.

II How Networked Individualism Works

5   Networked Relationships

Alarm spread in June 2006 when Miller McPherson, Lynn Smith-Lovin, and Mathew Brashears published "Social Isolation in America" in the *American Sociological Review*.[1] In this leading journal, the three scholars reported findings from the General Social Survey—the gold standard of American surveys—to the question: "Looking back over the last six months—who are the people with whom you discussed matters important to you?" Comparing Americans' answers in 2005 to answers in 1984, they found that the number of people with whom Americans reported discussing important matters had declined by 28 percent, from 2.9 to 2.1. Moreover, nearly one-quarter (23 percent) of Americans said they did not have any confidants with whom they could discuss important matters—not even their spouses. The nature of their confidants had also changed. There were fewer friends and neighbors in 2005 than in 1984 and more immediate kin and spouses. For example, the percentage of Americans with a friend as a confidant declined from three-quarters (73 percent) in 1984 to one-half (51 percent) in 2005.[2]

These depressing results raised an alarm that Americans had become more isolated. Although the researchers did not show that the internet was the cause of social isolation, the media speculated about this. *Toronto Globe and Mail* columnist Douglas Cornish sounded a common refrain when he wondered: "Will this glow [from the internet] produce a closed generation of socially challenged individuals, humans who are more comfortable with machines than anything else?"[3]

Anxieties about the withering of relationships are not new, but began many centuries before the coming of the internet. Every epoch experiences them. In past decades, they were tied to industrialization, bureaucratization, urbanization, socialism, and capitalism. Often, these alarms have been tied to the rise of technologies that connect people in new ways: from grumbling about nineteenth-century railroads spooking horses to more recent complaints about cars and telephones isolating people from in-person contact.[4]

The alarm is repetitive: Something is happening "now" to rend apart the supposedly supportive, fulfilling bonds of olden days—although in every generation the alarmists keep looking back approvingly to the previous generation. For example, in the now supposedly communal 1950s and 1960s, commentators were moaning that things were falling apart compared with the old days. They came up with a number of memes for it, such as "the lonely crowd," "mass society," and "the quest for community."[5] For example, here is Maurice Stein in *The Eclipse of Community*: "The old feeling of solidarity based on a sense that everyone in town belongs to common community gives way to sub-communities with hostile attitudes toward each other." He continues: "Community ties become increasingly dispensable, finally extending even into the nuclear family, and we are forced to watch children dispensing with their parents at an even earlier age in suburbia."[6]

Although such critics wrote before the proliferation of the internet, it has now became the scapegoat. The basic argument is that community is falling apart because internet use has led people to lose contact with authentic in-person relationships as they become ensnared online in weak simulacra of reality. As early as 1995, Texas radio commentator Jim Hightower warned, "While all this razzle-dazzle connects us electronically, it disconnects us from each other, having us 'interfacing' more with computers and TV screens than looking in the face of our fellow human beings."[7]

Social psychologist Robert Kraut and associates added to the unease in 1998 when major newspapers publicized his finding that newcomers to computing had decreased social involvement and psychological well-being. To their credit, Kraut and associates retracted their initial findings in 2002, when they found that as the newcomers became computing veterans, their negative symptoms disappeared. However, this got less media attention.[8]

The internet was also the force underlying social decay in William Gibson's science fiction novel *Neuromancer*, which portrayed people losing their real-world personas by "jacking in" to "cyberspace" (the latter being a word that Gibson coined for the novel).[9] More recently, social scientist Sherry Turkle has argued that people create separate selves as they immerse themselves in cyberspace and forget the real world. "People can get lost in virtual worlds," she warned in her 1996 *Wired* magazine article. Her 2011 book *Alone Together* continues the thread, bringing in a new techno-fear as added cause for alarm: connections with robots supplanting human interaction.[10] She also raised concerns about people being more preoccupied with the connections they make through mobile phones than with the real people who are standing mere inches away.

After the McPherson, Smith-Lovin, and Brashears article and ensuing commentary about technology's suspected baleful impact, network scholar Keith Hampton joined with Pew Internet to investigate how technology might be tied to social isolation and declining discussion networks. The resulting work showed the opposite: People who use ICTs (information and communication technologies) have larger and more diverse networks than others.[11] On average, a Pew Internet study showed, the size of people's discussion networks—those with whom they discuss important matters—is 12 percent larger among mobile phone users, 9 percent larger for individuals who share photos online, and 9 percent bigger for those who use instant messaging. The diversity of people's core networks—their closest and most significant confidants—tends to be 25 percent larger for mobile phone users, 15 percent larger for occasional internet users, and even larger for frequent internet users.

Contrary to some pundits' fears that the internet was drawing people away from local communities, Pew Internet research found that most internet activities have little relationship or a positive one to local activity. For instance, internet users are as likely as anyone else to visit with their neighbors in person. Mobile phone users, those who use the internet frequently at work, and bloggers are more likely to belong to a local volunteer association, such as a youth group or a charitable organization. Internet use does not pull people away from public places, but rather is associated with frequent visits to places such as parks, cafés, and restaurants—the kinds of locales where people are likely to encounter a wider array of people and diverse points of view.

Why do many commentators suspect that ICTs cause social woes? There are multiple traps in the notion that the internet is a separate, immersive medium:

• It assumes that people lead different "virtual" lives, distinct from their everyday real-world lives. As we showed in part I, this rarely is the case. With the partial exception of the intense gamers that Turkle has studied, online and in-person interactions—and lives—are intertwined.

• It assumes that in-person encounters are the only meaningful form of social connection, and it does not recognize that emails, text messages, Facebook posts, tweets, and the like are everyday tools that people routinely use to stay connected.

• It asserts the internet's limited capability for transmitting social cues such as facial expressions, smells, and body gestures. Internet encounters contain "less" social information and communication, and that might cause relationships to atrophy. Yet, people rarely interact with strangers over the internet. They have a strong sense of the others with whom they are online and internet encounters complement and increase the volume of communication among people, rather than substituting for richer in-person contact.

• It takes Marshall McLuhan's aphorism too seriously and *confuses the medium with the message*. In reality, people are not confusing the Facebook screen with the person at the other end of it, just as they have not confused the telephone receiver with the person with whom they were talking. Another McLuhan phrase seems more accurate: The media are "extensions of man" (in other words, people). When we send email to our spouse or look at a friend's Facebook updates, we do so with a strong understanding of the person with whom we are communicating.[12]

A large part of contemporary unease with technology stems from selective perception of the past and the superficial observation of other individuals. Many people think they are witnessing loneliness when they observe people walking or driving by themselves—not realizing they may be going to meet friends. They echo the Beatles: "All of the lonely people. Where do they all come from?"[13]

Yet, while people do not often open the door to strangers, they do drive, fly, and make internet phone calls over long distances to help their friends and relatives. People glance at Nelu Handa (chapter 4, figure 4.4) sitting by herself at her laptop and immersed in her iPhone chats and music, without realizing that she can also be interacting intensely with friends on the internet and the phone, as well as be available for in-person contact.

By contrast, tech enthusiasts have been excited about the positive possibilities of the internet for sociability. Their view has been that the internet would foster an enormous increase in cooperation by allowing far-flung people to interact. Rather than alienation and isolation, there would be more relationships, more long-distance relationships, and more connections among the members of a person's network. In the mid-1990s, John Perry Barlow was a leading enthusiast. The co-founder of the Electronic Frontier Foundation vividly prophesied that the Internet Revolution would bring about radical and positive social transformation: "With the development of the internet, and with the increasing pervasiveness of communication between networked computers, we are in the middle of the most transforming technological event since the capture of fire."[14]

Both sides of the debate—doomsters and enthusiasts—have been so excited by the internet that they can be too *presentist* and *parochial*: presentist, because they have rarely looked back to see if people had ever worried about relationships before the internet arose; parochial, because they have assumed that the internet's very existence would radically affect relationships. Social scientists call this sort of thinking "technological determinism," because it does not take into account how the use of ICTs is socially embedded and socially determined. This ignorance of context is why both the yeasayers and the naysayers have gone astray.

Their fixation on the internet has ignored nearly a century of research showing that technological changes before the internet—planes, trains, telephones, telegraphs, and cars—neither destroyed relationships and communities nor left them alone as remnants locked up in rural and urban villages. Fifty years of research have shown that people are in sizeable and supportive networks, both local and long-distance.[15] When asked, few people say that they, themselves, are living lives of lonely desperation, and they are aware that most of their friends, neighbors, relatives, and coworkers are also in supportive networks. Yet, even with these realizations, some people—and commentators—believe that they are the exceptions and that the masses around them are lonely, isolated, and fearful.

There is no reason to panic. The alarm that McPherson and associates sounded came from survey responses to only one narrow question. Looked at more broadly, a large body of evidence has shown that relationships and community and civic engagement thrive in social networks and that they are aided by the internet and mobile community. Take Robert Putnam's well-known book *Bowling Alone*, based on evidence from the middle to the end of the twentieth century. It argues that key reasons why involvement declined in community organizations such as bowling leagues is that people stayed home to watch television and many more women were doing paid work outside of their homes. But Putnam's own account shows that people are not bowling *alone*—despite the book's title—but in fact are bowling in networks of shifting sets of others who happen to be free that week.[16]

Research by Pew Internet, Toronto's NetLab, and others provides much evidence that that people have large and helpful networks. While the Internet and Mobile Revolutions have affected the nature of communities, they have transformed but not destroyed them for networked individuals in the networked operating system.

From Door-to-Door to Place-to-Place Networks

It helps to think about communities as fluid personal networks, rather than as static neighborhood or family groups. For too long, the model of community has been the preindustrial village where people walked door to door, and all knew, supported, and surveilled one another. These bygone village groups have largely transmuted into multiple, fragmented *personal networks* connected by the individuals and households at their centers. Figure 5.1 shows a typical network of close ties. For example, Wellman's early research found in 1968 that neighbors made up only 13 percent of Torontonians' core networks. Research elsewhere in North America confirmed this in Detroit, Los Angeles, and northern California. People find support and

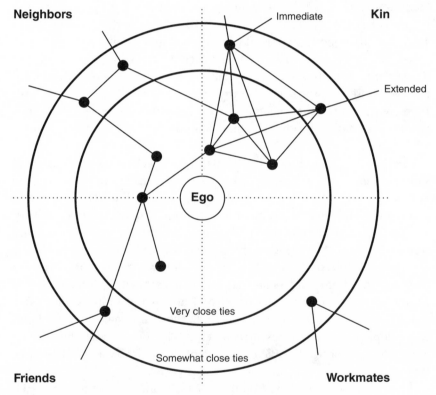

Figure 5.1
Typical personal network of close ties.
Note: Ego's ties to every network member omitted to reduce clutter.
Source: Barry Wellman. © 2004, used with permission.

sociability, but mostly with people who live outside of their neighborhoods and as often with friends as well as with kin. Rather than having a few go-to persons who provide a wide range of support.[17]

Although the move away from village groups did not happen instantly, it did happen after World War II, but before the Internet and Mobile Revolutions. The widespread abundance of cars, phones, and plane travel made "glocalization" possible (global + local connections). Social networks remained anchored in households, yet people often traveled substantial distances to get together with friends and relatives. Although neighboring remained, personal communities extended far beyond them. Wellman's awakening insight on this came when he was part of a "Save our Neighbourhood" meeting, intent on stopping the Spadina Expressway from knifing through downtown Toronto. The group was just like groups in other cities, fighting to preserve neighborhoods against cars. But as he looked around the room, he realized that many of that neighborhood's saviors did not even live there. They were not a little group of neighbors at all—they were a network of community activists who had come from all over Toronto.

Wellman's long-running research in Toronto has shown that although people continue to befriend neighbors, they have less connection with their neighborhoods than in preindustrial door-to-door times. Until the Mobile Revolution, phone calls came in by landlines to households— rather than wirelessly by mobile phones to specific people. Consequently, many interactions moved inside private homes—where much entertaining, phone calling, and internetting take place. At the same time, longer-distance connections proliferated. Both Wellman's first (1968) and second (1979) studies in the East York area of Toronto found that few strong ties were with neighbors. The more voluntary phone calls were stronger predictors of social closeness and support than in-person contact with neighbors and coworkers who might not have voluntarily chosen their relationships.[18] As such, people became connected *place to place*. They are aware of local contexts that they physically inhabit—especially home, work, bars, coffee shops, and airports—but they rarely know about the places in between them.

From Place-to-Place to Person-to-Person Networks

The personalized and mobile connectivity enhanced by the Triple Revolution and the weakening of group boundaries have helped relationships move from place-to-place networks to individualized person-to-person networks. Most have private internet connections and personal mobile phones, and their own cars. Lower numbers of children mean parents need to spend less time at home raising them. There are fewer children to keep parents housebound. The loosening of religious, occupational, and ethnic boundaries also encourages interpersonal free agentry.

Rather than ties between households or work groups, people connect as individuals to other individuals, in person-to-person networks. They maneuver through multiple sets of ties that shift in importance and contact by the day. Each person engages in multiple roles at home, with friends and relatives, and at work or school. Their networks are sparsely knit, with friends and relatives often loosely linked with each other. These loose linkages do not imply a complete untethering of social relations: There are only a few isolates "bowling alone." Most people are connecting in shifting networks rather than in solidary groups.[19] Such networks provide diversity, choice, and maneuverability at the probable cost of overall cohesion and long-term trust.

While place-to-place networks show how community has transcended local boundaries, person-to-person networks show how community has transcended group boundaries. It is the individual—and not the household, kinship group, or work group—that is the primary unit of connectivity. The shift puts people at the center of personal networks that can supply them with support, sociability, information, and a sense of belonging. People connect in person and via ICTs. Their networking activities shift as their needs shift. While network members relate to each other as persons, they often emphasize certain roles. They are bosses to their employees, husbands to their wives, friends to their friends, and so on—with somewhat different norms for each network.

Networked individualism means that people's involvement in multiple networks often limits their involvement in and commitment to any one network. It is not as if they are going to the village square every day to see the same crowd. Because people can maneuver among milieus, their multiple involvements decrease the control that each milieu has over their behavior. Yet limited involvements work both ways. If a person is only partially involved in a milieu, then the participants in that milieu often are not as committed to maintaining that person's well-being. Like corporations that segregate their activities into somewhat autonomous units, people are now in communities of "limited liability," to use the British legal term.[20]

The shift to person-to-person networks has profoundly affected how people relate. This is not a shift toward social isolation, but toward flexible autonomy. People have more freedom to tailor their interactions. They have increased opportunities about where—and with whom—to connect. As people maneuver through their days, lives, and networks, the nature of their ties varies from situation to situation. That means people are more selective about the people with whom they relate, because they no longer can be open to "the community." In the old days, people reportedly kept their outside doors unlocked and picked up their phones as soon as they rang. By contrast, a recent study showed that many Chicago homes, for example, are "islands of privacy." People practice selective concealment and disclosure. They don't open their doors readily—to avoid salespeople and religious proselytizers—and they use caller ID and voicemail to avoid phone contact with telemarketers, politicians, and others. Email is easily screened by software to remove most spam before viewing, and invitation-only Facebook offers preselected contacts.[21]

Norms are developing around these new social spaces. For instance, some teachers are now being encouraged not to become Facebook "friends" with their students. Moreover, Facebook and Twitter users control what information they disclose online. For example, neither Rainie nor Wellman discuss much of their personal lives on Twitter. Others provide code words to mask sensitive content, just as "partying" can mean sexual relations among teens. So far, texting and other mobile phone calls have been less of a problem because there are no public directories of their numbers.

Most people do not limit themselves to participation with just one or two groups. They gain advantages by having a diversified set of networks and knowing who has what to offer. That creates powerful social capital. For example, NetLab's Connected Lives research in the Toronto area of East York has found that people are apt to get hugs from their sisters, money from their parents, and sociability from their friends.[22]

Living in person-to-person networks has profound implications both for individuals and for the social milieus and overall societies that they are in. Networked individualism downloads the responsibility—and the burden—of maintaining personal networks on the individual. Networked individuals often have time binds, since they are constantly negotiating plans with disconnected sets of individuals within their expanding network. Active networking is more important than going along with the group. Acquiring resources depends substantially on personal skill, individual motivation, and maintaining the right connections.

What about our "self": that elusive concept of subjective identity that helps us to integrate our involvement in multiple social networks?[23] Are we the same person in different milieus, both online and offline? Sherry

Turkle has argued that our "second selves" online are different from our selves offline. Yet the research we present throughout this book shows that people's online and offline interactions are almost always integrated. However, Turkle rightly calls attention to the need for more research into how different aspects of the self get emphasized in different situations.[24]

We suggest it is useful to think of a *networked self:* a single self that gets reconfigured in different situations as people reach out, connect, and emphasize different aspects of themselves. Our working visual image of this is an amoeba, with both a core nucleus and constantly changing pseudopods.[25] While a small number of scholars have used a concept similar to the networked self, there has been little systematic research—or even theorizing. The most relevant discussion is conducted by Jay David Bolter and Richard Grusin, who talk about a networked self switching among a variety of media to make their social networks perform well. They point out that people are "constantly making and breaking connections, declaring allegiances and interests and then renouncing them—participating in a video conference while sorting through email or word processing at the same time."[26] However, they anchor the concept in communication media rather than in multiple roles in social networks, as we do.

Networked Relationships On- and Offline

With the shift to person-to-person networks, the gap between physical space and cyberspace—or for that matter, between writing and talking—is diminishing. For instance, a Pew Internet study found that American teens usually think of their texting as "conversations" rather than as "writing."[27] Teens are even more text-involved, checking for multiple Facebook updates and text messages from their "friends," who in fact range from close friends to distant acquaintances. Expressions such as "see you later" or references to conversations such as "she told me that" could as easily refer to in-person encounters, emails, tweets, texts, or Facebook postings. Technology-enabled interaction fits seamlessly into people's everyday lives and complements other practices.

When people think of the impact of the Internet and Mobile Revolutions on relationships and community, two contrasting images often come to mind. One is that of a world without borders and an endless amount of friendships and knowledge at people's fingertips—Marshall McLuhan's mythological global village come to life.[28] The contrasting image is of a lonely individual, hunched over a computer or smartphone screen, avoiding all human interaction. These two extreme examples are at odds, and the ambivalence has also been reflected in papal pronouncements. In June 2011, Pope Benedict XVI lauded the power and value of ICTs for spreading information, but warned that people need to get away from their computers and meet people in person:

The new technologies allow people to meet each other beyond the confines of space and of their own culture, creating in this way an entirely new world of potential friendships. This is a great opportunity, but it also requires greater attention to and awareness of possible risks. Who is my "neighbor" in this new world? Does the danger exist that we may be less present to those whom we encounter in our everyday life? Is there is a risk of being more distracted because our attention is fragmented and absorbed in a world "other" than the one in which we live? Do we have time to reflect critically on our choices and to foster human relationships which are truly deep and lasting? It is important always to remember that virtual contact cannot and must not take the place of direct human contact with people at every level of our lives.[29]

The Pope also tweets occasionally as PopeBenedictXVI.

It is appropriate that the pope recognized the importance of the Internet and Mobile Revolutions because in reality, people are positively embracing them. In July 2009, the Telus Canadians and Technology national survey found that more than half (55 percent) of Canadians aged thirteen and older agree that "the internet has improved my connections with friends and family." Only 15 percent disagree: a ratio of almost four to one. Moreover, 46 percent of the Canadians said, "the internet has improved the quality of my life": a ratio of nearly three to one. Almost as many (42 percent) go so far as to say, "I cannot live without access to the internet." Yet, the internet has not taken over completely, for only a minority say they spend more time interacting with friends and family online than in person.

Contrary to concerns that the internet would reduce other forms of contact, the evidence shows the opposite: the more internet contact, the more in-person and phone contact. These are not either/or relationships: People use the internet and mobile phones to keep in touch, to arrange get-togethers, and to follow up after they meet. Despite fears that the internet would curb relationships by luring people to the screen and away from in-person contact, the number of important relationships may even have grown. One survey found that Twitter users are more involved in social activities.[30] More broadly, the average number of friends whom American adults see in person at least weekly grew 20 percent in five years: from 9.4 in 2002 to 11.3 in 2007. Moreover, this does not include relatives unless the respondents consider them to be "friends." The same study shows that internet users have somewhat larger networks than non-users. Moreover, heavy internet users have had the biggest increase in their number of friends: a 38 percent average increase from 9.0 in 2002 to 12.4 in 2007 (figure 5.2). Similarly, a Pew Internet study found in 2004 that internet users have had 23 percent more active network members than non-users.

In short, being on the internet is associated with having both more friends and a greater increase in the number of friends over time. The number of friends has increased even for non-users, although not nearly as much. That non-users has increased their friendship contacts suggests two possibilities: The use of the word "friend" may have broadened between 2002 and 2007 as MySpace and then Facebook became popular, or the halo effect of the internet has created more opportunities for friendship because most of the friends of non-users undoubtedly are internet users.[31]

ICTs are about society as well as relationships. They support participation in traditional settings such as neighborhoods, voluntary groups, churches, and public spaces. They also support involvement in interest groups, whose membership might have been too small or spatially dispersed in pre-ICT days, to find one another and to get together in person. For example, communication scientist Nancy Baym has shown how the internet allows lovers of obscure indie bands to find each other online and becoming acquainted offline. Like rock parties, significant political

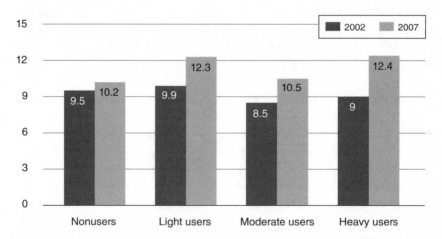

Figure 5.2
Change in average number of offline friends seen in person at least once per week. *Source*: Wang and Wellman, 2010 (see note 31).

organization begins on the internet, organizes via mobile phones, and then meets in person.[32]

As a result, North Americans are in more contact with the members of their social networks than ever before. For example, the Pew Internet's "The Strength of Internet Ties" study found that people who email the great majority of their core ties at least weekly are also in phone contact with more core ties than are non-emailers. Many people use the internet to keep up with their weaker ties. Computer science graduate student Sarita Yardi explains:

I use the [I]nternet for two reasons: First, to keep up with my family. I have 18 cousins, and most are married. Most have kids too and will often post pics. I've become closer—also in real world interactions—than I would otherwise be with all of them.

Second, I keep up with researchers in my community. For example, at the confer-ence I'm at, I see when people arrive, where they are going tonight, who wants to grab dinner, etc. Facebook is a little more manageable on a large scale than Twitter. One of the best benefits has been to see their work-life balance (most of them have a reasonable balance), and I see a mix of statuses and pics about kids, awards, travel, rants, updates about research, and it makes me confident that it is possible to do all that too.[33]

The more personal kinds of ICTs often intensify close relationships. Connected Lives participant Vamos values the personal autonomy he gets from using email. "If a friend sends me an email, I can respond—not immediately," he explains. "If I have something to do, I can say okay, I can send him an email after tomorrow when I have more time. Maybe [if he phoned] he can't understand that you can't speak with him for one hour, two hours. That's simpler on email."

Until recently, younger adults have been the most involved in the Internet and Mobile Revolutions. As Toronto student Nazia Shahrin recounts, "I find my mother and father value face-to-face communication a lot more than I do. To me, a phone call is good enough, while they really need to see my face. It creates a lot of arguments where I am screaming, 'I talk to you every day' and they are yelling, 'But I haven't seen you in two weeks.'"

Despite the ubiquity of the internet, the Center for the Digital Future's 2007 survey of Americans found that that only 23 percent of internet users have one or more "virtual friends" whom they have only met online. To be sure, the more people use the internet, the more virtual friends they are likely to have. Among those who have virtual friendships, heavy users (who use the internet at least three hours per day) report having an average of 8.7 online friends compared with only 1.3 for light internet users (online an hour or less per day). Moreover, just as in-person relations lead to more online contact, 20 percent of Americans have at least one relationship that started online migrate to in-person contact. Here, too, heavy internet users have more migrating friends (an average of 2.2) than do light users (0.5).[34]

While only a small percentage of people are heavily involved in virtual friendships, to some they are important—even consuming. Many of them are immersed in massively multiplayer online role-playing games (MMORPGs) that embrace thousands of players simultaneously, loosely organized as networked clans. But even in these, virtual friendships tend to "decay or grow inert without interaction," reports anthropologist Bonnie Nardi in her study of the World of Warcraft MMORPG. For example, there is no real group pressure to show up for clan activity, and people can switch clans easily. The games lack the rich ways that in-person relationships have to maintain connections.[35]

Still, neighbors and local concerns matter in both online and offline encounters. Communications scholar Keith Hampton spent considerable time looking at how people connect with neighbors online and offline. In the late 1990s, he and coauthor Wellman studied the pioneering "wired suburb" of "Netville" near Toronto, comparing residents who used the internet with those who did not. They found that as compared with non-internet users, internet-using neighbors had larger and wider-ranging local networks that socialized more with each other.[36] Further reflection suggests that the more active internet use resulted from the suburb setting up a local listserv that encouraged such interactions. Moreover, as settlers in a newly built suburb, the residents became part of the larger network of information—for example, where the dry cleaners were, who would baby sit, and efforts to press the area's developer to fix sinking driveways and leaky plumbing. The email list served to facilitate the flow of information regardless of physical proximity and according to the users' convenience. When such incentives for local internet connectivity are not present, neighbors interact less intensively. To help build local community, Hampton created a set of internet-based eNeighbors.org and iNeighbors .org sites across America to aid local connectivity.[37]

Despite the distance spanning of the internet, people are still much more apt to have friends, coworkers, and schoolmates who live a short walk or drive away, they use the internet and mobile phones between in-person encounters to share information, coordinate contact, provide support, and just socialize. In-person contact predominates in all neighborly interaction, but the amount of such contact may be declining. The Pew Internet "Neighbors Online" study found that while 46 percent of Americans talk face to face with their neighbors about community issues, only 21 percent discuss such issues over the phone. Even less—11 percent—read a blog about neighborhood issues, a mere 5 percent belong to a neighborhood listserv (such as Netville had), and only 9 percent have exchanged emails about neighborhood issues.[38] So, proximity matters to networked individuals, but for most, the neighborhood is not where their community lives are focused.

How Large Are Personal Networks?

The high level of friendship activity online and offline suggests that worries have been overstated that Americans have only an average of 2.1 close ties. Yet, the research on declining networks is based on a single question about people "discussing important matters" with others. But, that is only one kind of relationship in Americans' much larger core networks.

How large are people's personal networks? One widely known estimate by Oxford anthropologist Robin Dunbar argues that limits on people's cognitive information-processing capacity—what he calls their "social brain"—limits the maximum size of cohesive groups to 150. He bases his estimates principally on his studies of primates and villagers in less-developed societies and structured military organizations. Yet, as Dunbar himself points out, "The 150, as we understand it, is simply one of a series of layers of embedded relationships, and this seems to apply as much in the contemporary world as the ethnographic world."[39] The outer most layer, Dunbar explains, "demarcates those whom you know as individuals from those whom you recognize but only have casual relationships with." A social network "consists of four layers, the Circles of Acquaintanceship, which scale relative to each other by a factor of three (an inner core of five intimates, and then successive layers at 15, 50 and 150)."[40]

Does it matter if a personal network contains 150 or 1,000 people when most of these are undoubtedly weak ties—nodding acquaintances or people rarely in contact? The answer is "Yes" for many reasons. For example, the developers of social media want to know how much space to allocate for information about friends. They have eagerly seized upon what they call "Dunbar's number" because of their need to estimate the size of networks when they design social media such as Facebook—despite the fact that they are designing for less-bounded networked societies and not for village-like groups.[41] Likewise, policymakers want to know if people are lonely or connected, so that they can understand if they need extraordinary measures to build community. Even weak ties can provide a sense of community.[42] Social psychologists want to know about the origins of lonely people: Where do they all come from?[43] And epidemiologists want to know network size because many diseases, such as HIV/AIDS, come from human-to-human contact.[44]

Network size also matters because people can often reactivate latent ties when they travel to a place where they know people, or they rekindle a common interest. At the same time, when people move, they are able to retain some of their relationships in the places where they used to live.[45]

The larger the network, the more ties that can pass along information.[46] Moreover, people with more ties tend to connect to more networks. Larger, more diverse networks connect people to a greater variety of social milieus, providing a greater variety of information and social contacts.[47] There is a nice spin-off societal effect that sociologist Émile Durkheim first identified in the late nineteenth century as the "division of labor in society": When ties connect different social networks, their interconnections help to integrate these different milieus in an overall society, providing a social glue that can help hold a society together.[48]

The larger the network, the more health benefits. Larger networks provide more social support. As Peter and Trudy Johnson-Lenz found (see chapter 1), such support reduces psychological distress by providing more information, more goods and services, and a greater sense of connectedness. Moreover, larger in-person networks provide more immunity to serious infectious diseases by exposing people to a wider range of minor infections such as common cold viruses.[49]

Of course, the more people use the internet, the easier it is to connect online with large numbers of people.[50]

Thus, size matters. Although some might think that smaller networks will have higher-quality relationships—quality compensating for the lack of quantity—in fact, quantity goes along with quality. Not only do larger networks provide more support, but each person in a larger network is likely to be supportive.[51] We do not know why, but we suspect that social capital breeds more social capital in a positive feedback cycle. A large, active, specialized and resource-filled set of ties is an important resource in its own right.

Dunbar's number is set too low for most people in developed countries because their networks have many more than 150 members. Such higher numbers were found even before the advent of the internet because people have been moving among multiple sets of ties for generations. Moreover, social media such as Facebook have increased the carrying capacity of relationships: It takes little work to keep large numbers of hardly known (or long-lost) ties on your "friend" list. While many are weak ties at the moment, they can be called upon when needed. Networks are so large, segmented, and far-flung that many people are not in frequent contact with many members of their networks. This means that people may not remember many of those whom they know—unless they see them, see their names or pictures, or get another hint.

To deal with these complexities, researchers have used a variety of techniques to estimate network size. For example, one research team found that Americans can name an average of 290 persons as members of their personal networks when they asked them to spot names in a telephone book and identify first names they know.[52] Name identification is tricky, for people are more likely to remember a boy named Sue than a girl named Sue. When researchers more recently took into account the difficulties people have in recalling common first names, they found much larger networks: an average (or mean number) of about 611 members in of their networks with a median of about 470 people. The range in Americans' network size is vast, with 90 percent of the adult population knowing anywhere between 250 and 1,700 others, and half knowing between 400 and 800. Women know about 9 percent fewer people than men do.[53]

Scholars Keith Hampton and Lauren Sessions Goulet worked with Pew Internet researchers and a refined version of these name-recall methods to find that the average American has 634 social ties. Internet users, with an average of 669 ties, have more connections than nonusers, with an average of 506 ties. Moreover, heavy internet users have more ties than lighter users. At the same time, the average mobile phone user has 664 ties and the average user of a social networking site has 636 ties.[54]

But, even these larger numbers underestimate the number of people that each American adult knows—because they are all based on recalling names, and people will forget lots of others until they meet them or are otherwise reminded. As psychologists Melinda Blau and Karen Fingerman show in the well-named *Consequential Strangers,* people know many others whom they usually do not list in network surveys, such as the woman who runs the local variety store who smiles every weekday as she sells *The New York Times*.[55] All of these acquaintances embed people in society, provide useful services, sometimes open up new opportunities, and often give people a sense of belonging as they go through the day. The most accurate (and time-consuming) way to count these people is to follow someone around. Anthropologist Jeremy Boissevain did this in the 1970s when he followed two people in Malta for a year and had them keep records when he was not with them. Boissevain found the "true" average size of the networks in his small, intensive study to be more than 600, consistent with the estimates done by two recent research groups and much larger than Dunbar's number.[56]

Who Is in Personal Networks?

Personal networks tend to have roughly similar mixtures of people: friends, relatives, neighbors, and workmates (or schoolmates). Immediate family (parents, adult children, and siblings) and friends usually dominate the core of North American networks. For example, the Connected Lives study shows that half (50 percent) of very close ties were kin. The rest are with friends (41 percent), a handful of neighbors (4 percent) and work/school mates (5 percent) (see table 5.1). But in societies with monogamous marriages, people can have only a limited number of kin even if they get married more than once. In the 1950s, anthropologists estimated that the British had about fifty kin on average: Smaller families

Table 5.1
Percentage of Closeness

Role	Very Close	Ambiguously Very Close	Somewhat Close	All Close Ties
Immediate kin	44	20	6	22
Extended kin	6	10	14	11
All kin	*50*	*30*	*20*	*33*
Friends	37	50	53	47
Neighbors	4	7	9	7
Work/school mates	5	6	10	7
Organizational ties	0	0	4	2
Online-only friends	0	0	0	0
Other	4	7	4	4
All non-kin	*50*	*70*	*80*	*67*
TOTAL	100	100	100	100
Number of ties	348	229	462	1,039

Source: Pew Internet & American Life Project, "The Strength of Internet Ties," 2006.

may have made the average even lower now.[57] But there are no such limits on other types of relations; they are limited only by a person's carrying capacity for friendships, neighbors, workmates, and more distant relatives.

Any network of relations around an individual can be a personal network: be it one of emotional support, gift giving, or email exchanges. Thus, studying personal networks provides information about people's social worlds. Friends tend to outnumber relatives in personal networks. The larger the network, the higher the percentage (and number) of friends who are in it. Although the Connected Lives study shows that kin comprise 50 percent of very close ties, friends and other non-kin (neighbors, work-mates, etc.) comprise fully 80 percent of somewhat close ties. Using a somewhat more relaxed measure of closeness, Pew Internet research shows that Americans have twenty-three core ties in 2004 as well as twenty-seven other, but still significant ties: Most are friends and not kin.[58] Moreover, the average person's ten to fifty close ties are only in the core of their networks: Their other five hundred-plus ties are almost entirely with friends, acquaintances, and consequential strangers. The Connected Lives study does not show any close ties maintained solely via the internet; all meet in person at least once in a while.[59]

Sparsely Knit, Segmented, and Specialized Personal Communities

Networked individuals have "sparsely knit" personal communities, mean-ing that most network members are not directly connected with one another. As far back as 1968, the first Connected Lives study found that only one-third (33 percent) of an East Yorker's five socially close ties were linked with each other. Further research in 1979 showed that weaker ties are even more sparsely interconnected, with a density of 13 percent.[60] The larger the network, the less likely that two network members will be connected. We are not aware of more recent studies of the density of personal networks, although it is a good bet that the internet—especially Facebook, LinkedIn, Twitter, and email—enhances the density of inter-connections among a person's relatively close ties by allowing friends of friends to become aware of each other.

Personal communities are usually specialized, with different network members helping in various ways.[61] The exception is spouses who supply each other with many types of support.[62] Friends are valued as confidants and social companions. Neighbors and coworkers are conveniently suited for handling unexpected emergencies because their nearness enables them to react quickly with goods and services. Parents, adult children, and in-laws often provide emotional and long-term support: financial aid, emotional aid, large and small services such as childcare, health care, and home repairs. Similar to East Yorkers, Northern Californians name fifteen to nineteen network members who have helped them in up to ten different ways.[63]

Supportive people tend to have longer-lasting relationships.[64] Yet, networks do change over time. Friendships are not always forever; neither are some kinship ties. Breakups became more widely known as "unfriending" when the Facebook term "unfriend" became the Oxford University Press "word of the year" for 2009. However, there is not much research evidence about how friends break up, fade away, or become weaker ties. A preliminary study found that those who initiate friending requests on Facebook are more likely to be subsequently unfriended (disconnected) in the relationship than are those who receive the friendship requests—presumably because some friending requests were unwanted.[65] One small NetLab study, done before the advent of Facebook, suggests that changes in network membership are not gradual but sudden, triggered by changes in personal situations such as marriage, childbirth, and residential moves—a personal network version of what paleontologist Steven Jay Gould has called "punctuated equilibrium" on the global evolutionary scale.[66]

Core Networks Do More than Discuss Important Matters

We began this chapter with the alarm that Americans have only 2.1 people with whom they can discuss important matters, while a sizeable minority does not have any such discussion partners. Presumably these people are at the core of someone's personal network. But when we delved into the matter, we found that there was more to the core than discussion partners.[67] For one thing, the original survey did not ask about what "important matters" people discussed. When sociologists Peter Bearman and Paolo Parigi did, they discovered the variety of people's concerns. While some talked about war and peace or getting a job, others talked about eating less meat and cloning headless frogs.[68]

Not only is there variety in what people discuss, but their closeness comes from more ways than discussing important matters. Different people are close for different reasons, as sociologist Claude Fischer first documented in 1982.[69] For example, they could be *doing* things for each other (rather than *discussing)*; be *mutually enmeshed* in a broader kinship, friendship, or workplace network; *see* each other often at work or in the neighborhood; or *chat* frequently in person or on the internet. As new Connected Lives research is showing, the multiple ways in which people are socially close means that the core networks of close ties are much larger than the 2.1 persons whom the U.S. General Social Survey (GSS) reported discuss important matters.

To understand this better, the Connected Lives study interviewed 84 East Yorkers to learn about whom they felt close to in their personal communities—and why. The researchers asked about closeness in two different ways: by asking participants a direct question, and by asking them to place their network members on a series of concentric rings like a target, with the innermost ring indicating those who are "very close" (see figure 5.1). By only choosing those who are "very close" on both criteria, the researchers are more confident that they are studying ties that are very close. The Connected Lives study finds that the average Torontonian interviewed feels very close to 4.1 network members (answering "very close" on both measures) and pretty close to another 8.2. In short, they feel close to 12.3 people—not 2.1.[70]

But what does such closeness mean? Surprisingly, only 31 percent of the very close ties "discuss important matters" with each other: an average of 1.1 ties. The respondents also discuss important matters with 1.3 of their other somewhat less close ties. The total of 2.4 close ties who reportedly "discuss important matters" with the Connected Lives participants is more than the average of 2.1 found by the 2005 GSS but less than the 1984 GSS average of 2.9.[71]

If people do not discuss important matters with all of their very close ties, then what relationships connect them with their other very close ties? "Salami analysis"—cutting off and analyzing one chunk at a time—reveals that 20 percent of those who do not discuss important matters "chat about the day" with each other. Think of friends and relatives schmoozing. Another 12 percent of the very close ties neither discuss nor chat, but do provide various kinds of social support such as information about health, help with home renovations, and advice about computers.

What about the 37 percent of the very close ties who neither discuss important matters, nor chat about the day, nor exchange social support? Frequent contact seems to account for most of the rest: 13 percent see each other in person at least weekly, while 12 percent of the ties do not see each other in person but connect by email at least weekly. A few (4 percent) just keep in contact by talking on the phone at least weekly. The small number of remaining very close ties are almost equally divided among friends, neighbors, and workmates (4 percent) and parents and adult children (3 percent): These are ties with whom people feel very close, but contact infrequently.

These findings make it clear that "closeness" is not a one-dimensional phenomenon.[72] The variety of reasons for closeness shows that most ties in personal networks are specialized: People get different types of social support from different folks. Only when social closeness is measured exclusively by the "discuss important matters" criterion is there any evidence that North Americans have tiny and shrinking networks. As soon as multiple criteria for closeness are taken into account, there are larger supportive networks of strong, close ties. Toronto student Mirna Ghazarian put this nicely. "I would argue that close ties are not necessarily close friends," she writes. "For instance, I have a close tie with a lady I work with, with whom I discuss important political, environmental, and work-related matters, but I would not consider her a close friend. Why? Because I do not discuss my personal matters with her. I do not confide my personal problems as I would with my best friend."[73]

Despite the major changes in connectivity that ICTs have brought, the percentage of very close kin and friends in these networks is almost identical to what it was in 1979, when NetLab studied East York and found 48 percent were kin and 39 percent were friends, compared with 50 percent kin and 37 percent friends in 2005. However, friendships doubled between 1979 and 2005, from 24 percent to 53 percent, while the percentage of neighbors has dropped by half for both the very close and somewhat close ties. These changes suggest that ICTs help to expand friendships—especially with somewhat weaker ties—and diminish the importance of neighborly proximity.

Of course, styles vary with the stage of life. Marriage and early parenthood often entail high levels of commitment to kin, exerting strenuous demands on both time and energy for both spouses. Where singles use weekends for socializing with friends, married couples use weekends and weekday evenings for childcare and visits to their parents and in-laws. When working mothers are pressed for time, it is friendship that gives way and kinship that remains.[74]

Moreover, how men and women network is converging. In pre-internet days, women were most often responsible for keeping networks going, especially with kin, although husbands and wives often saw the same friends.[75] In the early days of the internet, men were more active than women. Now, on the one hand, there is less difference in what women and men do online. On the other hand, a study of American undergraduates still finds a traditional difference between men and women in their internet use. Women use the internet more to reinforce their existing core ties, while men are more apt to use the internet to develop new relationships.[76]

Networks in the Age of Facebook

Nothing has brought social networks more vividly to public awareness than the rise of social networking sites—first Friendster, then MySpace, and, most dramatically, Facebook. These sites have made social networks more salient and allowed networked individuals to share and capture more information about their friendships than has ever been possible. Moreover, this mutual exchange opens up countless avenues for dialogue and discussion among one's personal network, bringing to reality what mathematician Jon Kleinberg describes as "the visible conversations, the spikes and bursts of text, the controlled graffiti of tagging and commenting."[77] Social networking sites have become the dashboards of the internet for networked individuals. Half of all American adults (50 percent) now use such sites, according to Pew Internet work.[78] From early 2010 onward, the fastest growing user cohort for these sites has included individuals over age fifty (see figure 5.3).

Facebook, especially, has become a powerful stimulant to internet and mobile use. Some of the contours of the Facebook world and the visible conversations that take place there were captured in a Pew Internet survey

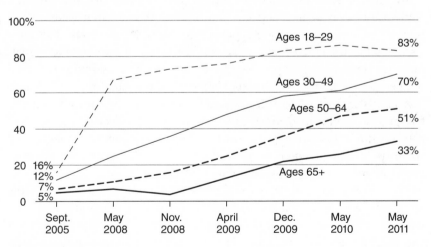

Figure 5.3
Growth in the percentage of adult internet users who use social networking sites.
Source: Pew Internet & American Life Project surveys.

in September 2010: Some 42 percent of all American adults (53 percent of internet users) are Facebook users.[79] Many have large and active networks on Facebook. The mean number of users' Facebook "friends" is 229, or 35 percent of the estimated size of Facebook users' overall social networks. Almost a third of the Facebook users (31 percent) say they check the site multiple times a day, and another 21 percent say they check in at least daily. And 15 percent say they change their profile at least once a day. The growing linkage between mobile connectivity and social networking is apparent in the study. Some 35 percent of those Facebook users access their profile pages from time to time with their mobile phones.

This same survey showed that 85 percent of the Facebook users comment on other people's status, wall, or links—and 21 percent do so every day. Similarly, 85 percent comment on other people's photos—and 19 percent do so every day. The survey shows that 78 percent use the "like" button to comment on others' status, wall, or links—and 25 percent say they do so every day. Also, 72 percent send private Facebook messages—and 10 percent do so every day.

Facebook has become so essential and appealing to networked individuals that it is consuming ever-increasing amounts of time. Nielsen Company figures show this (see table 5.2). The company reports that throughout the month of March 2011 the average internet user spent 6.5 hours on Facebook, compared with 21 minutes on Google, the most heavily trafficked site on the web that month.[80]

By engaging in these activities, networked individuals influence the content and flow of interpersonal information in ways that were unseen prior to the emergence of social networking sites. Figure 5.4 provides just a snapshot of the kind of personal information that networked individuals publicize on their online profiles. Nicole Soriano (a pseudonym) has filled out her Facebook profile with tidbits of personal information. For instance, just on this one page, Nicole has shared her location (Toronto), educational background (Political Science and Sociology at the University of Toronto), partnership status (in a relationship), languages (English, French, and Spanish), birthday (September 6), and religion (Catholicism). She provides links to her friends (also pseudonyms here), and has set up her social networking profile to indicate her favorite music, books, and movies. Nicole also shares a total of 921 photographs from her daily life and travel. Networked individuals on Facebook can share other details such as their current and previous work experience, favorite quotations, activities, interests, and contact information.

Table 5.2
Percent Using Top Ten Internet Sectors by Share of Time U.S. Internet Users Spend Online

Rank	Subcategory	% Share of Time June 2010	% Share of Time June 2009	% Change in Share of Time
1	Social networks	22.7	15.8	43
2	Online games	10.2	9.3	10
3	E-mail	8.3	11.5	−28
4	Portals	4.4	5.5	−19
5	Instant messaging	4.0	4.7	−15
6	Videos/movies	3.9	3.5	12
7	Search	3.5	3.4	1
8	Software manufacturers	3.3	3.3	−0
9	Multicategory entertainment	2.8	3.0	−7
10	Classifieds/ auctions	2.7	2.7	−2
	Other	34.3	37.3	−8

Source: The Nielsen Company. See note 80.

Although the award-winning 2010 movie about Facebook is called *The Social Network*, Facebook is mostly about groups rather than networks. Rather than making it easy to limit certain kinds of information to different types of people, Facebook's profiles are set up to default to the assumption that all people want to make all of their information available to all of their Facebook friends. This is a key part of Facebook founder Mark Zuckerberg's philosophy: "You have one identity. . . . The days of you having a different image for your work friends or co-workers and for the other people you know are probably coming to an end pretty quickly. . . . Having two identities for yourself is an example of a lack of integrity. . . . The level of transparency the world has now won't support having two identities for a person."[81]

So, Nicole's parents hear about her late-night partying, and her friends learn obscure details about her second cousins. Other social networking sites such as Google+ are trying to capitalize on this one-size-fits-all structure by allowing users to segment their networks and send different information and updates to those different segments.

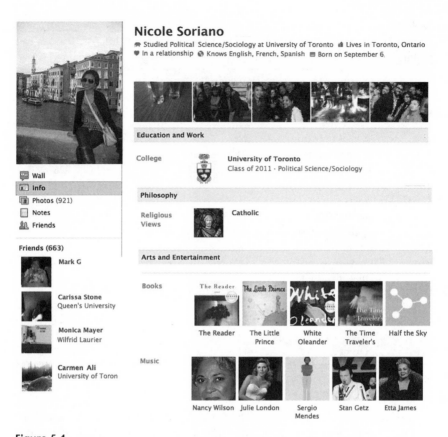

Figure 5.4
Screen shot of a networked individual's Facebook profile.
Source: © 2011, used with the Facebook user's permission.

Much of the information on Nicole's profile links to other pages within the social networking site itself and to external websites. For instance, the University of Toronto is a link to another page on Facebook that provides a description of the school as found on Wikipedia, related posts by Nicole's friends, and all the people who have also added this university to their profile. Similarly, the icon for her favorite book, *The Reader,* links to another Facebook page that gives a description of the book and shows how many other Facebook users like the novel. Thus, these links allow for a denser and broader network of information, not just about Nicole, but also about the things she likes and the other networks she is a part of.

Facebook news feeds update Nicole's friends with what is happening in her life. The feeds are neither random nor comprehensive: Facebook uses algorithms that try to tailor the information that each friend gets according to their interests. Thus, each friend gets a somewhat different picture of Nicole's life on their customized news feed. Some information is widely shared: When Nicole's status changed from "single" to "in a relationship," all her friends wanted to know "who?" and "why?"

What impact has the now-dominant Facebook had on networked relationships? It has clearly allowed more sustained contact with weaker ties. Even as people move, change jobs, and switch their attention zones, Facebook efficiently allows them to stay in touch with others, broadcast basic update messages, and receive similar updates from their friends. Facebook has also enabled reconnections. Long-lost friends can locate each other and reconnect with old school chums, onetime lovers, former coworkers, and former neighbors.[82]

Facebook promotes bridging as well as bonding: By following a chain of Facebook friends, people connect to other personal networks, providing potential access to other social milieus.[83] Mutual ties—both people are friends with the same third party—are especially important for forming new connections, as one friend validates the other.[84] As Toronto student Sharanpreet Kelley notes:

As I parted ways from my friends in high school offline, we maintained our relationship online. When I started university, my network swelled with new people. Facebook functioned unofficially alongside the university system, providing me with information on social events as well as on how my peers were doing. This open discussion played a key role in meeting people outside of my immediate network. I have depended on Facebook since high school, and it is difficult not to notice how dependent I am for social rituals, updates, and entertainment. Most of my friends and I do not see each other on a daily basis, so Facebook serves as a medium to continue light conversations and maintain our social ties.

Her story also shows how useful it is to be perpetually and pervasively aware of who is doing what with whom. Of course, this extreme transparency means that Facebook friends may learn unwanted things about one other—such as political leanings or sexual adventures—that may lead to unwanted attempts to control each other's behavior or may even rupture relationships.

Yet, the importance of Facebook goes beyond its role in connecting current and former friends. It has become a personal portal embodying the networked individual. Not only are there links to people, but to tastes—such as Nicole's books—and "likes" to even more books, music, and organizations. Corporations are now using Facebook pages extensively, so that if Nicole likes San Miguel Beer, she can link to the company's page and they will know about it. Facebook has become each person's "go to" page: their home base. It is why they stay on Facebook for so long. Just like the car has become the personal basis for transportation, the smartphone for personal communication, and Google for information, Facebook is becoming a key web in the social operating system—connecting each person to who and what they are interested in. At the same time, Facebook is amassing tons of information about the individual, the aggregated profiles of individuals (for example, young Canadian women with Chinese family names), and their social networks. Thus, Facebook is both the epitome of networked individualism—each person is an individual participant—*and* of the networked operating system as a whole.

The More, the Merrier

Critics used to worry that the internet would be an inadequate replacement for human contact because hugging a computer screen is less satisfying than hugging a friend. In fact, the evidence shows that ICTs supplement—rather than replace—human contact. People will make do with electronic contact if they cannot be together in person. A more anthropomorphic device is the mobile phone, which some people see as their third skin. But despite whispered endearments into the phones, the boundaries are clear even here.[85]

Do ICTs substitute for in-person communication, extend it, or transform it? The evidence for the *substitution* argument is almost nonexistent except for early studies of apprehensive newcomers to the internet. If anything was being substituted for, it was television.[86] Consider what happened when Toronto student Sharanpreet Kelley experimented with going off of Facebook and Twitter for two weeks in 2011. "As soon as I went offline, I wanted to check back immediately to see what I could have possibly missed," she says. "I had to distract myself with other activities, but my attention kept on going back to what was going on online. I felt like I was being isolated from my community. This was highly frustrating, because it was as if I had been exiled from my community."

Sharanpreet ended her cold-turkey experiment early: She could only handle her partial withdrawal from the network operating system for eight days instead of two weeks. There were events to plan and things to do. "FOMO"—fear of missing out—played a key role in her return: Her network was too individualized and spatially dispersed to keep in touch solely through in-person and telephone contact. Sharanpreet's experience partially supports the *extension* argument. Facebook, email, internet phoning (video and audio), mobile phoning, and texting are continuations of interpersonal conversations.

But, Sharanpreet says that things have gone beyond supplementary extension. ICTs have *transformed* communication, relationships, and community. They support rapid-fire exchanges among individuals—in pairs or groups—that would only be partially feasible in village pubs. Social media such as Facebook, Twitter, and email lists support "social neighborhoods" that may be as important as the physical neighborhood or workplace in providing frequent contact and information about others. Moreover, interconnected personal networks now aggregate so that the sum is more than the whole.

To what purpose? So far, systematic research has found ICT use to be more beneficial than harmful. This is true in city, suburb, and countryside.[87] The question is no longer the simple one of whether or not the number of relationships in personal networks is rising or falling in a hyperconnected world. Although earlier studies were ambiguous, it is now clear that they are rising in number and in the volume of contact.[88] Networks are larger, more diverse,[89] and supportive.[90] The question is not if but *how* ICTs intensify bonding and promote bridging. These happen both through casual interaction via email and Facebook, and through ad hoc support organized to help those in need. Susannah Fox reports this dimension of Pew Internet's research into how people support others with illnesses even when they have never met: "The most striking finding of the national survey is the extent of peer-to-peer help among people living with chronic conditions," she notes. "One in four internet users living with high blood pressure, diabetes, heart conditions, lung conditions, cancer, or some other chronic ailment (23 percent) say they have gone online to find others with similar health concerns. By contrast, 15 percent of internet users who report no chronic conditions have sought such help online.[91]

Fox summarizes that "people living with chronic disease who go online are finding resources that are more useful than the rest of the population."[92] Similarly, a Dutch study found that online communication stimulates teens' well-being,[93] while an American study showed that Facebook users provide social support. As one person in the Facebook study mused, "When you Google it, they just give you a list of medicines. You don't know if the medicine works or not. You talk to somebody else [on Facebook] who has a child and know that they gave it to their child."[94]

Networked relationships on and offline reinforce networked individualism. Both the internet and the mobile phone allow people to use their social switchboards to move between their social circles and to interconnect them. The internet and mobiles help people to bond within their circles by supplementing their in-person contacts. Further, their ease of use helps people to bridge networks as they never could before. They allow people to shop at specialized relational boutiques for support, similar to how Peter and Trudy Johnson-Lenz obtained diversified, often specialized, help from friends near and far in the story we recounted at the beginning of this book.

We have interviewed scores of networked individuals who use a panoply of gadgets and applications to orchestrate their lives. Theirs is a complicated dance through the networked operating system. They use email for certain kinds of networked communication; text messaging, Facebook posts, private Facebook messages, and Twitter posts for others; and phone calls for communication that requires more extensive conversation. Today individuals have more communications options than ever, and that means they have to work harder to figure out which gadget or mobile apps to use for which kinds of activities. Yet, segmenting their tools and messaging strategies allows them to handle different tasks across their segmented networks. It is common for multiple devices and applications to be running simultaneously in the network operating system. In many cases, ICTs are used to organize in-person contact.

The more people use the internet, the more friends they have, the more they see their friends, and the more socially diverse are their networks. The internet and mobile phones are both an outcome and a cause of larger networks. They help people get social support. They provide conduits for information, guides to services, and ways to seek and ask for help. The internet, especially, amplifies people's social capital—the resources they get from the ties that they draw upon for their needs and interests. As we have shown elsewhere in this chapter, the internet is especially good for connecting people with their weaker ties and with a broader diversity of people.

This chapter has described how personal networks have expanded, become more complex, and speeded up. Communities continue to exist, except as spacially dispersed and differentiated personal networks rather than as neighborhoods or densely knit groups. When we see individuals sitting alone, we should not assume they are isolated or lonely: With internet access and mobile phones they have community immediately at their fingertips. And when they need a real hug or material aid, transit, cars, and planes are often available.[95] People's lives offline and online are now integrated—it no longer makes sense to make a distinction.

6 Networked Families

The Triple Revolution—Social Network, Internet, and Mobile—has undermined the classic notion that people's homes are their castles: inviolate, defended households filled with family activity.[1] Rather, they are bases for reaching out and networking—with family members, friends and relatives, community groups, and work.

Hillary Clinton understood this in her book *It Takes a Village.* Despite the title, Clinton recognized in the text that families are not bound up in villages but are networked: "The networks of relationships we form and depend on are our modern-day villages, but they reach well beyond city limits."[2]

The evidence suggests Clinton can take her thesis further. No family is an island, and no house is a castle. They are multiply networked. The ways in which modern families are networked provide them with a great deal of individual discretion, abundant opportunities for communication, and flexibility in their togetherness. They spend less time physically together at home in the same room and even in the same house.[3] People network as individuals rather than within solidary family groups. Each household member operates as a semiautonomous individual, with her/his own agenda, using a multitude of transportation services and communication media to contact and coordinate with each other. But while structural changes in North America have centrifugally weakened the physical togetherness of families—for better or worse—multiple communication media links them. Families continue to be thickly connected at any time and anywhere, with in-person contact supplemented by mobile phones and the internet.

Although the trend to networked families began before the internet and mobile phone, the intrinsically *personal* nature of these technologies has encouraged the transmutation of households into networks. Where calls to wired (landline) home phones and visits to homes often were contacts with the entire household, new ICTs (information and communication technologies) foster individual person-to-person contact. Yet, this only tells part of the story, for in the network operating system, there are social and cultural changes in addition to technological changes. They include trends toward personal car ownership (rather than the one-for-all family car), women working outside of their homes, shifting family composition (smaller, with multiple marriages and parentage), and the substitution of paid services for the work formerly done by homemakers—such as lower-cost fast food and "family" restaurants.

The Way It Is in the Networked Age

To understand the networked nature of family life these days, consider how Tracy Kennedy operates. She's a single mom in southern Ontario with a teenage son, and their lives are oriented around ICTs:

I often start my weekday morning at the computer where I respond to emails and catch up on social networking sites. Much of my day takes places at my home office, and communicating via ICTs to friends, relatives and work peers is steady throughout the day. But not all of my online interactions are with people outside my home; I also connect with my teenager via ICTs when he is not at home and when he is at home. For example, just before lunch I receive a text message from my son while he is at school: "so bored in this class." Entertained, I text a pithy response to which he does not reply. In the afternoon he sends another text message saying he will be a few minutes late coming home because he has some work to do in the computer lab. I text him back letting him know that I am running errands and won't be home when he gets there. He sends me a text to let me know he has made it home—and to remind me to buy Coke.

Later in the day, dinner is almost ready and he's not responding when I shout up to him because he's listening to music on his headset. I send him an IM through Skype to let him know, and he immediately replies that he will be there right away. After dinner, we each grab our own Xbox to play the multiplayer *Modern Warfare* video game with each other. We also set up a game lobby with his school friends and my friends (all local) so that we can all chat on our headset during our game play. Later that evening, he sends me an email with a link to a laptop that he is really interested in buying for school use. We talk about it face-to-face before he goes to bed.

In my home, staying connected with my teenager throughout the day is vital as a single parent, and using ICTs gives me some peace of mind about his whereabouts and safety. More importantly I find that using these ICTs with my teenager is . . . engaging and entertaining when we are home together but not in the same room, and they also act as a generational bridge between parent and child. Our use of ICTs is individual, but we are connected and networked together as a family.[4]

This story points to how the increasingly ubiquitous nature of ICTs is embedded in the domestic lives of networked families. This is a dramatically different life from the one her parents experienced (table 6.1).

Changing Households: Size and Composition

North American families have changed substantially during the past several generations.[5] The percentage of married-couple households with children has steadily declined—that traditional model of the "nuclear family" of always-married mom and dad with kids, the *Fun with Dick and Jane* norm of American life. Between 1980 and 2005, the overall percentage of such households fell by one-quarter in the United States—from 31 percent to 23 percent. At the same time, the percentages of single-parent households and remarried-parent households have each increased.

Households have also become *smaller*. Part of this is attributed to women having fewer children. In the 1970s, only 10 percent of women in their childbearing years did not have a child compared with 20 percent in 2008. The percentage of family households containing children under the age of eighteen has declined from 52 percent in 1950 to 46 percent in 2008.[6] The result of this decrease means that there are now more childless (28 percent) and single-person (26 percent) households in the United States than married-couple-with-children households (figure 6.1).[7] Many households have *only one adult* because late marriage—with a median age of twenty-eight for men and twenty-six for women—provides more opportunities to develop separate networks as adults before marriage. Yet, while the percentage of one- or two-person households increased by nearly one-third during the same time period—from 46 percent to 60 percent—the percentage of households with five or more people dropped in half from 21 percent in 1970 to 10 percent in 2003.[8]

Households have become *less stable* in their composition. Fewer Americans between age thirty and forty-four are married than before: 84 percent of Americans 30–44 years were married in 1970 compared with 60 percent in 2007. American divorce rates in the United States are higher than they were a generation ago: increasing from 2.2 (per 1,000 people) in 1960 to 3.6 in 2005, while Canadian rates increased from 0.4 in 1960 to 2.2 in 2005.[9] Partnership—cohabitation without marriage—is becoming more normative. Whereas only 16 percent of American adults lived together before marriage in 1980, 41 percent did in 2000, and the 2010 numbers are expected to show more than 50 percent.[10]

Table 6.1
How Households Have Changed

	1950s to 1960s	2000s to 2010s
Mom	Homemaker	Paid worker outside home
Dad	Sole breadwinner	Largest earner
Marital status	Lifelong	Second marriage
Housework	Mom does almost all	Mom does more than dad
Children's play	Front/back yard, street, park	Baseball, ballet, scouts, piano
Mom contacts kids	Yell out window, call neighbor	Call kid's mobile phone
Joint family time	Watching TV	Showing & sharing at the computer
Number of cars	One per household	One per adult
Music	*American Bandstand*	iTunes
Mass communication	Radio, one TV controlled by dial, broadcasting	Multiple TVs controlled by remote, narrowcasting
Textual news	Daily (print) newspaper	Yahoo/Google News, RSS
Visual news	*Life* magazine, TV network news	YouTube, video links, blogs, tablets
Ads	Magazine, classifieds, Yellow Pages	Amazon, Craigslist, eBay, banners, AdWords
Spoken communications	One household phone	Personal mobile phones with Caller ID screening
Written communication	Letters, personalized stationery	Texting, email, Facebook
If not home	Call back	Leave voicemail
Spousal contact at work	Only in emergencies	Discreet email & text through the day
Household recreation	Charades, Monopoly, radio, TV	YouTube, downloads, video/online games
Movies	Movie theaters	Downloads, Netflix

Source: Table by Barry Wellman. © 2011, used with permission.

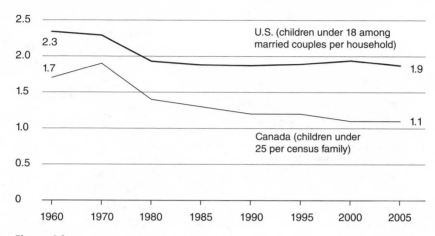

Figure 6.1
Average number of children per household in the United States and Canada.
Source: U.S. Census Bureau; Statistics Canada.

Shifting Family Roles

Family roles have changed, driven by the rise in the percentage of women who work outside their homes and the smaller number of children in households.[11] In most married couples, both wives and husbands go out to do paid work. While in 1960, 38 percent of American women were employed outside the home, that figure leapt to 59 percent in 2006. Similarly, in Canada, nearly three-quarters (73 percent) of the men and nearly two-thirds (62 percent) of the women were in the labor force in 2004. Dual-job households grew from 39 percent in 1970 to 53 percent in 2007. As a result, wives and husbands must negotiate multiple work and school schedules in addition to domestic work (such as cooking, cleaning, and maintenance), child care, family time, and social and leisure activities. Some of the biggest challenges emerge as the financial balance of power shifts in families. More spousal households in North America contain dual-earners (figure 6.2), and the gaps between the relative contribution of husbands and wives to the household income and the number of paid hours worked outside the home have narrowed over the past twenty years. For example, 42 percent of Canadian wives contributed at least 45 percent of the household income in 2007, as compared to 37 percent in 1997, with two-thirds (65 percent) of the wives working the same number of paid hours per week as their husbands.[12]

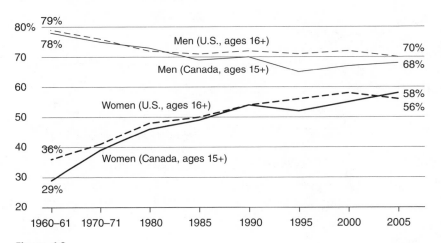

Figure 6.2
Percentage of employed workers by gender.
Source: Statistics Canada and U.S. Bureau of Labor Statistics (see note 12).

As women spend more time working outside the home, there have been big changes in how both men and women spend their time on child care, housework, and other activities. The amount of time mothers spend weekly on housework shrunk nearly in half between 1965 and 2005, from an average of 32 hours to an average of 19.6 hours, while the amount of time spent by men doubled from an average of 4.4 hours to 9.8 hours (figure 6.3). Fast food restaurants grew in number by 25 percent in the United States between 1998 and 2007 and by 36 percent in Canada between 1990 and 2007, reducing domestic time pressures on homemakers.[13]

Although mothers continue to spend more time than fathers caring for children, fathers have more than doubled the time they spend caring for their children—in addition to the increased time they spend doing domestic work. Men and women are spending more time in leisure activities, both outside of their homes (such as visiting friends and relatives and participating in voluntary associations, sports, and entertainment) and inside their homes (such as watching television, chatting online, playing online games, and downloading music and videos). Canadian women spend less time at home: 8.5 hours per day in 2005 as compared with 9.1 hours in 1992.

With wives doing more paid work and men doing more child care, husbands and wives have less specialized roles. In effect, this means that

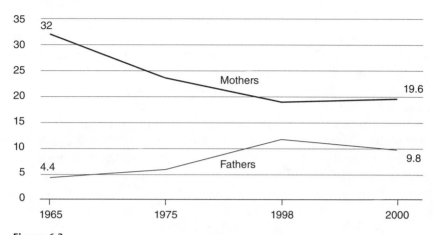

Figure 6.3
Mean number of hours of housework per week by parents in the United States.
Source: Sayer 2005, 2007 (see note 14); U.S. Bureau of Labor Statistics.

spouses need to perform as networked individuals to negotiate how they allocate household roles. Nevertheless, women are still primarily responsible for household work as well as for keeping up ties with neighbors, friends, and relatives.

The way couples apportion their time between work and home has also become more fluid as the boundary between home life and work life breaks down and the length of the work day increases. Some do all or part of their jobs at home, using the internet to communicate, access organizational databases, and find information. Some do part of their jobs on the fly, using their mobile devices to place calls, browse the internet, or access specialized apps that tie to their interests. Those who can control their schedules do best at avoiding stress and dealing with the often-competing demands of home and work.

Yet, the growth of computer-based work allows both men and women to spend more hours doing paid work than in the previous generation. In 1980, less than one-quarter (22 percent) of college-educated men worked a fifty-or-more-hour work week compared with almost one-third (31 percent) in 2005.[14]

Pressed for time, couples multitask throughout the day. They may have less time for each other or for their children. Some watch less TV, cut back on involvements with traditional volunteer organizations (such as the Scouts), or get together less often with neighbors and friends.[15] All of these trends have implications for how satisfied people are with the amount of time they have to spend with family and friends, and on leisure activities. The increase in both single-parent and dual-income families has resulted in both men and women devoting more hours to both paid and domestic work—and spending less time in person with their families.[16]

Household life has sped up and become more individualized. The time that adult Canadians spend alone increased 38 percent from 257 minutes per day in 1986 to 354 minutes per day in 1998. In 1986, men spent 215 minutes per day with their spouses and women spent 198 minutes. This had dropped by 1998 to 201 minutes for men (–16 percent) and 175 minutes for women (–11 percent) for women.[17] Moreover, smaller households mean fewer people at home to chat with.[18]

As people's routines have changed, so have the technologies they use to communicate with family, friends, and work peers. Over the last generation, home computers have become domesticated.[19] Pew Internet surveys in mid-2011 found that 55 percent of American adults own desktops and 57 percent own laptops and netbooks.[20] As personal computing tools have gotten smaller, desktop computer screens have gotten larger, growing from fourteen inches in the 1980s to at least twenty inches now. With more than twice the viewing area, these larger screens enable multiple programs to be viewed simultaneously and enable family members to use the internet jointly. Not all households are the same, and not all households use the internet and mobile phones in the same amounts or the same ways. Households with higher levels of education, higher income, and children still at home are more likely to use the internet at home. Many of today's households contain multiple ICTs. There are more technological toys in living rooms than a generation ago, such as digital cable boxes, gaming consoles, personal video recorders, and DVD or Blu-ray players. Much larger TV screens enhance the immersiveness—and potential inclusiveness—of the experience. Where a twenty-inch screen was normal in the 1970s, forty-two inches or more is the contemporary standard, providing more than 4.5 times the viewing area. [21]

Family Time, Network Time

When the University of Southern California's Center for the Digital Future issued a 2009 media release titled "Family Time Decreasing with Internet Use," hundreds of articles picked up on its premise. Such articles often portrayed internet use as a replacement rather a complement for in-person family contact, and they rarely saw the internet as a positive tool to help families cope with single parenthood, time-use challenges, or structured activities for their children.[22]

To be sure, social and economic changes have helped to reconfigure family life from solidary togetherness to networked semi-independence for family members. Nearly half of all American parents feel that they spend too little time with their children. A main reason is that greater involvement in paid work weakens parents' ability to respond flexibly to children's needs. Still, the majority of family members spend considerable time together, watching television, having family meals, and visiting with friends and family (figure 6.4). Although the time people spend with household members varies depending on the day's events and their schedules, they still find some time to be with their families. More than three-quarters of married parents (77 percent) report spending all or most of their non-work time with other household members.[23]

Family dinners have usually been a way for people to spend quality time with each other, chatting about how they spent the day, discussing things in more depth, and bonding with each other. So alarms rang in 2000 when Robert Putnam in *Bowling Alone* reported that the frequency of people saying they usually have family dinners together had declined by a third during the past twenty years, from about one-half to one-third. We wondered if the networking of family life means that spouses and children have less time to eat dinner together. The current situation is more positive

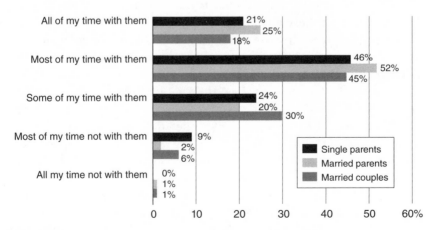

Figure 6.4
Percentage of time that adults spend with other household members.
Source: Pew Internet & American Life Project, Networked Family Survey 2008.

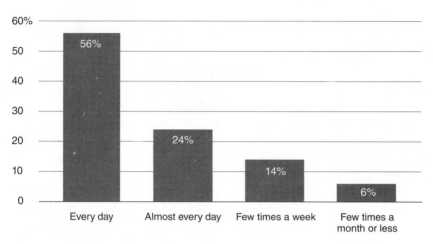

Figure 6.5
Frequency of dinners that adults share with other household members.
Source: Pew Internet & American Life Project, Networked Family Survey 2008.

than people might suspect. A Pew Internet survey on networked families shows that North Americans have dinner together more. Despite the demands of work, child care, school, and other activities, almost all (93 percent) of those American adults who live with a partner or a child have dinner with members of their household at least a few times per week. More than half (56 percent) have dinner daily with their families, while one-quarter (24 percent) have dinner together almost every day (figure 6.5). Only 6 percent have a family dinner just a few times a month.[24]

A generation ago, most people watched television every night. Hit shows were widespread bases for conversation around the TV set, at family dinners, in the neighborhood, and at work. Television continues to be frequently watched: Canadians, aged thirteen-plus, watch TV about fourteen hours per week (two hours a day), slightly less than the fifteen hours per week that they are on the internet. Yet, only three-quarters of Americans watch TV almost every day now, while a lower proportion—three-fifths, of young adults aged eighteen to twenty-nine—watch almost every day.

Television watching is a much different experience than it was a generation ago in the content of shows, the fragmented audiences, the technology people use to watch, and the time of day they watch. Fewer shows aim at the four traditional demographic segments of broadcasting: men and women, old and young. Instead, hundreds of stations now supply focused *narrowcasting* aimed at narrowly defined audience segments: "niche TV" with channels focused on such subjects as golf, cooking, pets, sex, old movies, and shopping. Choice has exploded, putting control of what people watch—and where—in the hands of individuals. There are more—and more flexible—ways to watch these shows. Many shows are streamed on the internet and excerpted on YouTube or Hulu. Netflix and most cable TV providers deliver movies and TV shows on demand; podcasts and personal video recorders (such as TiVo) allow people to watch shows according to their own schedules.

These changes allow for more flexible TV viewing tailored to personal interests, scheduling, and time constraints. In contrast to the previous generation's "must see" appointment TV, this has become an era of "my playlist" TV. The new era responds to smaller, more flexible households with networked family members going their separate, but connected, ways.[25]

Nevertheless, television watching is down. Where have all the watchers gone? The answer is the internet: Time-use analysis shows a good fit between less television watching and more internet use. For example, the Telus Canadians and Technology study shows that nearly half (44 percent) of all Canadian adults watched less TV in 2009 than they did in 2006. Moreover, nearly half (46 percent) of those who watched less TV say a main reason is that they are spending more time on the internet.[26]

Digital Technologies and Networked Households

Inside homes, computers have assumed a special place in communal spaces. As *New York Times* reporter Katie Hafner put it in 2003: "If the kitchen's warm, it may be the PC."[27] Evidence in the Pew Internet surveys and in ethnographic research shows that people reorganize their spaces at home to accommodate computer use. Most home computers are in shared space that is accessible to other household members most of the time. In Toronto's East York, 46 percent of those Connected Lives participants who use home computers have at least one in a public space such as a living room.

With computers and the internet so widely available, children can become proficient quickly. Maria Lianos-Carbone blogged about her five-year-old son Anthony:

I'm stunned that he learned how to turn my computer on, get into Internet Explorer and find his way to Nick Jr. so he can play video games. Watching his little fingers click on that mouse is simply astounding. Once he's bored of video games, he'll move to YouTube to watch the Wiggles online. When he's tired of singing along with Greg first in English and then in Spanish, he'll cross over to Starfall, a free educational website also used in his kindergarten class. . . . Did I mention that my husband is computer illiterate? My 5-year-old had to show him how to close a program. "See dad? All you have to do is to click on the X."[28]

Nor is Anthony a unique case. Even younger children grow up now becoming as unconsciously computer literate, just as many children start learning how to drive by watching their parents. *Wired* magazine editor Chris Anderson tweeted on September 11, 2009: "Our baby has a pull-along phone. There is nothing about it she recognizes as a phone dial: dial, handset, cord. But she chats on the TV remote" because it looks like a phone. [29]

These children are not necessarily prodigies. The Telus Canadians and Technology survey found that three-fifths (58 percent) of Canadian children have started using the internet by age seven, and one-quarter (23 percent) before they have entered kindergarten at age five. By age ten, the great majority (84 percent) are on the internet, using it at least four hours per week.[30] By that time, they have already learned a good deal about what to do, through watching their parents and older siblings, and playing around with their equipment.

Social scientists Gina Neff and Philip Howard sent us a picture of their eighteen-month-old twin boys Hammer and Gordon with "their favorite toys, mom's laptop and cell phone" (figure 6.6).

Figure 6.6
Eighteen-month-old boys playing with their favorite toys: their parents' laptop and iPhone.
Source: Gina Neff and Philip Howard. © 2010, used with permission.

As their homes fill up with ICTs, people make it a daily routine to use them to stay connected as they move about. Networked families use ICTs to keep their family together. Their ICTs enable them to communicate and coordinate despite their mobile, individual lifestyles. ICTs allow them to reach out to new information and new contacts, and then bring that back to the family. At home, their family spends quality time together showing and sharing web pages, online media, and email messages. They rely on blogs and other websites for advice from other parents and organizations, often getting a sense of connectivity with other, sometimes beleaguered, parents. All of this communication and content creation has helped family members as they operate as networked individuals using personal technology to navigate family life. Two-fifths (39 percent) of all American households have at least two computers; three-fifths (58 percent) of married-with-children families have at least two.

Not surprisingly, married couples with children stand out for owning more technology, including mobile phones and computers. Part of the reason is that these are the largest households. The more people in households, the more coordination and communication they need. For example, where a two-person household (married couple or single mom with child) has only two relationships to coordinate (one in each direction), a four-person household (mother, father, and two children) has twelve relationships to coordinate. Almost all (93 percent) American married-with-children families own a personal computer, and nearly six-tenths (58 percent) own two or more. In addition, more than one-third (37 percent) of such families (and 63 percent of those with multiple computers) have a wired or wireless home computer network, so that all can connect to the internet. Indeed, both spouses go online in three-quarters (76 percent) of these families as do an even higher five-sixths (84 percent) of their children aged seven to seventeen.

For example, Tanya (in the Connected Lives study) is a married woman in her late forties, a mother of two and a marathon runner. An engineer, she works full-time mostly at the office, although some of her evenings and weekends are also filled with work. Tanya has a tight schedule and coordinates all of the comings and goings of her family using the calendar program on her laptop. She uses her computer to plan all of her work, family, and social activities, and she sees it as a valuable asset that affects everything in her life. All of her family members have their own iPods, BlackBerrys, and computers, which are all linked and connected. They are in email contact with each other throughout the day. Tanya uses the internet all day and evening. She sees the internet as the ultimate resource. It is the newspaper, the phone book, and the encyclopedia combined. Whenever she needs information for her family—on healthy living, for example—she turns to the internet. For Tanya, as for almost all the people we interviewed, debates about whether ICTs help or hurt do not matter; she cares only about how these technologies let her get on with her life.

Families of married couples with children are also likely to own mobile phones according to Pew Internet data: 89 percent of such families own more than one mobile phone, while 47 percent own three or more. For parents, the patterns of mobile phone use are similar to those for internet use—both parents own a mobile phone in over three-quarters (78 percent) of such families. Children however, are less likely to own a mobile phone than they are to go online. For example, where the Telus survey found in 2009 that 84 percent of Canadian children aged seven to seventeen in married families go on the internet, only 57 percent have a mobile phone of their own. However, as costs have dropped for gadgets and monthly fees, mobile phone ownership for teens has increased: Pew Internet research showed that by mid-2011, exactly three-quarters of those between ages twelve and seventeen have a mobile phone, up from 45 percent in 2004.[31] Most teens get their mobile phones by the time they turn fifteen. It is a win-win situation for teens and parents: Teens value their ability to connect with their friends and to call home when in need, while parents have some comfort knowing they can connect with their kids at any time when they are apart.

Most children get their first mobile phone as a gift from their parents. Like most gifts, the phone comes with strings attached, in this case an obligation for teenagers to reciprocate by regularly checking in with their parents to maintain family ties while being physically apart. Parents use their mobile phones to make sure where their children are and to coordinate meeting spots. The paradox is that the Mobile Revolution makes children freer to be physically apart from their parents, because their mobile phones can keep them constantly accessible to their parents.

There is also pressure to add ICTs to the family so that each member can function independently. As Michelle told the Connected Lives study: "The whole reason why our three children got their own computers for Christmas the one year was because people were fighting over them. They'd all want to use them at the same time, of course. There's only so much time between after school and bedtime."

Helen, another Connected Lives participant, has two computers at home. Her personal computer is in her home office where she does her public relations business. Helen uses email to communicate with work people and friends. She too looks online for information to help her family. As was the situation for Peter and Trudy (described in chapter 1), ICTs are used in new and different ways when life disruptions occur. Helen recollects: "My father was diagnosed with cancer, so I went on the internet to find out everything I could about cancer, the cancer site and different things," she wrote. "And then, my mother once had a reaction to a medication and I went on the website to find out about that. I [also] use websites to find hockey camps for my son."[32]

Helen effectively uses ICTs to inform herself and those close to her. A single mom, she put the second computer for her children in the family room because she wants her kids to be close to her when they are using it. She keeps the door to her office open, unless her kids have friends over and it gets too loud. Her kids are computer whizzes; they have been using it since kindergarten. She draws on their skills if something needs attention. Her kids have no problems sharing one computer, as her fourteen-year-old son will use it after her nine-year-old daughter has gone to bed. Helen's son often uses the internet for homework, and also for learning guitar chords, downloading music, and following hockey. He also talks to his friends via email and IM. Her daughter plays games both on the computer and her mother's mobile phone.

More choreography is needed when families share a computer. Daniel and his wife, for example, have separate email accounts to keep their messages apart. These Connected Lives participants share a home computer that is in the basement, while Daniel also carries a laptop back and forth to work for his university research. He researches holidays, looks for genealogy information, and emails friends and family who live out of town. He likes to check his email in the morning, and might check it again after dinner. Sometimes when Daniel is at work he will forward email from his family out west to his wife at home—and she does the same. Although Daniel does not think of himself as an avid internet user, he goes online for travel and product information, buys tickets to local events, and more. He also looked up health information online once when his wife was misdiagnosed with a medical condition.

In such ways, networked households are working harder to keep in touch than households of the previous generation did. Where mom, dad, and the kids used to be in fixed locations all day and rarely communicated until they came home, on-the-go networked individuals are grabbing multiple means of communication to tell each other where they are and coordinate what they are doing. While sometimes they go online jointly, most often they are using personal communication media. They mostly dance solo but take part in a few duets and household ensembles.

The New Connectivity: Keeping in Touch with Spouses

Spouses use many different media to stay in touch with each other throughout the day. Those with children stay in touch more often than those without children, because the presence of children in a household increases responsibilities and workloads. There are more schedules to organize and more responsibility, and a definite need to stay connected instrumentally and emotionally.

Married couples—whether they have children or not—contact each other to schedule events and tasks, organize daily routines, and plan future events with friends and family. Spouses often communicate with each other when they are apart, just to say hello and chat. Despite the routines people keep, they still make time to talk. Much of their day is spent away from home: commuting and working, running errands, making social visits, picking up children at school, generally being busy and on the go. For example, Connected Lives participant Theresa lives in Toronto with her husband and three young children. She feels it is important to stay in touch with her husband at work throughout the day by email to make sure that important family organizational details don't slip through the cracks. For Theresa, these emails work as a helpful task list and reminder system—"we need to do this, this, this, and this"—to be discussed later when her husband comes home from work in between the "thousand things going on." Peter, with a wife and two young twins, tells a similar story. He emails his wife about sports schedules, scheduling pickups and check-ins: "Well, now I'm taking them to dentist: Put this in your schedule at work."

Spouses principally use mobile phones and landlines to stay in contact. Those with children connect with their partner by mobile phone at least once a day (a mean of 7.3 times per week), and several times a week by landline (a mean of 4.7 per week)—slightly more often than couples without children (figure 6.7).

Connecting with Children

Parents and children need to stay connected when they are not together, whether for socioemotional reasons, instrumental reasons, or for the children's security. While parents stay in touch with their spouses most

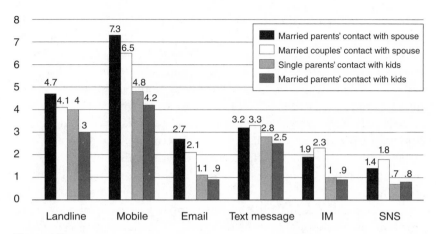

Figure 6.7
Mean number of daily ICT contacts with spouse and children.
Source: Pew Internet & American Life Project, Networked Family Survey 2008.

frequently, single parents connect with their children more often than married couples with children do. Some single parents may have more of a workload and increased responsibility because they are the sole domestic caregiver.[33]

Americans use the same media to connect with their children that they use to connect with their spouses: mobile phones and landlines. Single parents use mobile phones (a mean of 4.8 times per week), landlines (a mean of 4.1 times per week), and text messaging (a mean of 2.8 times per week) more often than couples with children. Not only are parents in frequent contact with their children, they also use a variety of ways to do so—what Caroline Haythornthwaite and coauthor Wellman have called "media multiplexity." This reflects the complexity of their lives, and how ICTs facilitate their needs.[34] While the danger of hovering "helicopter parents" has been overstated, many parents are fascinated with using ICTs to keep track of their children.[35]

James, a Connected Lives single father, regularly emails and IMs his son. His son especially enjoys IM-ing with his dad because he loves sending Yahoo's "emoticons" to him—like this winky ;-). James and his son often IM each other when they are at home together, but in separate rooms— they think it is fun. In Canada, mothers are the main communicators with children. Women use landlines, mobile phones, email, and IMs more than men do. Women contact children more than men do, especially using landlines and mobile phones.

James's experience with his son reflects broader social changes in how people communicate with their children, the time they spend with them, and ultimately how they parent them. Despite the concerns of parents about how they might navigate unchartered internet terrain, ICT use continues to become further embedded into daily home lives. Hence parents often become "digital parents," in order to work with their networked children who have grown up with ICTs, and who have different understandings of play, social interaction, leisure activities, and school work.[36] For better or worse, much of the teens' and children's time now is spent on the internet or with mobile phones. As researchers Caroline Haythorn-thwaite and Richard Andrew have found, when ICT use is tied into school, their use can enhance teaching and learning.[37] Moreover, playing video games habituates children into the routine ICT use that becomes an important part of their life and part of a skill set for jobs that have not even been created. And as communication scientists Hua Wang and Arvind Singhal have shown, games themselves can be designed to be educational—as well as enjoyable.[38]

Teens Texting

No other digital development highlights the new networked nature of family life better than short-message texting on mobile phones. As we noted, the rise of the mobile phone itself has changed household communication from a place-to-place basis to a person-to-person basis. Texting solidifies the transfer of activity from the group to the person. People use mobile phones to be a part of their social networks even when they are in the midst of their families—using their phones at the dinner table, riding in the family minivan, or sitting in front of the TV.

Teens are usually the most active networkers in households, especially girls. Mobile phones provide special power to teens because they allow them to conduct discreet private conversations without parental supervision. Pew Internet survey data collected in 2011 showed that three-quarters (75 percent) of those aged twelve to seventeen own mobile phones and 97 percent of that group use text messaging. Some 49 percent of those texters send fifty or more messages a day—1,500 per month—while a third send more than a hundred messages per day. Girls send or receive an average of eighty texts per day, compared with thirty for boys.

Teens in the Pew Internet focus groups also report that they like to text because it is asynchronous—two people don't have to be communicating at the same time. Teens can send messages and then simply await the answers. The persons receiving them can deal with the message as their situations allow. This gives them the possibility to interlace the communication into other parts of their lives without instant interruption. A young teen boy described this in the Pew Internet study: "I usually text my parents. Like, I guess although I'm not really supposed to in school, I'll just start texting them. I'll just be like, 'Hey mom come pick me up, this is happening,' Or just, 'Hey mom I forgot this can you drop it off?' I don't call that much."

Teens also text to cover their tracks. Unlike computer screens that often are open to parental view, teens texts privately from their mobile phones. As there is no sound when texting, teens prefer it to voice. For example, they can text their parents when the background noise of their location would give away too much information on their whereabouts. A high school boy described how his mother saw through this ruse: "Sometimes, I would text my mom, like, she, like, knows. She says to me, 'Sometimes I know you're doing something wrong if you're texting me.' She says she knows that, usually she's right if I tell her I'm supposed to be somewhere else. If she calls, she can hear the background. If she calls she says, 'Who are you with? Who are you talking to? Where are you?' If I'm texting, it doesn't give the location as much [because] she can't hear the background." Yet, mobile connections work both ways: Parents on the go can call their children to check in with them.

Texting can also be a buffer, keeping parents at an emotional as well as a physical distance. Since there is no synchronous interaction and since it is often more difficult for parents to construct a text message, teens use text messaging when they have to break bad news or make an uncomfortable request of their parents. It is not just parent-child communication patterns that have changed in networked families; child-parent communications have followed suit.[39] As one high school girl put it: "I usually text my mom whenever I want to ask her something or tell her something bad. That way I don't have to hear her yelling at me, like, give me a reason why I shouldn't go, or why she doesn't want me to go. 'Cause she doesn't text, she just writes short answers. I usually text her everything else I want her to know so I don't have to hear [her voice]."

Yet, teens described to Pew Internet how texting is an easy way to keep up with the flow of everyday life. Several teens noted that to call means that it is something that is important. However, if the teens are simply checking in with one another, texting is an easy way to touch base. A middle school boy said, "If I'm texting, it's just people I hang out with every day." And a high school girl said, "I text more than I talk . . . I call my family, but friends and stuff text me."

In short, texting is an emblematic activity for networked individuals—especially teens. It is personal. It can be customized to individual tastes and purposes. It allows people to be socially engaged with others outside the room. It provides ready access to multiple networks. It allows people to stay in touch with the flow of chatter that is coursing through their networks. It keeps parents at a distance, adding new ways for teens to respond to the previous generation's dialogue:

Parent: "Where did you go?"
Teen: "Out."
Parent: "What did you do?"
Teen: "Nothing."[40]

Netting Together

In the evocatively named *Alone Together,* social scientist Sherry Turkle argues that ICTs as a leisure pursuit may weaken the quality of time families spend together when individual family members spend it focusing on a screen instead of socializing with each another.[41] Some media depictions show fragmented families as heavy ICT users that text from the dinner table. Such commentators perceive ICT use as a solitary activity that is different from the TV experience of the 1960s, when families sat together in front of a single screen. For example, newspaper columnist Steve Collins writes:

There's no we in iPod. The TVs, computers and hand-held screens have . . . multiplied, enabling us to ignore each other any time and any place. Will we someday envy old-fashioned families who at least used to zone out in front of the same TV screen together, even as the Cassandras of the day prophesied looming social apocalypse? . . . Paradoxically, even as [the teenagers] seem unaware of each others' presence in the same room, they are interacting with each other online.[42]

Our evidence shows that people use ICTs to support, supplement, and enhance face-to-face interaction with family members. Networked families spend time netting together online: Most spouses have gone online together. Married childless couples spend more time online with their spouses than married parents. Just under half of them say they are "often" online with their spouses as compared with only one-third of married parents (figure 6.8).

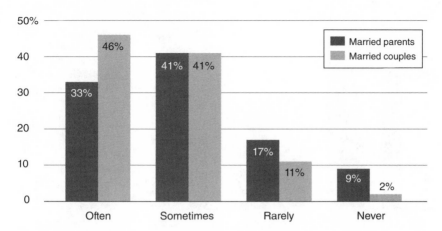

Figure 6.8
Frequency of spending time online with spouse.
Source: Pew Internet & American Life Project, Networked Family Survey 2008.

When couples go online together, their shared online activities are shaped by common interests, household needs, and talking to family members. For example, Connected Lives participant Theresa goes online with her husband to research things they need to make a decision about, such as buying cars. They also have been fans of "The Amazing Race" TV show. After they watch it, they go to the internet to check online clips for the next "fun things." Other participants go online jointly to plan family vacations. Olivia, a married mother in her 40s, told the Connected Lives researchers that she often goes online spontaneously with her husband when they want to follow up on information that comes up during conversations or while watching TV. "My husband looks at real estate online— all over the place," she said. "Just the other night we were both sitting and looking at condos in Mexico. 'Let's do it!'" Olivia laughs. "We were looking at property in Greece but it was ridiculous. On analyze.ca, you can look all over Canada on there, so we will often do that. So, we will sit together, mainly for that purpose—look at houses or something and dream."

Nearly 90 percent of American parents say they have gone online with their children. More than half of married mothers and one-third of married fathers go online "often" (figure 6.9). Single parents spend more time online with children than do married parents. Three-fifths of single mothers and nearly half of single fathers go online "often."

For example, Jennifer, a fortyish single Connected Lives participant looks forward to going online with seven-year-old Katie. For Jennifer, the

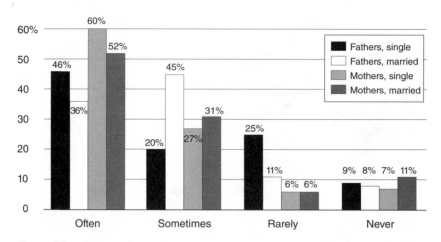

Figure 6.9
Frequency of spending time online with children.
Source: Pew Internet & American Life Project, Networked Family Survey 2008.

notion of netting together combines parent-child learning sessions, game playing, and boundary setting. Jennifer plays online games with her daughter Katie—choosing the games and showing her how to play—but feels her daughter is too young for an email address at this time, and she monitors what Katie is doing online. Although she knows that Katie knows how to search, her parental rule is that Katie has to ask and Jennifer has to be there.

Like Jennifer, many parents in the Connected Lives and Pew Internet studies worry about what their kids are doing online when they are not supervising them: what they are looking at, and who they are talking to. As a consequence, rules have emerged inside families for safety, netiquette, and metiquette. Beth Herina, in New Jersey, has a rule: "no texting at dinner" or on family outings: "not when it is family time."[43] In Toronto, some Connected Lives parents set themselves up as administrators of their children's computers to keep an eye on what software they are bringing in. They put their computers in public spaces, such as the den or kitchen, they ask their children about what they are seeing, and they randomly look over their shoulders at their screens.

Indeed, parental oversight of their children's ICT use is commonplace. Fully 85 percent of parents of American teens who use the internet have rules about the websites their children can visit; 85 percent have rules about the kind of information their children can share online with others, 69 percent have limits on how much time their kids can spend online, 53 percent have filters on family computers that keep users from visiting certain websites, and 45 percent have monitoring software to check up on what their children do online.[44] Similarly, a majority of parents impose rules about their children's use of mobile phones and use the phone as a monitoring and punishment tool: 64 percent of parents of teens who have mobile phones have looked at the contents of their child's phone such as the address book and log of calls or texts, 62 percent have taken away their child's phone as a punishment, 52 percent have imposed limits on the time of day when their child can use the phone, 48 percent use the phone to monitor their child's location, 46 percent limit the number of minutes their child can talk on the phone, and 28 percent limit the number of texts their child can send.

Many parents also embrace the internet with their children, going online with them to share interesting things, to find information, or simply to play. Connected Lives participant Tanya works with her husband and two children to use the internet throughout the day: doing everything from researching new toilets for home renovations to playing a computer game together to helping her kids do their schoolwork. Connected Lives participant Felicia also helps her son with schoolwork: "Oh, yes. Last year he was doing some work for a history project and we did some research on costumes and dress of the time." At other times, being online together is for fun. Henry, a married father of two, goes with his children to the Thomas the Tank Engine website because there are different little games, projects, and puzzles—his son loves to do internet puzzles. Sheena's sixteen-month-old daughter loves the games and music on the Treehouse TV children's website. More spontaneously, Greg, a married father of two, describes going online with his family when one of them finds something interesting online and calls, "Come and look at this!" Then all four family members crowd around the screen.

These shared online experiences show that home internet use does not have to be solitary. Showing and sharing on the internet can sometimes bring household members together, just as TV watching sometimes does. Although people may not have comfortable sofas at their desktops to accommodate an audience, they often have a laptop that can travel anywhere in the home, or their home computers are in places, such as living rooms, that allow several viewers.

Joint activity may redefine family boundaries as Facebook broadens from its original young adult base to encompass older adults. Parents who would never think of following their children in person often ask to "friend" them on Facebook. Pew Internet surveys in mid-2011 show that a substantial majority of parents use social networking sites and that 81 percent of the parents who use such sites are "friends" with their children.[45]

On the one hand, parent and child are sharing experiences. On the other hand, the Facebook link gives parents a good deal of information about what their children—and their children's friends—are doing. Some parents go even further, installing spyware on their children's computers to record the pages they view and the keystrokes they enter. Parents justify this to prevent misadventures and cyberbullying—reportedly experienced to some extent by about 30 percent of teens. However, parental surveillance of Facebook can become voyeuristic.[46]

Networked Families: In Connected Motion

Networked families have adapted to the Triple Revolution. They use ICTs to bridge barriers of time and space, weakening the boundaries between public and private life spaces. The mounting and interrelated changes in the composition of households—such as the life-cycle complexities of marriage and divorce and decisions to have children—mean that today's households are varied, complex, and evolving. Networked families use ICTs to mediate these complexities and adapt ICTs to their varied needs.

The findings from the Pew Internet and Connected Lives studies contribute to understanding of how networked individuals—who are perceived as technological nomads—operate within their daily activities and within households and families. Not only have families changed in size and composition, they have also changed in their lifestyles. ICTs have become thoroughly embedded in families' everyday lives, helping them stay connected and in motion. The internet and mobile phones connect family members as they move around, help them find each other, and bring them together for joint work and play. The result is that ICTs—often in conjunction with personal automobiles—have paradoxically provided household members with the ability to go their separate ways while at the same time keeping them more connected. Families have less face time, but more connected time, using mobile phones and the internet.

7   Networked Work

To see what white-collar work was like in the 1960s, watch the retro TV show, *Mad Men*.[1] It portrays life at a Madison Avenue (New York) advertising agency, where all the employees work nine to five. Men sit in separate offices or gather in small group meetings—each man with a young female secretary in the public space in front of their offices. Along with gender stratification, there is ethnic and religious homogeneity: All of the men are white Christians.

Except for the occasional phone call to other organizations, all the men work at desks, using pen and paper—only the secretaries type. To the men, multitasking is holding a cigarette, a glass of Scotch, and a notepad. They communicate only face to face. Work spills over into family life only when the lonely housewives object to their husbands working late at the office or accompany them to company parties. Work, itself, never comes home, while wives and children phone only in emergencies—and shock the system if they ever come to the office.

Although *Mad Men* is fiction, it fits well with William H. Whyte's 1956 bestseller, *The Organization Man*, which describes American organizational life after World War II. The book shows how loyal and conformist "organization men" (no women) fit into hierarchical structures with a single boss coupled to a stable work group whose members conformed to clear corporate-status pecking orders in conference rooms.[2]

Since the Industrial Revolution of the 1800s, the dominant images of work have been of employees in large factories or offices, in small groups such as retail shops, or as lone workers such as truck drivers. While there are still plenty of such factories, offices, shops, and lone workers, a host of forces has been transforming work from individual or group activities to networked activities. Many people work in a global economy where corporations deal agilely with turbulent market environments by fostering *networked work* as they participate in multiple teams often for multiple purposes. And they do so in *networked organizations* whose workers may well be physically and organizationally dispersed.

In this chapter, we examine the ways in which an organization's work has become distributed to different locations, the ways in which workers work in multiple teams on multiple projects, and the involvement of the Internet and Mobile Revolutions with these changes. We caution that our discussion is more tentative than in the preceding Networked Relationships and Networked Family chapters. That is because management gurus' assertions and advocacy about how networked organizations can—and should—operate currently outweigh evidence and analysis about how they actually do operate. We further caution that the turn to a networked operating system in workplaces is uneven. Neither all workers nor all organizations in North America have become connected. Moreover, the percentages of internet-using workers outside of North America are lower: A 2008 International Data Corporation (IDC) survey in seventeen mostly developed countries found that only 16 percent of the global workforce use multiple digital devices and new communications applications.[3] The figures have surely risen since then, but the larger point is true. This is decidedly a fractional minority of the global workforce.

Fostering the Turn to Networked Work in Networked Organizations

Five related trends have been encouraging the trend to networked work in networked organizations. First, the globalization of work, consumerism, and travel has expanded and diversified the reach and purview of corporations. As a rule, workers and companies are in contact with more colleagues and customers these days than they were in the past.

Second, there has been a shift in developed countries from growing, mining, making, and transporting things—*atom work* in the material economy—to selling, describing, and analyzing things via words and pictures—*bit work* in the information economy. The rising numbers of what urban scholar Richard Florida calls the "creative class" have fostered this trend toward bit work. The percentage of "creatives" in the American workforce has doubled in a generation: from 22 percent in 1960 to 43 percent in 2006 (see figure 2.11 in chapter 2). They are "people in science and engineering, architecture and design, education, arts, music and entertainment, whose economic function is to create new ideas, new technology, and/or new creative content."[4] In short, they are people who usually manipulate bits more than atoms.[5]

Third, although the shift to networked work and organizations began before the Internet and Mobile Revolutions, these two revolutions have accelerated the shift. The two revolutions allow "bit workers," who use computers to work with ideas and data, to have more ability to network—in multiple senses of that word—than "atom workers," who work with physical objects and often on assembly lines. For one thing, it is easier to use information and communication technologies (ICTs) to connect and collaborate when workers are pushing bits by calculating, searching, drawing, and writing. Some of the most ICT-connected workers have jobs built around creative effort rather than manufacturing or standardized paper pushing.

Fourth, the internet allows people to communicate and access shared information and databases at a distance—from publicly available libraries to secret corporate records[6]. Some workplaces can exist anywhere. As organizational scholars Paul Adler and Charles Heckscher put it:

The large corporation combining bureaucracy and loyalty-based community meets its limits in organizing the production of knowledge. Bureaucracy, as has been frequently documented, is very effective at organizing routinized production, but it does very poorly at these complex interactive tasks involving responsiveness and innovation. Under bureaucracy, knowledge is treated as a resource, and is therefore concentrated, along with the corresponding decision rights, in specialized functional units and at higher levels of the organization. However, in organizations that are competing primarily on their ability to respond and innovate, knowledge from all parts of the organization is crucial for success, and often subordinates know more than their superiors. Innovativeness and responsiveness cannot be rigorously preprogrammed, and the creative collaboration they require cannot be simply commanded.[7]

In other words, creative bit work may function better in networks than in the hierarchies of bureaucracies or the rugged individualism of markets. Of course, this is a bit of caricature as no hierarchies follow a strict tree structure—there are always side alliances and work-arounds. And no markets are truly atomized—market players are always making alliances and deals. But while networks permeate hierarchies and markets, they are more palpable now.[8]

The fifth and less pervasive shift has been enabled by the Mobile Revolution, which allows some bit workers to be productive with their laptops and smartphones while away from their desks. While the purchase of desktop computers have leveled off and of wired-in landline phones have declined, the purchase of smartphones, tablets, and laptops has soared. It is often as easy to push bits at home or in the coffee shop as it is in the office.

"Road warriors" do their work at home, offsite workplaces, hotels, planes, trains, and automobiles. To be sure, not all work is networked. Many people still work on assembly lines, sit in separate cubicles—white-collar assembly lines—or work alone in shops or trucks and taxis. Even so, the internet, tablets and smartphones have become pervasive at workplaces.

The Diffusion and Use of ICTs

There have always been long-distance communications within and between organizations. That is how Egyptian granaries fed Rome and the British East India Company ruled nineteenth-century India. That is how the smart use of alerting systems gave some nineteenth-century American companies a competitive advantage by being the first to know what ships were coming in and what they were carrying. Yet, communication this way was always difficult, slow, expensive, and intermittent, even when telegraphs, railroads, phones, cars, and planes came along. In essence, most communication at work was in person.[9] The Internet and Mobile Revolutions afford different opportunities for many networked workers to work from multiple locations, including their homes, and do so in less-hierarchical organizations.

At first, computers were giant mainframes, used principally for accounting and corporate services. It was only in the late 1980s that a few innovators and early adopters began to use truly *personal* computers, with even fewer having the ability to communicate online with coworkers. For example, the Cavecat and Telepresence projects at the University of Toronto spent several million dollars between 1989 and 1995 to build systems where a set of people could use their computers to text, talk, and video conference with one another. They had to design and build the hardware and software from scratch, and there was no user-friendly internet software.[10]

That all changed as computers became personal devices and then became networked. Management scholar Andrew McAfee identified what he dubbed "corporate America's ongoing love affair with geek gear." ICT spending in corporate America has grown greatly since the 1970s, outpacing other types of equipment expenditures, even after taking into account a growing workforce and the declining price of computer hardware and software. For instance, the nominal annual U.S. corporate IT investment from 1970 to 2008 increased from about $5 billion to almost $350 billion. Overall IT investment rose exponentially in the United States, from about $100 per employee in 1970 to $3,000 in 2008. While ICT expenditures made up less than 10 percent of the total corporate investment in equipment during the 1970s, they accounted for around one-third of total annual spending on fixed assets in the first decade of the twenty-first century. American industries became 5.5 times more ICT intensive between 1995 and 2008.[11]

At present, low-cost, portable, and increasingly usable software has become so diffused that most homes and workplaces have at least one personal computer with almost every computer connected to a network, either the publicly available internet or a private organizational network (see chapter 3, "The Internet Revolution"). The trend is striking. In its first survey in March 2000, Pew Internet found that 37 percent of full-time workers and 18 percent of part-time workers have internet access at work. By mid-2011, the figures grew to 76 percent of full-time workers and 52 percent of part-time workers who use the internet on the job. In Pew Internet's 2008 Networked Workers survey, 60 percent of workers use the internet every day at work, while only 28 percent say they never used computers. The same survey shows that ICT use at work varies significantly by the nature and prestige of the profession. About three-quarters of professionals, managers, and executives and half of clerical, office, and sales workers use the internet at work. By contrast, workers in service industries and skilled trades are less likely to use the ICTs at work.

Many American workers use a variety of ICTs, especially smartphones and personal computers. The Pew Internet Networked Workers survey showed that 77 percent of American workers use a desktop computer, 50 percent use a laptop, and of course, many workers use both. Other workers use ICTs that are not part of the internet: truckers, taxi drivers, and plumbers connected to their dispatchers and customers; cashiers and bank tellers at their terminals; consumers at self-service checkout and automatic teller machines. Workers also connect—overtly or covertly—via mobile phones. They are pervasive at American workplaces: Almost nine out of ten American workers had a mobile phone in 2008.

The same Pew Internet survey also found that almost all workers (87 percent) who have email at work check their work email at least daily (table 7.1). Half of all workers checked their phones several times an hour, with 37 percent checking constantly. One third (33 percent) of those who own a mobile phone regularly text with colleagues and two-fifths (39 percent) regularly text with friends and family. (The percentages are undoubtedly higher now.) In addition, more than three-quarters have personal email accounts and two-fifths (39 percent) of them check their account at least daily. The survey also found the first indications that social networking

Table 7.1
Frequency of Workers' Use of ICTs for Work and Personal Purposes

	Constantly	Several Times an Hour	Several Times a Day	About Once a Day	Every Few Days	Less Often	Never
Employed respondents who have personal email (n=751)							
Check your personal email	7%	4%	12%	16%	6%	9%	46%
Employed respondents who have work email (n=609)							
Check your work email	37	13	22	15	5	2	6
Employed respondents who use instant messaging (IM) (n=325)							
Send IMs to colleagues at work	8	5	13	5	5	8	55
Send IMs to friends or family	3	•	8	9	5	11	62
Employed respondents who text (n=455)							
Send texts to colleagues at work	3	2	6	6	6	11	67
Send texts to friends or family	3	2	11	12	11	16	44
Employed respondents using social networking sites (n=245)							
Communicate with colleagues using social networking sites	2	•	1	2	8	5	83
Communicate with friends or family using social networking sites	2	0	3	4	7	11	72

Source: Pew Internet & American Life Project, Networked Workers Survey 2008.

sites were being used at the workplace for work- and personally related communication.[12]

Not only are people more connected, but they are spending more time online. The mean amount of active internet use at work doubled from 4.6 hours per week in 2001 to 9.2 hours in 2010.[13] This means that those who have internet access at work are spending the equivalent of an entire workday per week using the internet. Of course, the actual numbers vary enormously for different individuals, and with broadband enabling computers to be on all workday, it is difficult to know what active use means. At the same time, ICTs make it easier for workers to relax at work. The Pew Internet Networked Worker survey found that while at work 22 percent shop online, 15 percent watch videos, 10 percent use online social or professional networking sites, while 3 percent play online games.[14]

What does all this ICT connectivity mean? The survey shows that despite diversions workers who use ICTs tend to be more productive, flexible, collaborative, and better connected. However, they also work longer hours, and are more distracted and stressed. For organizations, a McKinsey study reported that "technologically-enabled collaboration with external stakeholders [bridging ties] helps organizations gain market share from the competition."[15]

How Networked Workers Operate

Traditional work groups have featured densely interconnected relationships in physically compact spaces. People in such "fishbowl" offices work within clearly defined, ordered groups. Almost all communications are within the group and visible to all. In contrast to the walled silos of traditional workplaces, boundaries in networked organizations are permeable, fostering workers' interactions across departments. Compared with traditional "fishbowl" organizations, the "switchboard" is the descriptive metaphor for networked organizations, where each person plugs into a direct connection to his/her own colleagues. In a switchboard situation, the structure of networked organizations is more flexible, laterally coordinated, team based, and boundary spanning.[16] Many networked workers communicate frequently, easily, rapidly, and cheaply over great distances, using a battery of media—internet, phone, texts, emails, social network sites, blogs, instant messages, video encounters—whatever seems appropriate and is available. Increasingly, their tools and data are no longer stored on their own computers but are accessed from almost anywhere from the "cloud" on the internet. They link both within the organization and, if outward facing, to clients and suppliers.[17] Creativity and information gathering, especially, are encouraged by flexible arrangements with bosses, peers, and subordinates—pushing autonomy and authority onto networked individuals.

Yet, there are costs to all this networking, some inherent in the concept, some afforded by the ICT technologies, and some provided by the dynamics of labor-management relationships. For example, as networked workers move between teams—and places—they risk having less of a sense of "ownership" of a particular piece of work, and their attention and loyalty can be divided. When physically dispersed, workers may have difficulties in information sharing, coordination, and control, which means they may spend appreciably more time getting work done. These difficulties are increased when new teams are formed who do not know each other and do not share a common culture.[18]

The comparisons in table 7.2, although somewhat oversimplified, show how much traditional and networked office work can differ. Taken together, the comparisons suggest how a culture gap may develop between networked workers in networked organizations and workers doing repetitive work in bureaucratic organizations.

Networks at work may be larger than traditional work groups, as employees shift among teams locally and virtually. For example, the Pew Internet Networked Worker survey shows that four-fifths (80 percent) of American workers who use ICTs at work say these technologies have increased the size of their communication networks, with almost half of the workers (46 percent) reporting that these technologies have greatly increased their network size. The survey also shows that three-quarters (73 percent) of American workers who use ICTs at work say that ICTs have improved their ability to share ideas with colleagues.

As communication often crosses work group and organizational boundaries, information becomes a key asset and its flow becomes critical for success. In addition to standard leadership positions, strategic network positions contribute to superior individual performance, from those who can usefully connect various parts of the organization. Workers who have a higher level of "betweenness centrality"—they are in the shortest paths linking employees in the firm—tend to have better performance. One study has found different media-use patterns for the central "thought leaders" of an interorganizational scholarly network. Such leaders prefer giving advice in person while peripheral members tend to rely on email for giving and getting advice.[19] Knowing who knows who knows what and who knows whom becomes crucial for problem solving, leading to the growth of social media within organizations. One study of knowledge

Table 7.2

Traditional versus Networked Offices

Traditional "Fishbowl" Workgroups	"Switchboard" Networked Individuals
All work together in the same room	Each works separately
All visible to each other	Office doors closable for privacy
All have physical access to each other	Glass in doors indicate interruptibility
All can see when a person is interruptible	If door's locked, must knock; if door's open, request admission
All can see when one person is with another; neighbors have high visual and aural awareness	Difficult to learn if person is dealing with others, unless door is open
Limited number of group participants	Switch among large number of potential interactors
Densely knit: most are directly connected	Sparsely knit: most don't know each other or unaware of mutual contact; no detailed knowledge of indirect ties
Tightly bounded: most interactions are within a small group	Loosely bounded; many different people contacted in multiple workplaces; each person operates individually
Loyalty, motivation to the small work group	Self-motivation; loyalty to the career, profession, larger organization
Frequent contact; recurrent interactions	Variable, changing frequency of contact
Easy, swift to coordinate and control	Difficult, slower to coordinate and control
Long-duration ties	Switching between multiple ties
Work only at the office	Work at home: part-time, full-time
Sense of group solidarity: name, collective identity	Collective activities; transient, shifting sets
Mentoring by co-located workmates	Little mentoring: harder to learn tacit knowledge
Repetitive tasks, faster deskilling, lower wages	Multiple tasks foster multitasking; higher wages
Sense of ownership, loyalty to a work process	Commitment divided among multiple work processes
Little sense of the whole	More sense of how pieces fit together
Intranet/virtual private network: internal links only	Internet: links with outside organizations
Direct visual and aural social control by supervisor and group; plotting done off premises	Internalized normative control; electronic surveillance; plotting done by ICTs
Top of the hierarchy	Betweenness centrality

Source: Barry Wellman © 2011, used with permission.

workers in a large global information organization found that workers who are especially good in linking others in the organization generate more revenue for the firm.[20]

ICTs also facilitate friendships at work. For example, one study of computer scientists found that there is as much friendship contact at work as professional contact, and other research has shown that multiplex relationships (such as work plus friendship) help productive communication within and between organizations.[21] Observational data support this. When MIT organizational analyst Sandy Pentland put wearable "reality mining" sensors on workers, he discovered: "At more than a dozen companies . . . , social interaction has been found to be an important element in productivity. Staff who socialize trade information about how to do their jobs better. People who are cut off aren't wise in the ways of the company and don't have social support."[22]

When they chat about work or each other, workers often switch among different media, calculating which may be tactically important for the relationship and the situation. For example, Anabel Quan-Haase and coauthor Wellman found much media juggling at "KME," a high-tech Chicago company. The communication norm was to respond quickly to IMs from coworkers sitting at nearby cubicles. But coworkers switched to email when they had something long and complicated to communicate. When the subject got sensitive, they went out for face-to-face get-togethers at lunch or coffee.

People sometimes chose a platform because they do not want to communicate too much. As Andy at KME said: "I use IM a fair amount because there are times where I want to know something, but I don't want the other person to know how I am reacting or responding. And then, I can have a moment to think and compose myself, while figuring out how to respond."

KME's employees thought that getting up and walking to someone else's cubicle was too visible and too disruptive. For example, when an issue came up, Linda emailed Desmond asking him to meet, even though they sat across from each other. They emailed back and forth a couple of times to arrange a good time to have a coffee outside the building and talk.

Yet, despite the computerized connectivity at KME, organizational boundaries still mattered. Most communication stayed within a department, and the remainder usually went to KME employees elsewhere in the firm. Supervisors sent and received high volumes of messages and emails. Similarly, when a large Canadian telecommunications company set up a program for a few computer-based employees to "telework" at home, the structure of authority remained in place. The only major change was in work schedules, with more working hours replacing commuting. In short, organizational imperatives tended to outweigh the ease of cross-boundary communication. However, several people did become informal bridges carrying information and ideas between departments.[23]

Workplace design in many networked organizations favors open, fluid spatial arrangement that encourages interactions. Former CEO Eric Schmidt described Google as the "best example of a network-based organization." While self-congratulatory, Schmidt's remark points to Google's penchant for having small, focused, short-term work teams that are repeatedly reformed and refocused. Blogs disseminate information about who is working on—or thinking about—what. YouTube videos are created to share knowledge across teams. Frequent peer reviews foster quality control for networked workers' independent projects. Google employees work in the Googleplex—a campus-like environment designed to encourage interaction and discussions. These include cafeterias providing free lunches, game rooms, salsa dance clubs, and small shared offices instead of cubicles. Public spaces in the Googleplex aim to foster serendipitous contact among networked individuals beyond the traditional unanticipated meetings at the water cooler or photocopier.[24]

The Rise of Networked Organizations

Management analysts have suggested that flexible, networked organizations have several advantages in allocating people and resources to projects and at the same time can foster autonomy, flexibility, and decentralized control.

People often work in multiple projects with different teams. This allows firms to assemble ad hoc teams with diversified talents and perspectives. As workers shift among teams, they can develop cumulatively larger networks of expertise that are "glocal," with both local interactions and global connectivity. Instead of submitting to the traditional hierarchical mode of authority, workers have more discretion about the work they jointly accomplish. Networked organizations have advantages for boundary spanning, as employees work and network between work groups and organizations—and at times, between continents.

Networked workers in networked organizations are more likely to know and collaborate with colleagues outside of their immediate units. They can be knowledge brokers who receive and transmit diverse and novel knowledge and expertise within and between organizations, making them more likely to solve difficult problems and complete projects. At times, they become members of ongoing "communities of practice" within and between organizations, where they can access tacit knowledge and lore, earn professional reputations, and develop useful concepts. For example, organizational analyst Arent Greve has shown how boundary spanners in the Norwegian oil industry use their bridging ties to obtain information from other firms about important new technologies: "All successful firms had these links; the less successful did not." Yet, not all is boundary-spanning, as employees also call on their strong internal "bonding" ties for help in completing their projects.[25]

The organizational structure of networked organizations can be relatively flatter than traditional organizations, have less hierarchical reporting relationships, and be a more informal work culture. Moreover, networked organizations tend to enforce fewer bureaucratic rules. This encourages their employees to have multiple reporting relationships as well as open, fluid spatial arrangements. To make teams work, both hardware—such as knowledge and skills—and software—such as trust and commitment—are important. With less top-down supervision, team members need to obtain a shared understanding of the team's goals and actions. Moreover, the quality and the pace of knowledge transfer depend on the structure of the informal networks among employees. Therefore, networked organizations often encourage informal interaction and sociability in order to develop supportive networks of information exchange. Because networked organizations are often geographically distributed and internet based, they can offer advantages in the flexible organization of work, reduced real estate costs (fewer desks needed), more work time through less commuting, and rapid access to knowledge. Nevertheless, the thinness of the reporting structure necessitates active coordination to avoid duplication and different teams charging off in opposing directions.

Communication scientists Peter Monge and Noshir Contractor's review of the literature on this suggests that centralized organizations are more efficient for routine tasks and for facilitating the downward communication of knowledge and commands. They are less effective when tasks are not routine, as lower levels may lack the knowledge to create new knowledge, and they may lack the ability and incentives to transmit new ideas upward.[26]

By contrast, decentralized networks are more efficient for creativity and collaborative problem solving where people have more autonomy to find and use knowledge. Sociologist Ronald Burt argues that having connections to different work teams provides more diverse information because such inter-team networks help to find people who are interesting and think differently.[27] Similarly, organizational analysts Lynn Wu and associates demonstrate that the more "structurally diverse" the networks are in a large global information firm, the better the performance of both individuals and their work teams.[28]

Management experts argue that the process of shifting to a networked organization requires changes in the way work is scheduled and organized, as well as in how workers are managed. If early studies emphasized how technologies affect organization behavior and performance and ran the risk of technological determinism, more recent studies more carefully pinpoint the impact of ICTs on the bottom line. Networked organizations often require a workplace culture that encourages collaborative sharing of knowledge and expertise. Bill McDermott, president of SAP America, argues that using mobile tools "unleashes freedom. Because people who were formerly tethered to somebody else's rules, somebody else's processes, can now be free to manage their work life on their own terms."[29] Yet, such thoughts are essentially exhortation: Few studies besides that of Wu and associates have provided evidence about the conditions in which multiple teams actually work better than in fixed bureaucracies. As organizational analysts Steven Poltrock and Gloria Mark note, "A business case is easy to develop when a technology directly increases product sales or decreases development or manufacturing costs." However, they warn, "Working with people in our major divisions, we were able to develop a compelling story about likely benefits but could not identify specific, accessible budgets that would profit from the technology."[30]

In short, proponents claim and some evidence supports the argument that the networked/team approach has affected organizational operations by helping make problem solving more flexible and nimble. However, we do not want to sound like hype-mongering consultants claiming that most organizations have joined the networked bandwagon. Many organizations continue to be traditional bureaucratic hierarchies in whole or in part.

Working in Multiple Teams

When people work in multiple, fluid teams, they have more reporting relationships and fewer single-hierarchy relationships. The National Establishment Survey in 1992 showed that 55 percent of U.S. firms with fifty or more employees have adopted self-directed work teams to some extent.[31] More than a decade ago, sociologist David Knoke found that in more than

40 percent of the workplaces at least half of workers who are most directly involved in producing goods and services work in self-managed teams. Working in teams appears to have become even more widespread since then, especially in high-tech industries.[32]

One study of knowledge workers in a large American organization found that two-thirds (65 percent) of recent MBA graduates worked in more than one team at the same time, although most devoted almost half of their working time to a core team.[33] More broadly, the Pew Internet Networked Worker Survey found that two-thirds (64 percent) of American workers work in at least one team and two-fifths (41 percent) belong to multiple teams (figure 7.1). At one extreme, one-quarter (23 percent) of American workers are members of only one work group. At the other extreme, 15 percent of American workers are members of at least five teams. And what tends to be forgotten is that nearly one-third (30 percent) of Americans work alone as shopkeepers, truck drivers, and the like.

Intel is a good example of an organization filled with multiple teams of bit workers designing the chips, administering the company, marketing their products, and thinking about the future. (However, many Intel workers are on traditional—but high-tech—assembly lines making computer chips.) At the time of an evaluation in 2004, three-fifths (61 percent) of the workers studied used email, internet conferencing, and other means to work together in three or more teams, with each team consisting of two

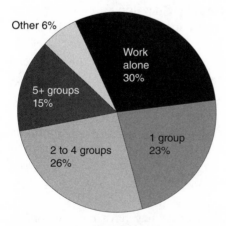

Figure 7.1
Percentage of workers who work alone, in a group, or in multiple groups. Sample size = 1,000 employed workers.
Source: Pew Internet & American Life Project Networked Workers Survey 2008.

to ten people. About half worked from home during normal business days at least once per month, and about half also regularly used mobile phones or laptops to keep in contact with colleagues.

Intel's teams span the globe. For example, approximately 70 percent of the Intel knowledge workers surveyed collaborate at least once a month via ICTs with colleagues whose first language is not English—although they all use English when they communicate. While Intel workers are generally comfortable with working in teams, the ICT-connected teams especially benefit. Seventy percent of team workers have never met in person. The researchers show that the spatial spread and different cultures of team members does not lead to lower performance, but using incompatible types of ICTs within a team does lower performance.[34]

At times, distributed work can mean physically disconnected work. As one administrator in another networked organization told us: "We have one guy who worked for the company for eight years before anyone ever met him face to face—he lives on a remote island. I was struck by the fact that people referred to him by his username rather than his real name."

Yet the employee got the work done.[35]

The blurring of boundaries can also cause problems. People may not know who to call to get a decision or to troubleshoot difficulties. This became apparent in the BP Gulf of Mexico oil spill disaster in the summer of 2010. As local official Billy Nungesser lamented: "If you asked me today, 'who was in charge: the Coast Guard, BP, or their subcontractors?' I couldn't look you in the eye and tell you who was making the decisions." When U.S. President Obama heard that, he told Nungesser to call him directly. Not everyone can call the president directly nor, we daresay, can Nungesser most of the time.[36]

Blurring the Home-Work Boundary

Along with the ability to enable work almost anywhere, ICTs at home afford individuals the opportunity—and the burden—to work at any time. Many workers stay connected beyond regular working hours, even during weekends and vacation time. Thanks to the Internet and Mobile Revolutions, some workers have become "teleworkers," performing their work in multiple locations: at home, clients' sites, cafés, airline lounges, hotel lobbies, on the go, or on the beach.

The Pew Internet Networked Worker survey showed that 60 percent of employed Americans do some work from home and 18 percent work from home almost every day. Other surveys also show high percentages:

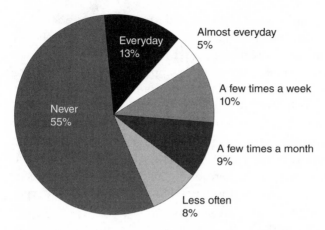

Figure 7.2
Frequency of internet use while working at home. Sample size = 1,000 employed workers.
Source: Pew Internet & American Life Project, Networked Workers Survey 2008.

The University of Southern California's Center for the Digital Future reports that more than three-quarters of Americans had gone online at home for work in 2008. Half (49 percent) of the Intel workers surveyed work at least part-time at home, 17 percent are mobile workers while traveling (airports, hotels, etc.), and 16 percent work at different Intel locations. While the share of teleworkers has been on the rise, few work entirely at home. Part-time and over-time work is more prevalent. In 2008, nearly half (45 percent) of American workers did some work at home (figure 7.2).

The Pew Internet Networked Worker study shows that about half of employed email users check their work emails on weekends, sick days, or before and after going to work for the day (table 7.3). More than one-third of employed email users check work email while on vacation. Phone calls also follow workers outside the normal working hours. Nearly two-thirds (63 percent) of American workers make or receive telephone calls related to work on weekends or when they are sick at home.[37]

And then there are the rebels who defy the new convention. When computer scientist Eleni Stroulia went off email during an eight-day holiday in February 2011, she found 470 emails piled up awaiting her return. Although she missed some opportunities, she says, "In the end I am terribly happy I stayed out of email. I had an amazingly relaxing vacation. I read a couple of books, I had some memorably inane conversations

Table 7.3
Frequency of Work-Related Emailing and Phoning beyond Regular Work Hours

	Often	Sometimes	Rarely	Never	No Email at Work
Respondents who check email related to work (n=807)					
On weekends	22%	18%	10%	41%	9%
On vacations	11	14	9	59	7
Before you go to work for the day	17	9	7	59	7
After you leave work for the day	19	16	9	48	8
When you are sick and cannot go to work	25	15	6	45	8
On the go, such as when you are commuting or shopping	7	6	5	75	8
Respondents who make calls related to work (n=1,000)					
On weekends	13	21	29	38	
On vacations	5	11	20	64	
Before you go to work for the day	10	17	24	49	
After you leave work for the day	11	25	25	39	
When you are sick and cannot go to work	14	21	27	37	
On the go, such as when you are commuting or shopping	12	17	18	53	

Source: Pew Internet & American Life Project, Networked Workers Survey 2008.

with both of my children, and my mailbox is not threatening any more this evening. And, maybe, I may start a different type of relationship with email. Where I do not respond immediately, but I let things sit for a week. It looks like email does not go bad in a week."[38]

About three-fifths (58 percent) of the respondents to the Pew Internet Networked Worker survey report that ICTs have allowed them to have a more flexible schedule and 24 percent say that ICTs have increased work flexibility a lot. Yet, teleworking varies greatly by job type, as different professions vary in the extent to which the nature of their jobs—and the nature of their employer—enables them to work outside of the worksite. Bit workers in professional and managerial occupations spend an average of about 10 percent of their working hours at home. By contrast, atom workers in occupations that require them to be at specific worksites (such as restaurants, health care, factories, mining, and construction) hardly ever telework.

While some teleworkers value the convenience as well as the elimination of commuting time, organizational analysts are trying to figure out the positive and negative impacts that telework has on cost efficiency, operations, workers' productivity, and commitment to the organization. Some companies encourage working at home to reduce real estate costs—a savings of over $10,000 per worker per year in major cities, one corporate executive told us. A meta-analysis of forty-six studies in 2007 found that, overall, telework increased workers' autonomy, job satisfaction, flexibility, and performance.

Most of the teleworkers interviewed by the Connected Lives researchers say that ICTs have improved their productivity and their quality of life. They say that telework reduces commuting stress, allows more flexible work schedules, and avoids interruptions in the office. Moreover, when a situation arises in which one parent must stay home with a child, telework helps many to deal with the situation easily rather than having to make awkward arrangements which might affect their jobs.[39] Yet, there are important variations. U.S. data shows that working at home aids multitasking, especially for those Creative Class workers who have some control over how they schedule work and other tasks. American women who telework have complex feelings about this, valuing the flexibility of taking care of household matters in between work activities but resenting the extra burden imposed by the prospect of contact at almost any hour. Male managers and professionals, with more authority and responsibility, are the most likely of all American men to report that work interferes with what is nominally their time off.[40]

Olivia, interviewed in the Connected Lives study, is a good example of a full-time teleworker. In her late 40s, she has worked full-time in public relations for two decades, doing corporate writing: speeches, press releases, and newsletters. She does not like working from nine to five at the office: "There was a lot of time in the office where I wasn't really making good use of my time." Getting a guaranteed workload of fifty-two hours a month from her primary client, she switched to home-based telework.

Many workers routinely cross the home–work boundary. When they do paid work at home, many workers try to tailor the focus, goals, and interpersonal styles that are appropriate to each side of the boundary. Because it is sometimes hard to maintain the boundary at home, they often have an off-limits room where children and spouses know that "Mommy is at work." Almost all of those in the Connected Lives study who work at home use ICTs to stay connected with spouses and children, who are out of the house, during "work" hours. Full-time home workers favor landlines while part-timers favor mobile phones. Yet, the more time people spend working at home, the more contact they have with their spouses via ICTs throughout the day—no matter where they are or what they are doing. Furthermore, the more time people spend working at home, the more they integrate their paid and unpaid work tasks in the home. For example, full-timers spend a mean of three hours per week more than part-timers doing chores and cleaning, cooking, and child care, but they also spend a mean of two hours more than part-timers per week with their children and spouses. In short, the more time people spend working at home, the more integrated is their work and family life.

Yet, telework can be both an obstacle and a solution to what family sociologists call "work–life balance." Many workers experience some negative impacts of ICT on their work life: prolonged working hours, increased availability outside of normal working hours, and distraction at work. The Pew Internet Networked Worker survey shows that almost half of workers who use ICTs say that blurred home-work boundaries increase working hours and stress. Moreover, teleworkers often pay for flexibility with longer, often uncompensated, hours. If circumstance allows, many teleworkers would not prefer to telework: Two-thirds work from home because of employers' requirements rather than individual preference.

The blurring boundaries of work and home may create intrusions when family members, especially young children, interfere with work. For instance, Olivia told the Connected Lives study that once she started working full-time at home, her family reduced the help they gave her with household chores: "As a general rule, I do it, and I think even more so now that I am not working outside of the home anymore. When I was working full-time [outside the home], I think there was a lot more chipping in."

Not only does her family expect her to do more household chores but Olivia herself also feels that she has to avoid getting preoccupied with domestic chores and focus on her work tasks. "I seem to be getting very good at this though: resisting the urge to do laundry or catch up on whatever – vacuuming or something like that," she says. "Fortunately, it's not something that I like to do. But I know when I first started working at home that was difficult. It was like, I'm home, so the house should be in better shape. My husband shouldn't come home and say, 'well, you've been home all day.'"

Teleworkers who work entirely from home may feel isolated due to the lack of face-to-face communication with colleagues. They fear that their careers could be limited because they are less visible at work. For instance, while Olivia is not too concerned about the isolation of working from home because she usually works as an individual, nonetheless, she makes sure that she goes to the office once or twice a month for meetings or just to stay in contact. In a similar vein, Homezilla's CEO Sandy Ward highlights maintaining visibility as a key issue for making remote workers happy and productive:

A remote worker can feel isolated and alone, which will greatly reduce productivity. Something as simple as them running a monthly meeting, having their manager over-communicate some of their achievements, and impromptu personal calls to chat (not check in) can all help. A big part of the visibility is connection to other team members; team members chatting in the hall over doughnuts doesn't happen to remote workers. . . . Impromptu personal calls go a long way: think of it as a 'hallway' chat for remote workers. Many people say IM is just as good but I disagree; a phone call builds relationships very differently than IM, texts, or emails.[41]

Net and Jet: Entrepreneurs Linking North America and China

Despite the reliance of distributed organizations on ICTs, "travel to trust" is often the watchword of such organizations.[42] As we saw in chapter 2, airline travel has soared—and along with it, business travel. As much as people depend on ICTs to communicate, networked organizations must have some in-person contact. That is the way they get a more multidimensional range of information about one another—how they look, talk, and smell; what the office gossip is, especially regarding their own standing within the firm and who is working on what. With such bonding, workers get tacit knowledge: the unwritten lore of organizational memories and know-how. As one entrepreneur told NetLab's Transnational Entrepreneurs project that "successful collaboration is a full-contact sport."

In-person contact is crucial for adding a human touch to glocalized networks.[43] Chinese-Canadian transnational entrepreneurs both "net and jet" to do business. They are jet setters in order to obtain high-value contacts in government and business, develop trust, and obtain deep knowledge of conditions in China. These Chinese-Canadian entrepreneurs have large and diverse business networks, with their travel helping them to tap into previously unattainable business partners or capital. Ties with government officials are a great advantage, but take a lot of time to build and maintain. "You must hook up with officials and state firms," Russell told us. As Taylor, another entrepreneur, said, "I search for partners on the internet. I found four of my joint venture partners on the Internet and the other half were recommended by friends. I send emails to people who I have never met before. I cannot imagine how I would do business without the Internet," He went on to say, "People ask me how I can trust someone who I have never met in person. Online contact is just the first step. You will have a lot communication through email. Yet, the most critical step is that you sit down together face to face. Then you must travel to China, meet him in person, and listen to his idea."

The entrepreneurs often find it hard to get tacit knowledge and fine-grained information online. They use face-to-face meetings to fill gaps. For instance, Stella does intensive market research on the internet when she needs to find new suppliers. However, she finds that "what you find on the internet, in particular in China, is often trading companies." As a trader herself, she is

not that inspired to deal with other trading companies because I do not want to have an additional middleman. I left for China to research what business we could do. I selected a few directions. I stayed two weeks in China and met about 20 people ranging from government officials to business owners: many of them were my friends or business contacts. I met with high-level officials in the National Committee of Development and Reform, the Ministry of Commerce, and the Ministry of Metallurgy. I met directors in imports and exports firms, insurance companies, and mining firms. As your contacts are high-level, they know very well the policy and the overall trend of the development. You learn a lot from chatting with them.[44]

The Distributed Designs of the Boeing 777 and 787 Airliners

The Boeing 777 and 787 airliners, designed by geographically distributed virtual teams, illustrate what such teams in a networked organization can—and cannot—achieve.[45] Where the networks of Chinese-Canadian transnational entrepreneurs were often interpersonal, Boeing's links were interorganizational. Given the competitive time pressure to enter service quickly, multiple international and interdisciplinary teams designed the 777 in more than a dozen countries. At the peak of the work, there were more than ten thousand team members. Analysts claim that the five-year time span from start to launch was 30 percent to 40 percent faster than working with comparable paper-based designs.

The 777's design process was greatly sped up by computer-assisted systems. It was the first plane in history built without physical mock-ups. Instead, the design teams worked on virtual mock-ups from 2,200 computer terminals connected to large mainframe computers in the United States, United Kingdom, and Japan. Key participants in the design process had immediate access to data and could track changes that other design teams made. A parts-tracking system monitored the delivery process of each plane part.

The Boeing 777 project focused on human networks as well as on computer networks. Boeing coordinated collaboration within and across cross-disciplinary design-build teams. Although Boeing had been an organization with strict hierarchical divisions between departments, it built a global network for the Boeing 777 project to support data exchange and process integration among suppliers, vendors, internal units and external customers worldwide. The entire plane was divided into specific areas of responsibility and each specific area was further broken down into subcomponents—such as the outboard fixed wing— with each run by a design-build team. To reduce within-organization rivalries and communication breakdowns, each of the two hundred teams was cross-disciplinary, bringing together members from engineering, finance, operations, manufacturing, customer support, suppliers, and customers.

Management made sure that virtual team members had considerable in-person interaction: Team members had to maintain participation, develop group cultures, and integrate with colleagues working in other places. Boeing worked to establish trust among dispersed team members by first trying to build good face-to-face relations, bringing design-build teams' members from a dozen countries to their headquarters near Seattle to work together for eighteen months. The teams also built collaborative relationships with major airlines and with foreign manufacturers. For instance, Japanese companies manufactured around 20 percent of the Boeing 777, with engineers from three major Japanese subcontractors working in the design-build teams.

Why did the process work so well, with reasonable speed and with all parts fitting together? Organizational analysts Arvind Malhotra and Ann Majchrzak concluded that Boeing's successful use of ICTs for far-flung virtual teams had four distinct dimensions. First, technology supported task coordination, in the sense that everyone knew what everyone else was working on. Second, ICTs supported the external connectivity needed to link the team with outsiders, whether to gain information or a new perspective on the project. Third, ICTs supported "distributed cognition," enabling team members to share and integrate their own diverse perspectives. Fourth, more prosaically but crucially, a variety of ICTs enabled the teams to interact in many ways, using the media they felt was most appropriate to the task at hand.[46]

Yet these kinds of collaborations through vast distributed networks can turn into nightmares. Although the Boeing 777 was a technical and commercial success, distributed work on the newer Boeing 787 Dreamliner has been a bumpy ride. The 787 project went further than the 777 project in its embrace of distributed work, passing out more of the manufacturing to more suppliers. A network of global suppliers is building four-fifths of the plane. Suppliers for the plane's structure span nine countries in four continents.[47]

However, the parts shipped in from around the world did not fit together well at first and sometimes did not ship on time: a major difficulty with interdependent manufacturing processes. Other problems included assembly instructions in different languages, variable quality standards between companies, and networked work going too far. Work was even more distributed when outsourcers further outsourced their work, far from Boeing's span of control. For example, one American company struggled to meet Boeing's standards in building fuselage sections, ran short of cash, and outsourced work to an Israeli company, creating a messy communication and accountability chain. The program fell three years behind schedule and billions of dollars beyond budget, with significant management upheavals. It appears that networked work and networked organizations are better at linking ideas—bits—than at linking parts—atoms.[48]

Networked Work On and Offline

Technology neither creates nor organizes work relations by itself. Instead, it affords possibilities and constraints. Even when computer-based communication developed in the 1980s and 1990s, the large size of the equipment and the low capacity of the communication links meant that almost all computer-based contact—if it existed at all—was confined to small groups. It is only with contemporary computers, smartphones, and the broadband networks to carry them that work has achieved the relative ease and low cost of distributed communication. Yet, how workers use ICTs and what they use them for are as important as the technological affordances of these technologies.

On the one hand, ICTs now play an enabling role in supporting interorganizational and intraorganizational communication and coordination among workers distributed in various locations. Workers are talking to each other, showing pictures and videos, and exchanging files that range from small documents to the huge databases that organizations from astronomy to insurance rely on. In many cases, ICT-using bit workers have been able to cut loose from working side by side—even if they are sometimes looking at each other face to face on video. And ICTs enable workers to have more contact with family members throughout the day.

On the other hand, ICTs convey fewer social cues than in-person contacts. That is why people travel to meet in person: to build trust, develop nuanced understanding, and exchange tacit knowledge. The success of collaboration in the network operating system often depends on a delicate balance of computer networks and human networks that provide the trust and incentives to share information and knowledge. As Lu Mei and her colleagues report, "a diverse repertoire of ICTs has created an environment where team members are often oblivious to other members' locations and backgrounds."[49]

Yet the need for in-person contact should not be overstated. Once people have established in-depth knowledge and trust, they can blend work via ICTs with in-person meetings. Widespread experience with using ICTs has created ease and comfort for long-distance communication.

Thus, the shift from organization men to networked individuals reflects fundamental changes in the organization of work. It makes organizations more flexible and more tenuous. More coordination, if not control, is needed. It increases problem-solving capabilities, if firms can get the attention of the teams and keep them focused. It diversifies information flows, at the cost of inefficiencies if networks are poorly connected. It increases

worker mobility at the cost of worker solidarity and, perhaps, worker solitude. It liberates workers from constant in-person supervision but tethers
them to constant communication channels and digital supervisions. It
often allows them to be casual in their dress code and punctuality, but puts
work–life balance at risk. It can make for happier, more relaxed workers
but it can distract and distress workers as they get torn between work and
family.

Most importantly, the time and space flexibility afforded by ICTs fits
the shift to networked individualism—at work as well as in the family and
communities. As IBM vice-president Irving Wladawsky-Berger argues,
"enterprises have had to become much more flexible and dynamic in order
to survive the intense competition they started to face from companies
around the world, large and small."[50]

8 Networked Creators

An old adage has it that you should never pick a fight with someone who buys ink by the barrel.[1] A new adage would be to never pick a fight with a networked individual with strong internet and mobile connections.

In the age of the Triple Revolution, anyone with an internet connection and a bit of digital literacy can create online content that has the potential to reach a wide audience. With all the different forms of creating content and the increasing ease with which people can do so, the boundaries between producers and consumers are becoming blurred in the network operating system—with noncredentialed amateurs participating in many of the arenas that were once limited to recognized and sanctioned experts.

Peter Maranci is one of the legions of digital creators and activists who have updated that old adage. He embraces ICTs, and tailors his use to satisfy his needs and accomplish his goals. In August 2003, for instance, he began a personal blog where he wrote a mixture of personal reflections, a chronicle of his young son's early years, political commentary and humor, thoughts on religion and atheism, information exchanges with fellow video gamers, stories of his experiences at work and play, and eventually ran a campaign against the Massachusetts Bay Transportation Authority. He had tried to contact local newspapers, rail officials, legislators, and the governor to voice his concerns over problems of the train system but had met with no response. So, he used his LiveJournal blog for several years to snipe at the problems he saw—dirty stations, stiflingly hot cars, late trains, and overcrowded trains.

In July 2007, a woman passed out on a packed, hot train. The aisles were so congested with standing riders that it was impossible for those near her to summon a conductor for help. An off-duty paramedic and a pregnant nurse tended to the woman until, in Maranci's estimation, a "clueless" conductor arrived several minutes later and eventually found a way to get the woman an ambulance.

The incident propelled Maranci to create a separate blog specifically focused on train problems, which he called "Charlie on the Commuter Line."[2] His first post was about the woman and the woeful conditions on the train: "Fortunately the ill woman's condition apparently wasn't critical. But if it had been, the delay caused by the overcrowding could have had serious consequences." He then joined a small cadre of other transit riders who were similarly blogging about their experiences. Their blogs linked to each other via blog rolls, lists of other blogs that are recommended by a blog writer. That helped to increase the visibility of all the blogs and to bring them new readership. Early in 2008, Maranci started adding photos of the crowds on the trains, the late arrivals of trains, and the conditions in train stations.

Over the following months Maranci could see how his postings were beginning to influence changes within the transit system. A conductor about whom he and other passengers complained was transferred off the train, the crumbling stairwells he documented were patched (albeit poorly), and he was contacted by a reporter from a free daily commuter paper, *BostonNow*, about the chronically crowded conditions on the trains. This led to a photo shoot and article. Maranci knows his public campaign was not the only reason for these changes, but as he explained in a personal interview: "It certainly brought more voices of protest to the attention of fellow riders, the press, and the officials who run these trains. It was really gratifying so many of the changes I'd been seeking were made."

While Maranci is not a celebrity or politician, he is able to act in the same environments they do and have influence on the issues that matter most to him. He would likely not have been very visible were it not for the affordances of ICTs. Still, content creation has its burdens and limits. On March 10, 2011, Maranci told his readers that he would blog much less often, because he had moved off the transit system to become a telecommuter, working from home and ending his four hours of daily travel. He said he had less to post, and his withdrawal shows that when unpaid networked creators lose interest or opportunity, they become less involved.

While he was at it, networked creator Maranci joined the roughly two-thirds of adult internet users in America who have created online material, taking advantage of these new technological affordances. The act of creating media—text, photos, audio, artwork, or videos—can serve many purposes. For many digital creators, content creation is a simple act of documenting memories, the way traditional photo albums are constructed. For others, making and sharing media is a signpost of friendship and communication. Then, there are those who create material so that they can learn and explore. Finally, there are content creators who want to advertise themselves, reach out to strangers, or show their technical skills.

With rapid technological advancement, it becomes difficult to keep pace with all of the ways people can create content online. Nevertheless, Pew Internet research has tracked some of the clearly identifiable online activities of networked individuals. It has found that the number of content creation activities and the age of networked individuals engaging in such activities are increasing over time as is reflected in the following percentages from surveys in 2011:[3]

• writing material on a social networking site such as Facebook: 65 percent of internet users do this
• sharing photos: 55 percent
• contributing rankings and reviews of products or services: 37 percent
• creating tags of content: 33 percent
• posting comments on third-party websites or blogs: 26 percent
• taking online material and remixing it into a new creation: 15 percent of internet users do this with photos, video, audio, or text
• creating or working on a blog: 14 percent
• using Twitter: 13 percent

Pew Internet's studies show that networked individuals often do not confine themselves to one type of content creation activity, but are involved in multiple types. For instance, along with his activism as a train rider, Maranci, like a third of internet users, reviews products and services. He contributes to Yelp—a website that allows users to rate and review local businesses such as barber shops and restaurants—where he mixes praise of good businesses with withering reviews of firms that fall short of his standards. He's also similar to the 15 percent of internet users who combine material previously created by others to create original material—remixing or mashing it up. For example, he created a "separated at birth" video of Democrat John Kerry and a zombie from a cult movie, attracting several thousand viewers.[4]

Maranci's history as an online contributor is but one example of how content creation affords networked individuals opportunities to solve problems, make decisions, increase their social standing, and gain political support. Networked creators, such as Maranci, are the focus of this chapter. We examine how the rise of ICTs has changed the landscape of creation, providing networked individuals with an increasingly accessible vehicle by which to display their creativity to a larger audience. Armed with their internet connections and mobile phones, they can document and report on what is happening in their surroundings, providing an alternative source of news. They can produce online content in ways that not only expand their social network but also elevate their social standing.

Creating in the Age of the Triple Revolution

Media making in the network operating system is a participatory act. Even old-style media events turn into social media "conversations" in the age of the Triple Revolution. After President Obama's State of the Union address in 2011, Americans still could not respond to their television sets and be heard, but networked individuals created simple and original YouTube videos to voice their concerns. A week later, Obama spent forty-five minutes responding to a few of the videos by answering questions on a number of topics including policies in Iraq and Afghanistan, the war on drugs, and even his favorite Occidental College course.[5] A similar phenomenon also happened the year before when more than fourteen thousand videos were generated in response to Obama's 2010 State of the Union speech.[6] Of course, these numbers are not representative of the entire population of America and creating a video did not ensure a response from Obama. Nevertheless, ICTs had opened up the possibility for greater dialogue, especially with a public figure who is nearly impossible for an ordinary person to reach.

"Making is connecting," asserts sociologist David Gauntlett. ICTs not only enable people to create their own original work with the material available on and offline, they also encourage networked individuals to do so collaboratively. As Gauntlett emphasizes, "Sites such as YouTube, eBay, Facebook, Flickr, Craigslist and Wikipedia only exist and have value because people use and contribute to them, and they are clearly better the more people are using and contributing to them."[7] Similarly, social analyst Sebastien Paquet argues that it is "ridiculously easy" to create a group in this environment because the transaction costs of finding and connecting with others are so miniscule.[8] Paths where only recognized experts would once tread are now open to noncredentialed amateurs. As media commentator Douglas Rushkoff puts it: "The people have crashed the gates of professionalism, penetrating the formerly sacrosanct boundaries protecting elites of all industries from challenges from below."[9]

Thus, networked individuals are voluntarily creating content every day in tandem with other networked individuals within and outside of their own personal networks, and in ways that can expand and enrich collective knowledge and solve problems. Digital tools have helped networked individuals reconfigure the structure of their social networks by extending their reach and their potential for influence, blurring the lines between producers and consumers. With costs to creating and disseminating technology lowering, many more people are telling stories, giving personal testimonies, contributing their ideas, and interacting with others.

Collaborative Content Creation on Wikipedia

Wikipedia relies almost entirely on the collective, creative efforts of networked individuals who create, edit, and manage the content of the more than 3.6 million articles currently stored in its English-language version.[10] Although the online encyclopedia allows any registered user to create and edit articles: A small number of the 2.8 million do much of the editing.[11] At times, the process of contributing is simple and uncontested. At other times, it is protracted and contentious as contributors engage in "edit wars" over facts, interpretations, and sourcing, all professing to uphold Wikipedia's primary principles of reliable authorities and a neutral point of view. After all, the fact that "editors have diverse motivations, have never worked together before, and [are all] free to participate as little or as much as they want" makes collaboration on something as intricate as an online encyclopedia quite a difficult task.[12]

Wikipedians sprung to work right after the devastating northeastern Japan earthquake and tsunami of March 11, 2011, and the subsequent nuclear reactor failures. Some Wikipedians updated existing articles on topics such as "Sendai," the principal city in the area, and "boiling water reactor," adding discussions of the reactors' inadequacies. Wikipedians also created new articles, such as "Tohoku earthquake and tsunami" and "Timeline of the Fukushima 1 nuclear accidents." Between March 11 (0600 UTC, the approximate time the quake began) and March 25 (0000), 2,871 different editors contributed 13,175 revisions to 75 relevant articles.[13]

Figure 8.1 shows links between Wikipedia editors and the key articles they contributed to.[14] The principal articles are the larger rectangles and the small dots are the editors. The halo—or if you like, peacock feathers—around the principal nodes show those editors who only contributed to one article. Yet, there are bridges suggesting that some editors contributed to multiple articles, such as the clump at the bottom between "Fukushima 1 Nuclear Power Plant" and "Nuclear 1 nuclear accidents." The cluster of

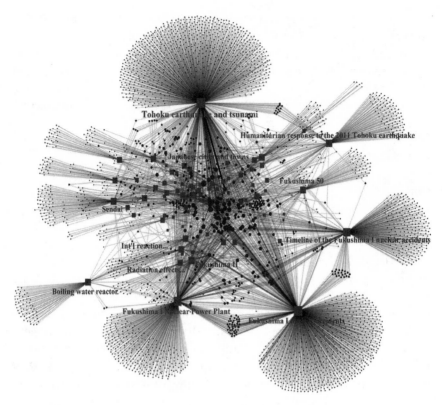

Figure 8.1
Wikipedia links between articles and editors for the 2011 Japanese disasters.
Source: Brian Keegan. © 2011, used with permission.

small squares running left to right in the middle of the network shows the main links among the twenty-nine articles related to the destroyed cities and towns. And the cluster of dots in the center—looking like a cluster of stars at the heart of a galaxy—shows that many editors were at the core of these revisions, contributing to many articles. The graph shows one strength of Wikipedia: It itself provides networked information, as hyperlinks connect articles with one mouse click.

The graph shows an interconnectivity of articles, ideas, and people: what sociologist Ronald Breiger calls a "duality of persons and groups." People link groups, but groups also link people.[15] For example, there were more than three hundred edits alone on March 14—just to the "Fukushima 1 Nuclear Power Plant" article, many made minutes apart, and some repeatedly made by the same people. A "discussion page" accompanies every Wikipedia article for editors to discuss issues. Twenty-five items appeared here on March 14 along with a structured vote about whether the article should be merged with another one about the accident itself. While Wikipedians may argue with and edit each other a lot, the result in all the places we have looked appears to be steady progress toward creating informative and reliable encyclopedia articles. Indeed, we have judiciously relied on Wikipedia in writing this book.

In addition to the drive of Wikipedia contributors to chronicle major news in real time, there is a more fundamental urge to participate in knowledge creation of all kinds—even with the frequent stresses that are part of that process. One Wikipedia contributor, "Willowaye" (a pseudonym of the real alias), told us about his involvement during the 2008 U.S. presidential campaign, where content creation between networked individuals on Wikipedia became a heated process.[16] Willowaye had made over a thousand edits to existing and new articles by that time, which made Wikipedia a natural space for him to contribute to the election campaign. He shied away from the main articles about Barack Obama and Republican candidate John McCain because they were too filled with warring contributors. Instead, he focused on Obama's mother, Ann Dunham, and stepfather, Lolo Soetoro. He put these articles on his Wikipedia watch list, receiving alerts every time anyone changed the text.

When it came to editing the Ann Dunham article, Willowaye encountered vulgar insults about her sexual exploits and choice in men that were not based on any evidence. Although he had to deal with these allegations on a recurring basis, they were easy to fix with a one-key-click deletion (what Wikipedia calls "reversion"), using a tool called Twinkle. Another issue with the Dunham article was more subtle. Certain editors wanted to emphasize her supposedly atheistic beliefs. However, several editors, including Willowaye, were able to show that Dunham was broadly supportive of the humanist aspect of religion. The editing battle continued behind the scenes on "talk pages"—separate from the main Wikipedia pages—until a compromise was made and the section was renamed from "Religious Beliefs" to "Spiritual Beliefs," although further editing has renamed it again: "Personal Beliefs" (as of December 25, 2011, 2210 UTC).

Meanwhile, the Wikipedia pages on Obama's stepfather, Lolo Soetoro, posed another challenge. Several editors wanted to make Soetoro's Muslim religion a main focus of the article, which Willowaye thought might well affect how people thought about Obama. During the time of the campaign, recurring misrepresentations asserted that Obama was a Muslim when he was, in fact, a Christian.

The interactions that Willowaye experienced while editing these articles illustrate how networked creators collaborate to produce a collection of valuable information. In the debates over what to include and exclude for the Wikipedia articles, disagreements were phrased according to Wikipedian norms such as verifiability, civility, and neutral point of view: "the article would be tighter without that"; "you're providing undue emphasis"; "we need to make the article more complete"; and "please document your facts." Without such outwardly civil discourse, cooperative editing could not take place.

With Wikipedia, ordinary networked individuals who are committed to contributing their time and effort have the power to create and amend articles that contain valuable information. The content on Wikipedia is not necessarily introduced by professional experts. It is almost always edited after it is created on the website rather than being edited and fact-checked by third parties ahead of time. The articles are not usually controlled by the administrators who act as the arbiters of disputed writing. In other words, many of the traditional processes of creating reliable information are abandoned in favor of a procedure based on interpersonal exchanges among networked individuals during their leisure time.

This does not mean that the information on Wikipedia is always the most accurate, although it usually is the most handy. Willowaye's experience demonstrates just how protracted content creation can become as many different networked individuals attempt to push their interpretation of information online. Nevertheless, the rise of Wikipedia illustrates how power dynamics have changed with the coming of ICTs. Professional elite producers no longer hold a monopoly on content creation and dissemination. Wikipedia, an online encyclopedia mostly edited by amateurs, is a primary example of how knowledge is crowdsourced—that is, produced by groups of people who are interested, motivated, and have internet access. It has become the most widely referenced work on the internet.[17]

Tag, You're It!

Institutional actors are recognizing the collaborative potential of networked creation and integrating it into their own work processes. For instance, the U.S. Library of Congress posted a message on its website in January 2008 and sent an email to its list of "friends" asking them for help. A small team of the library's staff managed by Michelle Springer was in the process of placing approximately three thousand historic photographs on

Table 8.1
Library of Congress Project Statistics for 6,416 images in Flickr Commons

26.3 million	Views of the photos
130,033	Tags added by 3,507 unique Flickr accounts
29,348 comments	Left by 7,889 unique Flickr accounts
Fewer than 25	Instances of user-generated content removed as inappropriate

Source: Library of Congress (2011).

the popular photo-sharing website Flickr. They needed people who were members of Flickr to help provide more information about the photos by commenting and adding labels, also known as "tags," to the photos.[18] The august Library of Congress was giving up some control over the creation of information to amateurs.

Many library officials were stunned in the ensuing weeks at the outpouring of interest and helpful public contributions to the project. They were most surprised by the unforeseen outcomes: the pace by which new tags and comments piled up over time, the willingness of Flickr members to expend effort on history detective work, the sourcing of new information through links to newspaper archives and highly specialized websites, and the desire of other institutions to launch similar efforts. While cynics predicted that the project would yield little payoff or that troll users would disfigure their photos or overwhelm the site, the Library of Congress team noted that the response from Flickr members and observers was "overwhelmingly positive and beneficial" (table 8.1).[19]

Thousands of Flickr users made tags and added comments to the photos, creating a resource rich with information. At times, comments would lead to further discussions of a wide range of topics including people's memories of farming practices, recollections of their grandparents' lives, women's roles in World War II, and the changing landscape of local neighborhoods. One particular photo that drew extended interactions was labeled Weavers at Work (figure 8.2). Networked individuals created tags to indicate that the women in the photo were blind and that they were making rugs at the New York Association for the Blind. One networked individual even recognized the photo as one taken by his great grandfather, Percy Byron (prompted by the logo at the bottom of the photo).

Moreover, the Library of Congress team decided to enable Flickr's "blog this" functionality, which allowed the photos to be incorporated

Figure 8.2
Weavers at Work: blind women weaving rugs in Byron, New York, ca. 1910–1915.
Source: Flickr Commons Project, 2008.

into the personal blogs and websites of networked individuals. At the onset, the team worried that the photos would lose context and authority if allowed to migrate to different online venues. Instead, they found that networked individuals used the photos in creative—and not abusive—ways. One photo of World War II nurses was used to illustrate a blog post on giving blood, an image of a half-built skyscraper was embedded in a blog post announcing the blogger's upcoming trip to New York City, and a picture of a two-story-high stack of paper outside a 1940s paper mill was accompanied by a blog post mocking a judicial order regarding access to data. Networked individuals were remixing these old photographs from the early 1900s in a way that made sense in their contemporary context. Thus, in releasing the historic photos, the Library of Congress opened the doors for people to integrate the past with the present through their own creations and with their own interpretations.

Similar to Wikipedia, then, the library's material had been enriched, expanded, and made more accurate and accessible by its invitation to people to contribute their own insights and engage in their own conversations about the photos. Instead of asking professional historians to fill the missing gaps of information about its artifacts, the Library of Congress witnessed networked individuals creating their own information taxonomies through their tags. More importantly, it spurred the creation of original tools and resources from public institutions and networked individuals alike. Other public institutions have been quick to grasp the potential of collaborative creation present in the library's pilot project. By December 2011, fifty-six additional museums, libraries, and archives were participating in the Flickr Commons initiative, a project compiling artifacts from the past and crowdsourcing them for modern interpretation. These dedicated virtual volunteers have created a resource for history enthusiasts to share and discuss the collections available in the Commons. In addition, they have also served as a programming resource for some of the participating institutions as well as pushing for more innovative creations.[20]

The Egyptian Revolt, On and Offline

Social media—Facebook, Twitter, and email—plus mobile phones played a major part in the "Arab spring" of protests and rebellion against authoritarian regimes in the Middle East and North Africa throughout 2011. The activity of networked individuals in Tunisia, Egypt, and other states was a prime example of how online content creation and community building, in tandem with offline gatherings and backstage maneuvering, can aid mass mobilizations.[21] Although only 21 percent of Egyptians subscribed to the internet in 2010, these were predominantly the young adult Cairo men who were at the heart of the revolt. They also were new kinds of influencers who could shape a movement that did not emerge in the traditional way of relying on a formal, hierarchical opposition organization. Moreover, the internet in general—and Facebook in particular—were important metaphors for the revolt. The young men pronounced themselves as "the Facebook generation," signifying that they were no longer the nonmodern Egyptians of the past.[22]

The movement toward revolt did not happen overnight, though it may have seemed so for the many people around the world who relied on traditional media for the news: newspapers, radio, and TV. There had been serious preparatory links both online and offline between Egyptian activists and other like-minded activists in other countries. In 2008, Ahmed Mahar and his friends had created the "April 6 Youth Movement" as a Facebook group to promote and coordinate a nationwide general labor strike. A few months later, young networked individuals in Tunisia created the group "Progressive Youth of Tunisia," which became a link for communication between activists in the two countries.[23] The Egyptian activists also communicated with activist networks elsewhere, including the Centre for Applied Nonviolent Action and Strategies (Canvas), an offshoot of the Serbian youth movement Otpor, which had participated in overthrowing the Slobodan Milosevic regime in 2000. Members of the April 6 Youth Movement had traveled to Belgrade to learn how to organize peaceful protests, and Serbian activists had similarly traveled to Egypt to train protest organizers.[24]

Digital tools also helped Egyptians to build networks within the country. When Khaled Said, a twenty-eight-year-old Egyptian businessman, was beaten to death by police officers in June 2010, cries of police brutality and public outrage erupted. Then-anonymous networked individuals created a Facebook page called "We Are All Khaled Said," posted photos taken from a mobile phone of Said's beaten face, and created YouTube videos depicting Said as a happy individual prior to his death. One of the Facebook group administrators, Wael Ghonim, a marketing executive at Google, worked closely with Maher of the April 6 Youth Movement and other activist networks, using knowledge he gained from those networks to help mobilize those drawn to the new Facebook page.[25]

The Egyptian movement was galvanized in early 2011 after large-scale protests in Tunisia toppled the regime of President Ben Ali on January 14. The Tunisian uprising had special resonance in Egypt because it was prompted by incidents of police corruption and viral social media condemnation of them. When news of the Tunisian revolt's success spread, networked individuals in Egypt began to plan protests for January 25. The "We Are All Khaled Said" Facebook page was used to actively recruit people to join the protest with more than one hundred thousand reportedly signing up (figure 8.3).[26] With information widely disseminated online, the revolt moved from cyberspace to physical place as protestors occupied Tahrir Square on January 25 and the days to come. Facebook became a platform where discontented citizens could voice their frustrations, share relevant expertise, and overcome the fear that comes with living under an oppressive regime.[27] Moreover, where activists of the past had relied almost exclusively on direct access to friends to get the word out about a cause, these networked individuals simply convened with those who were online, even if they were initially strangers to one another. A sense of shared purpose arose and communities of mutual trust and assistance developed.

Figure 8.3
Facebook T-shirts commerating the January 25, 2011, Egyptian uprising for sale in Tahrir Square, Cairo, May 2011.
Source: Zeynep Tufekci. © 2011, used with permission.

While protesting, organizers leveraged their networks and resources. They continued to communicate with the experienced activists from Tunisia and Serbia who provided practical advice, such as sniffing lemons, onions, and vinegar for relief from tear gas, using spray paint to cover windshields of police cars, and shielding their bodies with plastic bottles or cardboard.[28] They relied on traditional media, such as flyers, to mobilize people who might not have access to the internet.[29] And of course, they coordinated their efforts online, announcing demonstration sites and sharing instructions, and tactics with one another. As Egyptian blogger Mahmoud Salem said, the internet created a "parallel Egypt" through which networked individuals could communicate.[30]

Mobile phones played an important role because many more people owned them than personal computers. Those in Tahrir Square at the center of the revolt relied on mobile phones to learn about fast-paced events that were unfolding around them and then share those stories via the internet. They—and others—sent bulletins by Twitter and text, and uploaded protest (and repression) videos to YouTube, thereby spreading the flow of information worldwide. In fact, after Egyptian authorities closed down Al Jazeera's office in Cairo, the pan-Arab broadcasting network called on people to send blog posts and videos of what was

Figure 8.4
Egyptian activists recharging their mobile phones during the revolt.
Source: Karim Marold. © 2011, used with permission.

happening on the ground to expand coverage of the protests.[31] Mobile phones were so essential to the group that they ingeniously kept them charged by tapping into streetlamp wires to obtain electricity (figure 8.4).

Despite the importance of the Internet and Mobile Revolutions, other factors were indispensible for the revolt's success. The participants were not a random group of Cairenes. As in almost all social movements, the revolt was built upon established networks of friends and political groups such as the Muslim Brotherhood.[32] Moreover, international pressure played a role in the outcome of the revolt.[33] Most significantly, the Egyptian military leaders decisively did not intervene violently in the protests. They sought to remain powerful after the revolt and were influenced by the American government, which has been providing sizeable funds to them.[34] Their nonintervention contrasted with the Libyan experience a few weeks later when most of the military initially remained loyal to the established Qaddafi regime.

How All This User-Generated Content Is Changing the Media Landscape

In addition to serving the needs of networked individuals, the rise of social media has also changed the character of the overall media environment. The most dramatic evidence of this comes in the content analysis research of the Pew Research Center's Project for Excellence in Journalism (PEJ).[35] Starting in early 2009, on January 19 the PEJ produced a News Coverage Index (NCI) of the major news stories covered by the traditional news media and a separate New Media Index (NMI) of the top topics that were discussed in the social media universe.[36] The NMI looks separately at blogs, Twitter posts, and YouTube videos.

Side by side, these indexes show strikingly divergent universes of coverage and commentary. Even individual social-media channels seem to have their own news agenda and personality. A PEJ analysis of a year's worth of material in each index showed little overlap among the channels in the stories that gained prominence.[37] For instance, PEJ analysts noted that bloggers gravitated toward stories that elicited emotion, concerned individual or group rights, or triggered ideological passion. Often these were stories that people could personalize and then share in the social forum—at times in highly partisan language. And unlike in some other types of media, the partisanship of the blogosphere as a whole did not lean strongly to one side or the other. Both conservative and liberal voices come through equally strongly even in stories on contentious topics such as the Tea Party protests and public support for President Barack Obama.

On Twitter, by contrast, technology itself is a major topic of posts—with a heavy focus on Twitter itself—while politics often plays a much smaller role. "The mission of the short-form messaging site is primarily about passing along important—often breaking—information in a way that unifies or assumes shared values within the Twitter community," the PEJ report stated. And the breaking news that trumped all else across Twitter in 2009 focused on the protests after the Iranian election. It led as the top news story on Twitter for seven weeks in a row—a feat not reached by any other news story on any of the platforms studied.

YouTube stands apart from other social media channels as well. PEJ found: "Users don't often add comments or additional insights but instead take part by selecting from millions of videos and sharing. Partly as a result, the most watched videos have a strong sense of serendipity. They pique interest and curiosity with a strong visual appeal. The 'Hey you've got to see this,' mentality rings strong. Users also gravitate toward a much broader

Table 8.2
News Topics across Media Platforms, January 19, 2009–January 15, 2010*

	Blogs (Percent of Stories)	Twitter (Percent of Stories)*	YouTube (Percent of Videos)	Traditional Press (Percent of Stories
Politics/government	17%	6%	21%	15%
Foreign events (non-U.S.)	12	13	26	9
Economy	7	1	1	9
Technology	8	43	1	1
Health and medicine	7	4	6	11

*Twitter was tracked from June 15, 2009, to January 15, 2010.
Source: Pew Research Center's Project for Excellence in Journalism.

international mix here as videos transcend language barriers in a way that written text cannot." Table 8.2 shows the top five news topics in each of the channels.

Overall, PEJ found that blogs shared the same lead story with traditional media in just thirteen of the forty-nine weeks studied. Twitter was even less likely to share the traditional media agenda with the lead story matching that of the mainstream press in just four weeks of the forty-nine weeks studied. On YouTube, the top stories overlapped with traditional media eight out of forty-nine weeks. Moreover, even when the topics aligned, the treatment of them was very different. News media organizations focused on new developments in their stories while social media creators did several more personal and visceral things in their creations. Often, they proselytized about the meaning of those developments, as when President Obama's massive stimulus spending package was being debated. Other times, social media creators gave personal testimonies in reaction to the developments, especially events like the death of pop music star Michael Jackson.

The Week of March 30 to April 5

A more detailed examination of one of the weeks when the coverage diverged strikingly illustrates some of the main differences between social media and traditional media. In the week of March 30, 2009, three major economic stories dominated traditional news media coverage: an economic summit among developed nations in London that was aimed at coordinating global policies to recover from the financial meltdown; continued problems in U.S. banks that were highlighted when Federal Reserve Board Chairman Ben Bernanke spoke of his reluctant support for the 2008 bailout of investment house Bear Stearns and insurance conglomerate AIG, and the problems with the U.S. auto industry that were highlighted when the White House forced the dismissal of General Motors CEO Rick Wagoner.[38] The fourth major story for traditional media involved a shooting rampage at an immigration center in Binghamton, New York, that left fourteen people dead, including the assailant. And the fifth story involved President Obama's attempts to gain support among America's NATO allies to provide more troops for the war in Afghanistan. Of course, the economic summit was a set-piece news event that drew the highest-ranking journalists among the broadcast networks—all of the anchors were on scene to report events. The formality of the venue also drew news-coverage seekers to the event. Antiglobalization protestors gained a fair amount of coverage. In addition, sidebar stories to the event, such as Michelle Obama's reputation in Europe, were part of the coverage of the news entourages who descended on London.

By contrast, the blogosphere and other social media outlets did not much care about the summit or any of the other stories on the mainstream media list. Bloggers and other social media creators are not "on scene" and obliged to cover topics. They are distributed and more distant observers of news. This gives them more room to range over subjects and choose where they want to link and comment.

The most discussed and linked-to story of this same week on the New Media Index was not even a real story—or an American story. As an April Fool's prank, the *Guardian,* a British newspaper, said it would end its print edition and use the popular online communications site Twitter to draw attention to its stories.[39] While bloggers got the joke, it gained attention because some felt the phony Twitter story offered genuine insight into the huge economic and technological changes transforming the news business. The attention given to this story also highlighted a special trait of social media creators: They love practical jokes. Earlier in the year, the New Media Index had registered high levels of linking to a report in Foxnews.com about hackers in Texas who broke into a traffic-control room and digitally altered a road sign so that it warned of a "zombie attack." The index also had high scores for a small BBC report about a British lad who painted a sixty-foot penis on the roof of his parents' house that had gone unnoticed for a year. Writing about the penis stunt, Yasha at *Heeb Magazine,* an online journal that permits user contributions, explained: "It's these little things that make life's hiccups—a bleak economy, climate change and missing an episode of Gossip Girl – just a bit more bearable." Yasha said she had been sent a link to this story, underlining one of the common traits of stories that take off among social media creators. They are passed around a lot and gain velocity after a critical mass of internet users finds them funny or otherwise valuable.

That same week, the second-largest story in the New Media Index questioned the effectiveness of torture as a technique for gaining intelligence information. Bloggers, especially liberals, focused on a March 29 *Washington Post* report that the harsh interrogation techniques used on al-Qaeda suspect Abu Zubaida had yielded no useful information. The article gave fodder to those who opposed the use of such methods by the United States. This highlights a common element of stories that gain high levels of attention in social media: If they address hot-button issues that matter to a portion of the blogosphere, they are found and passed around quickly. Generally, those in like-minded tribes easily share information. Often, it is well-trafficked bloggers who provide the spark as they link to—or their posts then become link bait—for viral pass-arounds of stories. In this case, liberal media blogger Dan Gillmor's favorable blog report on Abu Zubaida's story was one of the sparks of its eventual popularity.

The third most linked-to story of the week was a mix of Hollywood and politics. It also represented the mirror of the previous story because it was especially circulated among conservatives. Actress Angie Harmon, in an interview for Fox News, said she was tired of having to defend herself against charges of racism because she opposed President Obama. This story got a special lift among conservatives when the Sarah Palin blog cited it thusly: "Support Angie Harmon. She is smart, beautiful, talented, and not afraid to stand up for her beliefs! Angie Harmon is an endangered species —A Republican in Hollywood."[40] Citations like this from key influencers are often the drivers that take a story into the highest reaches of the content creation world.

In sum, the stories in social media often differ from stories in traditional media because they frequently cover different subjects, have a different narrative sensibility, and have different pathways to capture the attention of their audience. Because social media creators are in a participatory frame of mind, their material also has a more powerful impact on their sense of community and efficacy. Pew Internet studies of the role of social media in health care situations, political activities, and news media show that social media users are more engaged with their topics and aware of the world around them.[41]

Networked Stars are Found and Mashed-up

Beyond the news, networked individuals can also experiment with ICTs to create in other ways. Their networked creations also have the potential to reach a wider audience. This, in turn, allows them to build their networks. In doing so, they receive validation, reputation enhancement, feedback, and crowdsourced social support.

Since its launch in 2005, YouTube has become a hub for aspiring singers and musicians. Talent managers look to it to find budding pop stars. When teen idol singing star Justin Bieber uploaded his first YouTube video in 2008, he was, without a doubt, an amateur. He was a thirteen-year-old boy singing a cover of Chris Brown's song "With You" in his living room with a low-quality camera. And yet, the video went viral with tens of millions of people watching the clip. Found and signed by a talent scout, he has become a best-selling singer with three CDs, a movie, a book, and the 2010 Artist of the Year award from the American Music Awards.[42]

Mashups, mixed and modified digital material, are becoming widespread. Although the most widely known are Google maps combined with street views, a Pew Internet study found that in 2010, 15 percent of adult Americans made mashups: taking digital material they found online such as songs, texts, or images and remixing them into their own creations. This activity was a particular province of younger users. One-fifth of online teens (21 percent) had created remixes, compared with 13 percent of internet users aged thirty and up.

Kutiman (Ophir Kutiel) has built his career this way. In early 2009, Kutiman downloaded a YouTube clip of American drummer Bernard Purdie and began to play guitar along with the recorded percussion.[43] "All it needed was some bass and guitar . . . I loved the idea that I was playing along with him and he didn't even know it," he told *Wired* magazine.[44] Then it occurred to him that "maybe I can find someone else on YouTube to play with him." He spent the next two months scanning YouTube videos and building a widely acclaimed work he called "The Mother of All Funk Chords." In this video, he combines sounds he gathered from musical clips that are in the same key but from a wide array of do-it-yourself videos made by similar networked individuals—a guitar instructor showing viewers how to play an E-9th chord ("the mother of all funk chords"), a teen practicing trumpet, a middle-aged man playing harmonica and wailing the blues, a man playing theremin in a music store, a church organist, a Game Boy player, and so on. In all, Kutiman used twenty-two different clips to pull together the right brass section, percussion, and reed instrumentation for the work.

After assembling this video, Kutiman spent months creating enough clips and remixing them into seven songs he calls his "ThruYou" project.[45] He posted the material online in March 2008 and sent links to twenty friends. A week later, ThruYou had been viewed a million times. When Kutiman woke up the morning after uploading the videos, he says he "opened up my MySpace page and saw all these new friend requests and messages. People had found it, and after that, so many people tried to look at the site that the whole thing crashed."[46] From there, it was a short step to laudatory coverage in the mainstream media such as National Public Radio.

These acts of networked creation by networked individuals were best encapsulated by Kutiman himself when he said: "I think people are caught by the same thing that caught my attention: seeing people sitting in their small flat with a webcam, filming a song, hoping for something to happen."[47]

Identity Creation and Reputation Management

In addition to expressing their artistry, there are people who use the internet and smartphones to advertise themselves and build networks that reach out to strangers. They can brand themselves, using ICTs to become celebrities of a kind. One typical example studied by scholar Patricia Lange concerns a mother named Janet and her daughter Maddie who in 2006 made a "channel" on YouTube they call "Beyond Reality." Their videos summarize and discuss reality TV shows such as *Survivor, Big Brother, Beauty and the Geek,* and *Top Chef.*[48] Maddie, a teenager when the channel was created, originally characterized herself as a "future filmmaker" on her YouTube channel while her mother joined in to encourage her daughter and spend some time together.[49] By the start of 2010, they had created over three hundred videos that they uploaded to YouTube, which had collectively been viewed more than 3.8 million times.

Lange noted that Maddie and Janet use "branding" hallmarks in their programs. For instance, every video created the by daughter-mother team begins with an opening clip of their show's stylized title "Beyond Reality with Maddie and Janet," along with a sound bite. Always sitting in front of a graphic with the name of the show they are discussing, they banter in the style of celebrity-show news anchors. As Lange described it, Maddie and Janet watch a reality show together, each taking notes, and then discuss how they would describe and critique the TV show. After each video was posted, Maddie behaved as a networked individual would, alerting others on YouTube, posting bulletins on MySpace, and notifying friends via instant messaging.

Maddie and Janet gained an audience without the help of a professional producer or director. They were able to amass more than five thousand subscribers to Maddie's channel who wanted to be notified every time a new video was posted, gaining influence among reality TV viewers. As one viewer commented on a video recapping *The Bachelorette*, "I don't even watch The Bachelorette. I just watch your recaps." Maddie often subscribed to the channels of popular YouTube members in the hopes that other people would see her channel icon and view her work. Her YouTube portfolio brought her a step closer to her goal of becoming a filmmaker when she was accepted to the New York University film program. By mid-2011 it was mostly Janet's show, although their YouTube channel says that Maddie "will continue to do some recaps & college life vlogs" [video logs].[50]

With ICTs acting as doors that could lead to relative fame and increased social standing, some networked individuals are becoming more deliberate in building their reputations online.[51] They have to think in new ways about their identities, creations, and the degree to which their personal information is disclosed and archived. The result may be the erosion of the distinction between creating for a network of friends and broadcasting to the general public. We discuss this more in chapter 9, Networked Information.

Why Become A Networked Creator?

What advantages do networked individuals enjoy by engaging in networked creation?

A form of self-expression: Like all other ways of making content, online creation provides networked individuals with an outlet to express themselves. That motivated Peter Maranci to document his life; Kutiman, the creator of the musical YouTube mashups, is another example. He says he "didn't expect it to blow up like this, even in my wildest dreams." Instead, he was motivated by the music and the act of creating: "I lost track of night and day. I'd just pass out and wake up on the computer. I was fascinated by the idea. It was magical."[52] One study of top political bloggers found that the main reasons they blogged were to let off steam, keep track of thoughts, and formulate new ideas. As one such blogger explained: "I am a writer by nature. I have a lot of things to say because my mind never stops working. Blogging allows me to express myself and prevents me from being trapped under a mountain of half-formed ideas."[53]

An opportunity to learn: Networked creation provides powerful informal learning opportunities for people to acquire and share knowledge. For instance, reality TV commentators Maddie and Janet produced better videos over time in terms of both their performance and their technical production. Part of this was inspired by feedback that was sometimes brutal in the comments section of their YouTube channel. Moreover, a MacArthur Foundation study of teenagers and young adults also highlights many examples of youth who learned the ropes of ICTs from friends who taught them how to create profiles, write elementary computer code, edit and remix material, and then upload their creations online.[54] Creating online material then gives networked individuals the opportunity to expand skills that may prove valuable in other circumstances.

A space for collaboration: Content creation provides milieus in which networked individuals interact and negotiate with one another. Reciprocal exchanges of information and social interaction are the norm among those who use ICTs. As they are creating content, users are often intimately involved in bargaining with their peers over social niceties. Consider the Wikipedian Willowaye. His editorial experience on Wikipedia during the Obama presidential campaign required that he interact with fellow editors to produce articles about Obama's parents. This often meant that there were back-and-forth discussions of what statements should be included, omitted, enhanced, or downplayed, especially when it came to Lolo Soetoro's religion. Creating online material not only gives networked creators a sense of teamwork, but also may lead to new forms of innovation.

A place to connect with community: More than the fulfillment that comes with self-expression, networked creation produces spaces where people can build their social networks among friends and among others who share their interests, even if at first those people are strangers. Pew Internet researcher Susannah Fox has studied patient support groups whose members often start out as strangers helping strangers and turn into friends supporting friends. She calls them "just-in-time, just-like-me" communities. They are built around content creation and network sharing.

One emblematic example is Karen Parles, a research librarian at a New York art museum until January 1998, when she learned she had advanced lung cancer. Frustrated at how scattered information online was, she found a support group on the Lung-Onc mailing list. After this network of new friends helped her make a hard choice about proceeding with surgery, she recovered and devoted herself to helping others. "My membership in the group provided instant access to the wisdom and experience of hundreds of other lung cancer patients," she said, "and I wanted to pass that along." Parles and her network built a large portal called Lung Cancer Online that offered everything from highly specialized medical advice to wellsprings of social and emotional support for cancer patients. Tens of thousands of them have found help there. Parles said: "I treasure the emails I receive, thanking me and saying how much I've helped. When people say, 'If it weren't for you, I'd be dead—or severely depressed'—well, that's gratitude on a whole different level."[55]

Parles eventually succumbed to lung cancer in early 2009, eleven years after she was given less than a year to live. Many of those she helped posted their memories of her and their thanks on a memorial website. And a common theme permeated their comments: how important it was that Parles's story or other stories on her site so closely paralleled their own. This is the special power of just-in-time, just-like-me connections. When people are suffering or searching, they strongly prefer connections with those whose circumstances are most similar to theirs, rather than those who have general empathy. People attach a singular authority and appeal to those who have gone through precisely the same circumstances or whose experiences match theirs. Networked individuals using ICTs provide that kind of community.

A sense of empowerment: Though networked individuals may begin blogging or creating other forms of online content with internal, expressive motivations, as they continue to create, they become empowered by the act of creating and the potential influence it wields.[56] Egyptians, for instance, first created online material as a means of self-expression. The initial uploading of images of Khaled Said's brutally beaten face and the creation of videos about his life were ways for disillusioned networked individuals to express their thoughts and opinions on police brutality and the repressive regime. After the Facebook page created in Khaled Said's honor amassed many thousands of members, they felt more networked and empowered—perhaps in great measure because the volume of traffic on the Facebook page made them aware of how many other Egyptians felt as they did. As Ghonim, the Google executive behind the page expressed, this form of networked creation "gave us the impression that—Wow, I'm not alone! There are a lot of people who are frustrated. There are lots of people who actually share the same dream."[57]

Other networked creators have felt similarly empowered not just by the support they received from other networked individuals, but also by the influence they gained through creation. Citizen journalists, for instance, have the potential to shape political discourse if their blogs or videos gain enough traction. As one political blogger attested, "I saw that they were capable of moving information around the traditional media bottlenecks and decided that it was something I could contribute to."[58] The rewards that come with wielding influence are thus an important motivation for networked creators.

The prelude to greater glory: Networked creations are a prelude to greater glory as some networked individuals see their creations, and sometimes even themselves, become popular and even relatively famous. Kutiman's "The Mother of All Funk Chords" mashup received critical acclaim and even led to an album of more of his original creations. Maddie's YouTube videos enhanced her portfolio to help her gain admission to the New York University media program. And once-unknown singer Justin Bieber has reached international stardom. Although networked creations usually do not provide material rewards in themselves, they can lead to bigger things, turning amateurs into professionals.

Everybody Wants To Get Into The Act

A major impact of this democratization of media participation is that it enables a new breed of media creators to step onto the cultural stage. This reshuffles the relationship between experts and amateurs and reconfigures the ways that people can exert influence in the world. Those who have things to say have new opportunities to pitch their voices into the information commons and gain a following. At their best, these networked individuals can work together to create collective goods—the way many Flickr members provided additional information about photos from a bygone time. Networked creators can also provide support and information for one another as the networked individuals in Egypt did when they created and participated in Facebook groups. They can break stories as the Egyptian protestors did. Finally, networked individuals can gain relative popularity for themselves and for their creations like Maddie and her mother's YouTube channel and Kutiman's mashups.

Of course there is a dark side to affording everyone the opportunity to be a networked creator. In Elizabeth Eisenstein's accounting, the 15th-century invention of the printing press gave new life to charlatans, quacks, alchemists, disseminators of ridiculous folk "wisdom," propagandists, and other assorted evildoers. The printing press probably created more junk information in history, and Eisenstein argues that it took generations of Enlightenment insight and research to cleanse the information ecosystem of the problems.[59]

Digital technology has been roundly challenged for having the same impact. But many creators somehow find ways to make it work for them in the network operating system. Often, they do this by using the same digital applications to provide their own counternarrative to those who are spinning information incorrectly. Just as not every violinist makes it to Carnegie Hall, not every networked creator becomes a star—or is even heard by more than their friends and relatives. Yet, their creations contribute to the expanding universe of networked information that we will describe in chapter 9.

9 Networked Information

Information not only wants to be free, it also wants to be networked.[1] There is no way to overstress the importance of this insight, first articulated by John Seely Brown and Paul Duguid, that information has a social life.[2] The first reader who scribbled a note in the margin of a tract understood this. The first scholar who added a footnote to her writing to highlight a primary source or a complementary thought understood this. The sixteenth-century Viennese printer who published the Talmud as a series of concentric boxes of commentary around the original text understood this.[3]

Vannevar Bush, the pioneering American scientist and science advocate, definitely understood this. His idea of the "memex" was meant to help librarians aid scientists in their search for material that existed in an environment where information overload had become a daunting problem. He envisioned it to be a desk-like device into which people would enter large amounts of microfilm that could be tagged by users; then data in disparate documents could be associated through "trails" that ran between them. In the July 1945 issue of *The Atlantic Monthly*, Bush wrote: "Thereafter, at any time, when one of these items is in view, the other can be instantly recalled. It is exactly as though the physical items had been gathered together from widely separated sources and bound together to form a new book. It is more than this, for any item can be joined into numerous trails."[4]

Bush's idea of linking and physically associating information became a passion for some computer geeks in the following decades, even more than it was embraced by librarians or other scientists. The leading proponent of linking information was a computer scientist named Ted Nelson who introduced the idea of "hypertext," which is the text on electronic devices with embedded links.[5] It is this idea that inspired Tim Berners-Lee in his creation of the World Wide Web, which then led to the raw material—the billions upon billions of links—that would ultimately fuel Google's search algorithms.

Information has a special nature now that it has become computerized. Because it is composed of bits, information can be easily produced, reproduced, remixed, and disseminated. In ways that were never possible when a person's encounters with text on a page were linear experiences, digital material can be directly and effortlessly connected to related material. It can be added to and amended by anyone with access to it. In short, information is unleashed when it takes digital form. More obviously than in the past, the social life of information can be uncovered and understood as part of the network operating system.

Getting information has become so much more interactive than the twentieth-century way of passively receiving it from printed sources such as books and newspapers and from one-way broadcast media such as TV, radio, and movies. Take, for instance, the social ways by which we gathered information for this chapter. We drew on our memories, to be sure: the human "memex." But we also consulted files of papers that we had archived online and in traditional printed paper. We followed up on many links that friends gave us via Twitter and we used the Google search engine to find material and exact references. We used the Endnote textbase to get references right. When we were stuck, we emailed and tweeted friends, asking for help in finding the right information. Although we drew on the resources curated by University of Toronto—their online catalogues and their ejournals—we never set foot in a physical library.

This chapter examines the number of ways in which the information and media ecology has changed with the Triple Revolution and how these changes have affected networked individuals. The growth and acceleration of information flows have made it important for networked individuals to develop new skills to manage the wealth of information that is available—both the institutional information provided by corporations, organizations, and governments as well as the interpersonal information gathered from one's own network. More than that, networked information now also includes what was once considered intimate, personal information, giving rise to serious privacy concerns.

The News Then and Now

One way to see how the networking of information changes people's experience of information is to consider the dimensions of a news story then and now. Coauthor Rainie's colleague Tom Rosenstiel at the Pew Research

Center's Project for Excellence in Journalism has worked with him to compile a list of the elements of news stories in printed format compared to the elements in digital format. In the print-dominant era of news, news stories could have a handful of elements: headlines, narrative texts, photos, graphics, sidebar stories, and "pull quotes" that featured people cited in the article. In the digital age, the number of features of a news story could rise to over fifty items as websites could contain links to other stories and primary resources, spaces for readers to add their own comments, tags and pictures, links to archives of stories and timelines, full transcripts of interviews, audio material, video clips, background material from the reporter about the process of gathering the story, photo albums, details about the reporter such as a biography and an archive of her previous work. In other words, web treatment of news provides fuller context than print media because of the associations that can be built into a story such as links to background material, other stories, archives of past coverage, as well as newsmakers and organizations mentioned. Among other things, the digital, linked format invites browsing and "horizontal" reading through links, rather than linear "vertical" reading.[6]

This display of digital material also invites challenge, amplification, and adjustment by users of the news site. Networked individuals can now respond to stories more easily and in more ways than they ever could in the "Letters to the Editor" sections of newspapers. With commenting features embedded within news stories, readers can immediately post their thoughts and opinions—not only for the editorial team to see but also for anyone else who happens to be reading that same article. With links to the writers' email addresses or Twitter accounts, readers can communicate directly with journalists and may sometimes receive a response with greater speed than they would have in the days when readers would mail in their comments and await their publication—if they even made it to publication. Online follow-up chat sessions also give readers the opportunity to discuss matters directly with the journalists in real time.

Of course, these interactive features do not guarantee that readers will receive a response from journalists. In fact, the increasingly easy ability to contact journalists can work to a disadvantage as it can overload them with so much information that they cannot read—let alone respond to—all the comments from their readers. Plus, many of these responses are not very well thought out and are often vulgar and hate-filled. Nevertheless, news in its digital form is unique in that it provides more opportunities for networked individuals to interact with journalists and express their views as compared to its print counterpart. And even if journalists do not respond, networked individuals can share with one another through the comments sections or pass it along to their friends who may have more thoughts to add to the discussion, which all together may add a potentially rich supplement to the original text.

Compared to the print environment, then, data in the digital environment are denser, broader, and deeper. The digitalization of news thus offers the potential for richer coverage and therefore deeper understanding. Moreover, decisions about the structure and hierarchy of content found online, on how to allocate attention, and on how to respond are now likely to rest in the hands of both the traditional editorial professionals and ordinary networked individuals. But while all this applies especially to the news, it is also more broadly true of the creation and consumption of all kinds of information in its digital form.

The Changing Information and Media Ecology

The Triple Revolution—Social Network, Internet, and Mobile—has created a new information and media ecology that is distinct from the past. The process of creating, collecting, assessing, and distributing information is increasingly becoming networked through social processes and is very much tied to the rise of networked individualism. Key technological changes have given rise to new affordances that shape the everyday lives of networked individuals as well as their decisions and their behavior.

The dramatic growth of information: More information is generated and circulated now. Social scientists Hal Varian and Peter Lyman calculated in 2002 that the total amount of new information stored on paper, film, magnetic, and optical media had doubled in the past three years. This pace may well have increased, as Google has become a major consumer of electricity to run its storage sites.[7] The marketing research firm IDC projected that by 2020, the "Digital Universe"—a term IDC uses to describe the digital information that is created and replicated—will grow to 35 trillion gigabytes.[8] Much of this growth may be attributed to the spread of digital media in amateur and professional hands; growing adoption of computers, cameras, and audio equipment; increases in the online display of creations and information; and the development of large server facilities to store the data.

The differentiation of information use: As all this material becomes displayed and distributed online, and as mobile devices and cloud computing allow internet users access to media and data from any place and any time, users exercise more choice in their media consumption. Sometimes this means simply adding more sources to a person's information arsenal, and at other times it means gravitating toward new and different material and away from more mass-oriented products. Internet analyst Chris Anderson found that as the volume of information grows, people focus less on "hits" in any media domain—best-selling books, hot music acts, or blockbuster movies—and focus more on the particular things that interest them, no matter how small and quirky the niche is.[9]

The greater variety of information: On the one hand, as communication scholar Pablo Boczkowski has shown, fewer organizations than before control traditional media, making it less diverse. For example, the News Corporation now owns the *Wall Street Journal,* the (London) *Times,* the *New York Post,* and the *Australian* as well as *HarperCollins* (books), *20th-Century Fox* (movies), and *Fox News* (radio and TV). And in some ways, the internet makes the news less diverse because online availability and rapid dissemination affords both traditional media and bloggers more opportunities to imitate each other. There are also studies suggesting that news organizations' surveillance of competitors—especially through the social media posts of correspondents—has resulted in greater similarity of coverage in the digital age, rather than differentiation.[10]

On the other hand, counterbalancing this, the internet and mobile news apps enhance people's capacity to obtain diverse information, as does their involvement in multiple social networks. People bump into more information in more ways when they browse the internet and exchange electronic communications. Search and discovery are now kissing cousins because when networked individuals search for a specific query, they sometimes discover other information that they were not explicitly looking for. In the case of news, for instance, people encounter new information by accident when they are online for purposes other than finding out the news. In a survey after the 2006 election, Pew Internet discovered that 36 percent of internet users had run across campaign news and information online while they were pursuing other unrelated material on the internet.[11] Moreover, in 2004, Pew Internet found that the most active internet users are also the most aware of all kinds of political arguments. They are especially attracted to material that conforms to their point of view. Yet, the people with the most capacity to filter and customize information online—the people who use the internet the most and have the most sophisticated connections and digital gadgets—are also the most likely to be aware of points of view that challenge theirs.[12] Thus, the internet makes available a wide variety of information that encompasses a diverse range of perspectives.

The acceleration of information flows: With high-speed broadband and always accessible mobile devices, information flows through people's lives more quickly than at any time since most people lived in small villages. In 2008, one set of researchers estimated that the average American was receiving thirty-four gigabytes (100,500 words) of information on an average day. Per capita hours of information consumption had increased from 7.4 hours in 1980 to 11.8 in 2008. Thanks to the Internet and Mobile Revolutions, they were receiving about one-third of the words they read interactively—and more than that of the bytes they looked at (mostly pictures). Indeed, reading has tripled since 1980 as a way of receiving information: on computer and smartphone screens.[13] With the explosive growth of texting, the phone—traditionally a device for exchanging information by voice—has become a device for reading.

The pace of communication—emails, updates from social networking profiles, tweets, text messages, mobile phone calls—has accelerated and intensified. All of this connectivity means that news and updates about people and institutions cascade through the internet and mobile devices, bringing a range of insights and developments into people's lives. Sometimes, big news travels more quickly and disseminates more widely than in the past. Half-humorously, some commentators believe that wired individuals develop FOMO: "fear of missing out" on the latest development.[14] Moreover, minor developments and information nuggets from niche worlds spread more rapidly and to broader audiences than in the past. Co-authors Rainie and Wellman both have had live bloggers and Twitter users write about them in real time as they were speaking at conferences. Their PowerPoint slides often appear on conference websites. Photos from these events have also been captured with mobile phones and instantly uploaded online using photo-sharing services. These efforts expand the audience to many times the size of the crowd that is physically present and spreads their thoughts far more rapidly than was possible in the days when the content of their speeches never left the confines of conference halls.

Finding relevant information with greater ease: The convenience of Google, Wikipedia, and other easily accessible sources gives the networked individual the instant ability to pull together masses of information directly related to every query of the moment. Search engines are instrumental in this quest for information. They have consistently shown up in Pew Internet data as the second most popular online activity, behind email but catching up to it over time. In mid-2011, Pew Internet surveys found that 92 percent of online Americans were search engine users and that on any given day, 59 percent of internet users had consulted a search engine.[15] On a typical day, online searches are done by more people than any other online activity—except email. The vast majority of internet users are satisfied with the results of their search queries, especially when it comes to news, health information, material about products and services, government information, and the people they are looking for online.[16] It also helps that people can set up alerts and aggregate material for their personalized *"Daily Me"* newspaper to receive information that is relevant to individual preferences and tastes.[17]

Emergence of new signposts of credible and trustworthy sources: In the old days, people would cite only print publications such as the *New York Times* or the Toronto *Globe and Mail* as the source for their news and information. But with the rise of networked information, there is the need for new markers of trustworthiness and credibility. One such new marker is the increasing popularity of user ranking, rating, commenting, and tagging systems. A 2008 Pew Internet survey shows that more than 33 percent of internet users had tagged online content with descriptive information (name, location, etc.), while 32 percent had rated a person, product, or service online.[18] At one level, this shapes the credibility of information because thoughtful reviews or comments on material can help viewers assess what they are reading. At another level, the volume of ranking and rating has its own effect as a large number of reviews sometimes signal that a seller has a lot of customers. Conversations are a powerful currency in the marketplace of goods, services, ideas, and learning for networked individuals dealing with networked information.

Facebook's pervasive "like" or "recommend" buttons, which have expanded beyond the social networking site into external websites, serve a similar function of establishing trust and credibility. These buttons can exist anywhere online: news articles, blog posts, photos, videos, or entire websites. Once clicked, a notification on one's Facebook profile will appear showing that the user has "liked" or "recommended" that item. This can be seen by anyone within that individual's personal network, opening up the possibility that those in the network will at least click on that link and even "like" it as well. Capitalizing on the trust that traditionally exists in friendship relations, these Facebook buttons have established a new signpost for pushing out credible and relevant information.

A separate measure of information's relevance and worthiness comes from the aggregation of user data. Amazon.com has built a successful recommender system with its feature "Customers Who Bought This Item Also Bought." With this feature, businesses have moved networked information from word of mouth directly to their websites. News websites and others commonly point out which stories garner the most traffic and which have been passed along most as links in emails. This does not directly address trust, but it does allow networking behavior to be some kind of indicator of worthiness or entertainment value. Sometimes the giant aggregation of user data helps computer systems move toward appropriate decisions about the best and most relevant information. Google algorithms, for instance, give extra weight in rendering search results to links and websites that users seem to find useful.

Beyond the direct contributions of users helping to validate information, there is one last way that the value of information emerges from its "networked" character. Computers can analyze patterns of information and recognize similarities. One popular example is Pandora Radio, a U.S. online music service that offers listeners the chance to build playlists by declaring favorite songs or artists. Once the service has learned a user's preferences, it scans the "music genome" to find attributes in other songs that match these preferences. It begins to play music it believes are similar to the user's choices and the user then indicates whether the songs Pandora chooses should be added to the playlist. Thus, the powerful collection and analysis of data by computers to provide individuals with information that may suit their needs and tastes has opened up new signposts for credible sources.

The intermixing of information and communication: The ancient Greek philosopher Socrates reportedly argued against writing because of its dehumanizing effects. Among other drawbacks, Socrates reasoned that text written on a page cannot respond to questions and can be twisted and manipulated without being able to defend itself since it is external from the person who first wrote it.[19]

This is not necessarily the case with the internet. It allows writers to respond to the comments and criticisms of their readers in a variety of ways: with another comment, a live-chat session, or a follow-up article or blog post. The *New York Times* writer Nicholas Kristof, for instance, wrote a column about the "Do-It-Yourself Foreign Aid Revolution" that featured the efforts of young Americans starting their own grassroots organizations in developing countries.[20] This spurred discussion from his readers in the comments section of the article and even on their own blogs as many criticized the sustainability, accountability, and overall effectiveness of such projects. After a week, Kristof responded by writing a follow-up post on his own blog, addressing specific complaints and defending his arguments.[21] The original article, honoring the networked form, had a link to the archive of comments as well as Kristof's blog response to those comments. Thus, information and communication are tightly bound together in digital form, providing richer coverage and the possibility for deeper understanding.

The social experience that underlies the consumption of networked information also deeply connects it to communication. In early 2010, Pew Internet research found that half of all American news consumers say they rely to some degree on people around them to tell them the news they need to know. In the online environment, the social experience is widespread: three-quarters (75 percent) of online news consumers say they get news forwarded through email or posts on social networking sites and half (52 percent) say they share links to news with others via those means.[22]

The intermixing of information and communication also deepens the interplay of information flows: the feedback process between institutional information and interpersonal information. This is different from the original two-step flow of information, in which people received information from the mass media and then discussed it with their friends and families.[23] Nowadays, Connected Lives research has shown that people often obtain information first from their friends and family—in person or via ICTs—and then go to the internet to check it and amplify it. Instead of a two-step flow, there is often a multistep process, with people checking back and forth with their social networks and institutional sources on the internet. People also often discuss in person with fellow network members the news and developments they discover online. Because they are often uncertain about whom and what to trust, there is this continuous cycle between institutional information and people in their networks (both in person and reached via ICTs) to pin down and assess information.[24]

For instance, in late February 2011, Wellman picked up disconnected rumors on Twitter about discontent in China that stemmed largely from the Middle Eastern revolutions that were occurring at the same time. Curious to learn more, he emailed a student with connections in China who in turn found out from a friend that the key word being used on the ground was "Jasmine Revolution." After searching on Google, Wellman found 1,610 hits on the phrase, but after realizing that almost all of these were speculative and from Western sources, he continued his search. With nothing on his *Associated Press* news feed, *Yahoo!* news, or on the radio, Wellman decided to check the *New York Times* online where he found a short article on the developing story that confirmed the original rumors that he heard on Twitter. Finally, he sent a link to the article to a few friends who he knew would be interested in the topic.

This sort of cyclical behavior of weaving institutional information with interpersonal information means that networked individuals exploit the networked information available to them according to their assessments of what is most beneficial and efficient for their needs. They use a number of different media and human sources to collect and verify information. Connections between clusters of friends, websites, professional experts, and books and other print media satisfy people's information sources. In most key information searches that are taken en route to decision making, people have told the Pew Internet project that they have consulted multiple sources and used a combination of searches of written material and conversations with smart friends or professionals.

TMI: Too Much Information

Perhaps the biggest complaint that people have in the era of networked information is that there is just too much information to monitor and digest. Still, the feeling of being overwhelmed by information did not start with the Internet and Mobile Revolutions. Historian Ann Blair has found scholars complaining as early as 1550 about a "confusing and harmful abundance of books."[25] However, what is unique is that ICTs provide not only more information but also more channels that connect people to this information and feed it to them. The unprecedented abundance of information that permeates the networked individual's life can often be difficult and stressful to manage. As one Connected Lives participant complained: "Searching [the] internet for information can be quite tedious, time consuming and not quite successful. You can spend hours trying to find one stupid fact and there is too much information, which is really hard to sort out."

To deal with TMI, networked individuals employ a number of strategies that range in complexity to cope with and manage the information overload. They rely on search engines, bookmarks, and tags. Moreover, people develop ways to alert them to new information about issues that matter to them. Pew Internet data show that two-fifths of Americans have set up news alerts through Google, Yahoo, news services, financial sites, and sports operations to update them every time a subject is mentioned on the web. Close to two-thirds get online newsletters related to work or hobbies. Some 37 percent have set up customized web pages to display information on subjects they care about—virtually all of which get up-to-date news of one kind or another about the people and the subjects that the creators have designated.[26]

Microblogging sites such as Twitter have led to yet another new way of managing information flows. With Twitter, networked individuals have the power to choose the people they want to follow and receive curated information from. Unlike Facebook, Twitter is asymmetric, so people can follow more (or fewer) people than follow them.[27] As entrepreneur Mark Suster blogged, "I follow really smart people from a wide variety of backgrounds and interests [and] they tell me what to read. . . . I pay attention to people I trust & respect and let them be my guides."[28]

Still, these strategies only begin to scratch the surface because they only involve gathering and disseminating information. Information assessment is another issue. With a deluge of information cascading from a variety of different sources, networked individuals must actively develop the skills to critically assess the institutional information they find and what they receive from their personal networks. The ability to balance these two information sources is a key for networked individuals as they cope with information overload.

Pew Internet conducted research in mid-2009 aimed at understanding the new patterns of engagement people had with their social networks and with media in a particularly important context. Through a national telephone survey, online interviews, and in-depth phone interviews, Pew Internet researchers asked people about the ways they were getting information and advice about the 2008 recession and its lingering effects.[29] There was a clear sense in the survey data that people were trying to make sense of complex economic problems that were not easily explained by traditional economic theories. As such, Pew Internet asked about five specific sources of information and support that may help networked individuals to better understand the general economy and their own personal finances (table 9.1). For information about the general economy, a great majority of Americans were most likely to seek out traditional news sources in broadcast and print media (84 percent and 64 percent, respectively). When it came to the two-thirds (64 percent) of the sample who had home broadband internet connection at the time, the internet became more prominent as an information source about both the general economy and one's personal finances. Significantly, use of the internet did not displace people's reliance on interpersonal networks of friends and family.

The survey showed that during this time of uncertainty, networked individuals balanced both institutional and interpersonal information. People did not either talk to others or consult a single media platform.

Table 9.1

Sources Used for Information about the Economy and Personal Finance in the United States

Sources of Information	General Population		Those with Broadband at Home (64% of Sample)	
	General Economy	Personal Finance	General Economy	Personal Finance
Television and radio	84%	45%	85%	46%
Newspapers, magazines, books	64	44	67	43
Internet	48	38	67	52
Friends and family	40	37	45	40
Financial professional	17	24	21	28
None of these sources	6	20	5	18

Source: Pew Internet & American Life Project.

Rather, they foraged among sources and communicated with a range of people. As the recession took hold, the average American used two or three sources of information to make sense of what was happening and to plan personal coping strategies. People talked to other people, sought updates from media sources like newspapers and broadcast media, and actively searched for insights that would help them explain what was happening to the economy and how they might adjust to those changes.

The behavior of networked individuals permeates the survey results. One typical example is Sharon Hockensmith, the sixty-six-year-old wife of a retired Air Force officer. Sharon was attuned to market vibes in part because of her nature as an information omnivore. During the time of the recession, she and her husband signed up to receive email newsletters and online alerts from several financial companies. They also subscribed to some financial blogs with a free market emphasis. Moreover, they began to watch cable TV financial shows like *Fast Money* and *The Kudlow Report*. In short, Sharon juggled networked information the way a networked individual would: "We research companies online; check with our invest- ment adviser; exchange ideas with a couple of financially-savvy friends, and when we know enough, we make investments that, more than ever before, take into account the way the political wind is blowing."

Another example of a networked individual balancing between institutional and interpersonal information sources was a Pew Internet respondent who was looking to buy a new home:

We had been in the market to buy a home for a little less than a year. As the housing bubble burst our neighborhood (Park Slope, Brooklyn) saw a stagnation in housing prices. We felt with our rental lease up we should pounce. We used NYTimes.com to check out open houses. That came after a year in which we had talked to a lot of brokers to get a sense of what we needed to ask. We used online forums (Brooklynian.com and others) to find out about the area and get recommendations for our lawyer. We talked with our parents and siblings (home owners) and we talked with friends working in finance to determine if it really did make sense to buy. After all this work, we decided it did. We found a place being sold by someone not using a broker. We got approved for the mortgage and got the place.

Thus, networked individuals use a number of strategies to help manage the abundance of information that is available to them both online and offline and they exploit both institutional and interpersonal information to help them in their everyday decisions.

The "Veillance" of Personal Information

Beyond general information and news, however, much of the content that is being unleashed in the digital world is that of networked individuals' personal information. The increasing popularity of social networking sites, especially Facebook, has resulted in the public sharing of sensitive personal information such as one's location, marital status, workplace, contact information, and many other details. Users are at least partially aware of some of the information about them that is available online. In a national survey in September 2009, Pew Internet found that 57 percent of internet users had used search engines or other search strategies to see if there was material about them online and 63 percent of them had found at least something about themselves.[30] Among all internet users:

- 42 percent know a picture of them is available online
- 33 percent know their birth date is listed online
- 31 percent know their email address is listed online
- 26 percent know their home address is listed online
- 23 percent know that something they have written is listed online

• 22 percent know the groups or organizations they belong to are listed online
• 21 percent know their home phone number is listed online
• 12 percent know their political affiliation is listed online
• 10 percent know a video of them is available online
• 44 percent of employed internet users know the name of their employer is listed online
• 12 percent of the internet users who have cell phones know their cell number is listed online

When shared online, personal minutiae can be offered by networked individuals to build trust and make online interactions more efficient. Though this brings benefits, it also impacts people's privacy. "Surveillance" is a commonly used word, adapted from the French. But the rise of networked information in the Internet Revolution has enhanced two other forms of "veillance," or "monitoring": peer-to-peer "coveillance" and "sousveillance" by ordinary people of those in authority.

Surveillance
The social life of digital information has opened up the doors to new means of surveillance by government and organizations. Through monitoring social media, governments have found a new way to keep an eye on the behaviors and actions of their citizens. In China, for instance, the Ministry of Public Security has developed an extensive and sophisticated system of surveillance that limits access to information that Chinese leaders believe may disrupt the state's stability or undermine security. In the words of Greg Walton, a researcher for the International Centre for Human Rights and Democratic Development: "Old style censorship is being replaced with a massive, ubiquitous architecture of surveillance . . . [and] the aim is to integrate a gigantic online database with an all-encompassing surveillance network—incorporating speech and face recognition, closed-circuit television, smart cards, credit records, and internet surveillance technologies."[31]

To take one important example, Chinese regulations require all internet service providers to record for at least sixty days the identities of their users, the websites they access, the time they spend on those websites, and any other online activities.[32] That information is handed over to government officials when requested. More than monitoring the websites accessed, the government also scrutinizes the electronic communications of its citizens. Thus the government works with TOM-Skype, the Chinese version of Skype, to gather users' private voice, video, and text conversations. They regularly scan chat messages for specific keywords that are deemed offensive or politically sensitive.[33]

China is far from being the only state exercising such surveillance practices. Western democracies, including the United States, also take part in such activities. For example, following the September 11, 2001 terror attacks that gave rise to Americans' safety concerns, the U.S. government instituted an eavesdropping program to collect both domestic and international communications. Corporations, too, are taking advantage of technological advances to gather information about consumer habits, behaviors, and interests as a means of turning a profit.[34] Writer Evgeny Morozov quips: "The only difference between the two is that one system learns everything about us to show us more relevant advertisements, while the other one learns everything about us to ban us from accessing relevant pages."[35]

Ironically, corporations are exploiting the very same systems of aggregation of user data that were noted earlier as creating new signposts of credible and trustworthy sources. They are using internet-tracking technologies to collect information about networked individuals' online activities, behaviors, attitudes, buying habits, and interactions. For instance, the *Wall Street Journal* found that America's top fifty websites installed an average of sixty-four pieces of tracking technology onto computers resulting in a total of 3,180 tracking files.[36] These data are often commodified and sold to the highest bidder to help businesses market services and products. One telling example discovered by the *Journal:* A tracking "cookie" surreptitiously installed by Lotame Solutions on Ashley Hayes-Beaty's computer that consists of a single code—4c812db292272995e 5416a323e79bd37—accurately identified her as a twenty-six-year-old female in Nashville who has searched for information about the movies *The Princess Bride* and *50 First Dates*. The cookie knows that she has watched the *Sex and the City* TV series, browses entertainment news, and likes to take quizzes.[37]

Hackers or criminals can also gather personal information such as location tagging or status updates about one's daily activities. To demonstrate this, Dave Marcus, director of security research and communications at McAfee Labs, has followed people through the location tagging of their tweets and documenting their schedules, place of work and home—many of which were unwittingly provided online by the individuals themselves.[38] The website PleaseRobMe.com, established in 2010, also brought attention to the dangers of providing such information online. Aggregating and live-streaming publicly shared check-ins via foursquare and Twitter,

PleaseRobMe.com showed when people left their homes: pointing out just how easy it is for technologically savvy and determined criminals to accomplish their goals.[39]

Coveillance: We Watch Each Other

Ordinary citizens now frequently engage in practices of "coveillance," which people use so they can observe each other.[40] Search engines and social networking sites are the primary sources people use to find out more about both known and unknown individuals.[41] While they may be unhappy about others checking up on them, Americans are quite willing to exploit the internet to check up on others. In a 2009 Pew Internet survey, 69 percent of internet users reported searching for someone online, up from 30 percent in 2001 and 53 percent in 2006 when similar surveys were conducted. The later survey found that:

• 46 percent of internet users had searched for someone from their past or someone they had lost touch with
• 44 percent had searched for someone whose services or advice they were seeking in a professional capacity like a doctor, lawyer, or plumber
• 38 percent had searched for friends
• 30 percent had searched for family members
• 26 percent had searched for coworkers, professional colleagues, or business competitors
• 19 percent had searched for neighbors or people in their community
• 19 percent had searched about someone they just met or someone they were about to meet for the first time
• 16 percent had searched for people they were dating

What do they search for? Contact information (69 percent), social network site profile information (48 percent), photos (43 percent), information about professional or career accomplishments (36 percent), personal background information (27 percent), public records related to things like real estate transactions, divorce proceedings or bankruptcies (27 percent), and whether someone is single or in a relationship (17 percent).

Tracking others in this manner can seem creepy. Indeed, the terms "Facebook stalking" and "creeping" have been coined to describe the act of using Facebook to find out more information about those within or even outside of one's personal network.[42] As one student says of this endemic practice: "There's only so much you can learn when you first meet someone. By Facebook stalking, I can learn more about the person, like, who they're dating, their interests, any common friends that we may have and random tidbits of information you wouldn't get on your first encounter."

Scholarly research agrees: Facebook stalking has become so prevalent in the lives of teens and young adults that a Facebook page called "Facebook Stalking. . . . Admit it, you do it" has more than 820,200 "likes" in August 2011.[43] More than that, a Google search of the keywords "Facebook stalking tips" pulls up a number of blogs and articles of do's and don'ts, anecdotes, and best practices. The increasing popularity of online dating websites has also resulted in more coveillance as individuals looking for romantic partners must assess the credibility and reduce the inevitable uncertainties of encountering others online. In 2010, nearly one-quarter (23%) of American online daters engaged in information-based triangulation: checking public records and cross-referencing and comparing profiles on multiple websites.[44]

There is coveillance in other realms of life. In interviews tied to its 2009 survey on reputation management, Pew Internet heard from a large number of those who learned important things in their searches about others. One woman who was adopted as an infant wrote:

I used the internet to trace my birth father through a search. He gave me the name of my birth mother. Through a combination of in-person research and online queries, I patched together the history of my birth mother through property records, birth records, divorce records, and genealogical records (especially a family history placed online by a birth-great uncle). [T]hrough all of this, I found my birth mother, who refused contact.

A few years later I found my birth sister. She is now one of my best friends. We look alike and are alike in many ways. It turns out she vaguely knew about me and looked on the internet for me and came close. We both are very happy to have met each other and it would probably have never happened without the internet.[45]

Other respondents to the Pew Internet survey told of tracking down damaging personal information on a pastoral candidate who was being recruited by their church; described how a search about a physician giving important, intimate advice provided details that he is a transsexual; revealed that a man who used to date a Pew respondent's sister was an avid participant in "furry fandom" events where people dress as animals that exhibit human personality traits; that a dentist who had been wrongfully billing the Pew respondent had also been overbilling other clients; that a boss had quietly accepted a job at a competing firm; and that a would-be tenant was a convicted pedophile. Many described tracking down old flames and long-lost friends. Some described learning too much information about the sexual adventures of younger relatives.

Some of the most riveting stories came from Karyl Chastain Beal about the website she runs for those who want to memorialize suicide victims, suicidegrief.com.

A woman named Melissa submitted her own name for our website's memorial wall. I only knew she lived in Illinois. I searched and found where she lived and I contacted the police. They got to her house before she died. She was getting ready to take an overdose of pills. . . . In another similar case, a woman sent a note to one of the suicide-watch groups I run. She sent a suicide note through the group to me. I used Google to track her down in Canada and called the Royal Canadian Mounted Police. They found her on the floor; she had already taken the pills, but fortunately they got her to the hospital in time.

One of the consequences of all this self-monitoring and tracking of others online is that it increases people's awareness of very weak ties—and that likely changes the way they mobilize our networks. It gives people a better sense of the potential power of their networks and the specific people who might help them address a problem—whether that problem is finding a cancer specialist or a new job. The disclosures and revelations that would have previously been shared with only a small, intimate network of family and friends (or not at all) become valuable indicators of the professional and personal competencies among networked individuals. Such enhanced awareness also gives networked individuals more information than they might otherwise have about such things as the political views, the cultural tastes, the friendship circles, the basic lifestyle preferences, and even the daily activities of those in their networks.

Sousveillance: Watching the More Powerful

In direct opposition to panopticon surveillance where organizations observe people from on high, "sousveillance" is the observation from below of more powerful organizations and people. Steve Mann invented the term when he decided to watch the watchers by video blogging his interactions in department and chain stores, including the surveillance cameras on their ceilings.[46] But most of the sousveillance action is now on the internet, where networked individuals can now find information that has the potential to destabilize power relations.

Wikileaks.org has undertaken the most controversial and publicized sousveillance. It is an organization that releases confidential governmental information online from anonymous news sources who submit sensitive material to an electronic drop box. Its motto is: "We help you safely get the truth out." In October 2010, the organization leaked approximately four hundred thousand private and classified documents that became known as the Iraq War Logs. Following that, in November 2010, Wikileaks began to release and publish a total of 251,287 secret diplomatic cables from U.S. embassies around the world dating from 1966 to February 2010. The release of the cables revealed controversial foreign strategies, such as a secret intelligence campaign where information such as passwords, credit card numbers, email addresses, and even biographic and biometric data of United Nations leaders were collected.[47] Although proclaiming its openness to multiple contributors, the controversial nature of the site—and the probability of government surveillance of its contributors—has reduced contributions. Indeed, the Wikileaks site would not even open on April 9, 2011, and the Wikipedia's "Wikileaks" article reports that "the wikileaks.org domain redirected to mirror .wikileaks.info."[48]

Surveillers can also be sousveilled, as when the OpenNet Initiative works to "investigate, expose and analyze internet filtering and surveillance practices" of seventy countries.[49] It provides information about their filtering and censorship practices as well as the legal, technical, and administrative tools they use.[50] Its sister project, the Information Warfare Monitor, uncovered the GhostNet cyber-espionage network emanating from within China that targeted the computers of the Tibetan community in 2009, including the private office of the Dalai Lama.[51]

The different manifestations and levels of "veillance" show that networked information, more specifically personal information, is bound to privacy concerns. A variety of actors can now more easily exploit the wealth of information available online in ways that fulfill their respective needs. For governments, this means watching over citizens to ensure the "stability" and "security" of the state, while for businesses, this means collecting data about consumer behaviors to find new ways of making a profit. Ordinary citizens may also watch each other side by side in an attempt to find more nuanced information about family, friends, acquaintances, employees and employers, prospective romantic relationships, and even strangers. And of course, sometimes the purpose may be to challenge the surveillance practices of authorities, essentially "surveilling the surveillers." In this way, networked information lives a social life that is deeply complicated and heavily layered.

Dealing with the "Zero Privacy"

The ever-thorny and increasingly salient issue of privacy has been brought to the forefront of popular discourse as networked individuals share information about themselves and as governments, organizations, businesses, and individuals have more power to watch over one another. Former CEO of Sun Microsystems Scott McNealy famously said, "You have zero privacy . . . get over it."[52] Facebook founder and CEO Mark Zuckerberg claims that "People have really gotten comfortable not only sharing more information and different kinds, but more openly and with more people. That social norm is just something that has evolved over time."[53] His company followed that belief by giving third-party application developers access to each member's data.[54]

As surveillance, coveillance, and sousveillance proliferate, people have become more aware of the issue of privacy. The evidence shows that networked individuals, both adults and youth alike, would like to keep the norm of privacy alive by controlling the information going out both to their networks and the wider public. In May 2011, a Pew Internet study found that 58 percent of the adult users of social networking sites set their accounts so that only friends can see what they post and another 19 percent use settings to make their account partially private. A quarter of those who restrict access to only their friends (26 percent) have taken the further step of limiting what their friends can see. Another strategy people use to control their identity is to use different profiles on multiple social networking sites: 42 percent of social networking site users have profiles on at least two sites and another 8 percent have more than one profile on the single site they use. Moreover, despite concerns that sites such as Facebook make it hard for users to adjust their privacy controls, 79 percent of social networking site users said they found the privacy-setting systems not difficult or not too difficult to use.[56] These strategies show that adult internet users are trying to manage their identity to some extent.

danah boyd and Eszter Hargittai found that in response to these issues of surveillance, young people are actively seeking ways to protect themselves and control the information they release to the public. For example, about one-quarter (24 percent) of all Facebook users changed their privacy settings four or more times in 2009, with this number increasing to more than half (51 percent) in 2010.[57] Similarly, Pew Internet surveys have found that 66 percent of all teens with an online social networking profile have restricted access by making the profiles private, adding password protection, hiding them entirely from others, or even taking them offline.[58]

Moreover, tech-savvy youth are creating specialized sublists of friends to further manage information flows.[59] One student explained, "I put a lot of my family members, especially the older ones like aunts and uncles, on my Limited Profile list so that they don't see my pics or my status updates. Sometimes I have pics at parties and swear on my updates and they just don't need to see that."

Other social media users are even removing friends with whom they no longer want or feel obliged to stay in touch.[60] As Pew researchers Amanda Lenhart and Mary Madden conclude: "For teens, all personal information is not created equal." They filter the personal information that they share with particular others, controlling who is able to see what on the basis of the nature of the tie and the particular circumstances.

The evidence for both adults and youth show that there is at least some deliberate attempt at controlling what personal information is released on the internet and for whom specifically. Networked individuals are aware of the costs that come with giving unfettered access to their personal information online and thus adjust their online behavior accordingly. Of course, there are still the covert methods of surveillance practiced by governments and organizations. But in the face of these mounting challenges against privacy, networked individuals are trying to find ways of adapting to surveillance and coveillance.

Lifestyles of the Rich in Networked Information

The relationship of networked individuals to networked information is a key part of their lifestyle. Consumption of information for them can be—and often is—a networking experience as they contribute to and share it as part of their effort to enrich their network relationships and build their reputations. Similarly, social networks help people find the information they want and to understand its meaning. It is commonplace now for a point to come up in conversation, and for participants to use their smartphones or wireless laptops to search for more information. Coauthor Wellman notes that he cannot get away with an assertion in a lecture any more without having his audience check it out.

That is the networking side of the equation. Yet there is an individualism side as well. In the digital world, people's consumption of information is increasingly tailored to their personal tastes and they are more in charge of what they consume, how they consume it, and where they consume it. Twitter and Facebook have become early-warning providers of information that sometimes is as prosaically personal as a change in someone's "relationship status." Other times, social media provide material that is earth-shakingly important, such as breaking developments about Egyptian upheavals in 2011.

Vannevar Bush was trying to think through a "library problem" when he concocted the idea of the memex in 1945. He wanted to craft a way that librarians could stay on top of things in a world where information was growing at a staggering pace. Nowadays, people often have the library come to them rather than actually visiting the library. And what search engines cannot find for them, their online friends often can. Bush was really envisioning the major social change that would unfold as information proliferated and became networked. Ever since the Internet and Mobile Revolutions took off and gave new power to the network operating system, the processes of creating and sharing information have become more vividly a networking activity—with all the joys of sharing and the perils of disclosure.

Interlude: The Conversation Never Ends

Weekend social organizing has a very different flavor with very different com-munication logistics in the age of hyperconnection. We asked one of Wellman's students, Justine Abigail Yu, to describe the process as it plays out in her life.

With Reading Week fast approaching, the past few days have been a flurry of planning as my friends and I have been busy coordinating with one another in an attempt to make the best out of our short, but much-needed break from the stress of university life. Many of the plans have been ongoing for weeks but it is only now as the date draws nearer do we finally start filling in details. It is always a mad influx of phone calls, texts, Face-book messages, and tweets during the days and even hours prior to an event as we must continually clarify what exactly our plans are . . .

Thursday

2:00 pm   I just finished my tutorial and now have a four-hour break ahead of me. I call Nadia to see where she and the rest of my girl friends are and if they are available for the afternoon. I meet her and Sally at Sidney Smith Hall where we make ourselves comfortable for a long day of "studying." I also send a text to Vic and Abby to tell them where we are and invite them to join us.

2:30 pm   After awhile, I walk around the building looking for something to snack on and I run into Anthony, another good friend of mine. We get totally sidetracked and talk for about an hour or so catching up on what has been going on in our lives and discussing the birthday party we are both invited to on Friday. We try sorting out the details as to when and where it's taking place but we're not entirely sure. This plan was made about two weeks ago through a Facebook event but at the time, no venue or time was indicated on the invitation; that information was simply noted as "TBA" (to be announced).

3:30 pm   More than an hour has passed and all I've managed to do so far is socialize so I decide to pack up my things. I need to isolate myself to write an essay that's due the following week . . . these things don't write themselves, you know. So I head on over to the 11th floor of Robarts Library where I know no one can disturb me. I need a computer to do my research and after a few minutes of concentration, I need a little break. I log on to my Gmail, Twitter, and Facebook accounts and it all goes down-hill from there . . .

I get an email from my mom with her flight details asking me if I can pick her up from the airport on Monday. I respond, "of course." On the Twitter front, I respond to tweets asking me about an upcoming "tweet-up" that I'm organizing at the end of the month. Finally, Facebook is where I spend the bulk of my time. I respond to a private message thread from my high school friends. They want to go watch a movie on Saturday, but I've already made plans so I try to negotiate another day during the week.

At the same time, I check who's online and see that Andrew and Sarah are available to chat. They were also invited to the birthday party on Friday, and I want to find out if they're attending or not. Sarah is out of town and Andrew first wants to know who else is going before he makes his decision. So I text our mutual friends Silvia, Heather, and Isaac to find out if they're coming. Isaac can't make it; Silvia and Heather can. I relay the message back to Andrew. Meanwhile, Sarah is upset that she won't be able to make it, but her birthday is next week and she's throwing her own party with essentially the same people invited. We look online for various places around the Toronto area where she can hold her event, consulting websites such as Toronto.com, martiniboys.com, and blogTO.com. She seems pretty set on this one bar but is perplexed because there are quite a few bad reviews online. This goes on for about another hour or so with back and forth emails, tweets, chat messages, and text messages killing my produc-tivity time. There are too many conversations happening all at once and it's starting to get just a bit overwhelming . . .

5:00 pm   I have written just a single sentence for my essay and I am start-ing to feel lonely here in the library. I text Cass and Abby to see if they can meet up before my 6 pm class.

5:15 pm   I meet Cass downstairs and we grab a bite to eat. We spend most of our time brainstorming nearby restaurants for our lunch on Saturday with a few other friends. We have been planning this over a Facebook private message thread for quite some time now but as usual, the concrete details are only being worked out a couple of days beforehand.

6:00 pm Class again. Today should be interesting as we have a guest lecturer who will speak to us via Skype.

6:15 pm Never mind, not interesting at all. I use the TweetDeck application on my iPhone and start tweeting with some people. Thank God for data plans!

9:30 pm I finally arrive home from a long day on campus and automatically turn on my computer. Just like earlier today, I log into my Gmail, Facebook, and Twitter and the whole cycle begins all over again! I check the Facebook event for tomorrow's party and finally, Samantha, the birthday girl and hostess, has provided more details about the location.

12:00 am I am finally able to pry myself away from the computer and get ready for bed . . .

Friday

9:00 am I'm up early and ready to start another day! Is it sick that the first thing I do every morning is reach for my phone to check my email, Twitter, and Facebook? Maybe I should change my morning routine . . .

10:00 am I spend the day working on my essay . . . really this time. I am miraculously able to keep myself away from the internet and my phone.

1:15 pm My friend Abby gives me a call telling me that she has made reservations for our lunch tomorrow and asks if I can pick her up. We agree on a time and I quickly get back to writing. It's a busy weekend so I need to get as much done as I can now.

3:20 pm A sudden flood of texts from my friends arrive, asking me where we are going to meet for tonight's party. It's a back and forth negotiation with each of them that goes on for about 20 minutes. We all finally agree to meet at 9:00 pm at a specific bar for pre-drinks before we all go to the party together at around 10:30 pm.

5:00 pm I start getting ready for the party when I get a call from Silvia. We discuss how we will get home tonight and decide on taking a cab and just splitting the cost.

7:10 pm I am on my way out of the house when I send Silvia a text notifying her that I'm about to leave. I do this so that she can time her departure in such a way that will ensure that we arrive at the bar at about the same time.

8:40 pm   While I walk towards the bar, I send a text to Heather to see if she is there already. She's not. I text Silvia and she responds saying that she is about 10 minutes away and that I should save us a spot as it will be a packed night tonight.

9:00 pm   Everyone has arrived and on time too . . . that's unusual for this group of friends! I "check in" to the bar, using my foursquare application, a location-based social networking site. This particular mobile service allows users to give tips on what to order or what to look for at any given location. Many users suggested ordering a pitcher of the beer instead of a pint to save money so I thought I'd follow their advice. I also tweeted about where I was and what I was up to.

I disconnected myself from technology the rest of the night and instead immersed myself fully in the real-life conversation and laughter.

Saturday

10:30 am   Morning ritual: I grab my phone, check my email, Facebook, and Twitter. My high school friends have responded to my message on Facebook suggesting that we watch the movie on Wednesday instead. I respond to say that I'm free and I block the evening off on my iCalendar. I also receive a Facebook invite to Sarah's birthday party next Saturday and it looks like she has decided to go with the bar that we were talking about the other day. I click on "attending" in the RSVP box and block that night off on my iCalendar as well.

11:15 am   I'm almost ready to pick up Abby for our lunch, but before I head out of the house I give her a call to make sure that she will be ready by the time that I get there.

11:30 am   When I arrive there, I call to tell her that I am waiting outside. It doesn't take her long to come out and we drive to the restaurant to meet everyone else.

12:00 pm   As we wait for the others to arrive, I "check in" using the four-square application on my iPhone again. I've been to this place many times before, and I add a tip for other users to order the calamari as it is the best I've ever had. Just like last night, I tweet about where I am, who I'm with, and what I'm doing.

12:10 pm   Text from Nicole. She's running late, and we should just order without her.

12:30 pm   When our food arrives, I tweet a photo of all the goodies before us.

2:30 pm   Back in my own home, I receive a text message from Nicole saying that she had a good time and that we should do it again sometime soon. I respond with similar sentiments. Abby texts me as well thanking me for the lift to and from the restaurant.

2:45 pm   I log into my Facebook account to find myself tagged in a number of photos from last night's party. I comment on a few of them and write on Silvia's wall reminding her of an inside joke that had us laughing for what seemed like hours yesterday. I upload photos from my own camera as well. The day after a big party almost always means that no work will be accomplished as I usually spend hours looking through our pictures online and reliving the night.

3:30 pm   Realizing that I don't really have anything else to do for the rest of the day, I give one of my high school friends, Carmen, a call to see if she's still up for watching a movie. She makes a few phone calls to see if our other friends are available.

3:45 pm   Carmen calls me back to let me know that everyone is free. While chatting on the phone, I go online to tribute.ca to check movie times and listings. We decide to catch the 4:30 pm show and I send a text to those who are coming with us to let them know about these new developments.

4:15 pm   I arrive at the movie theatre but I'm the only one there. As I use the machine to buy my ticket, I make a few phone calls to find out where everyone is.

4:40 PM   Movie time!

6:00 PM   I'm back home and don't intend to do anything else for the rest of the weekend. I'm pooped from all this planning and coordinating so I relax in front of (can you guess it?) my computer. I have new emails and Facebook messages with friends asking what my plans are for the coming week . . . here we go again!

Reflections

Though I hardly ever think twice about it, there is an incredibly complex interplay of mobile, online, and face-to-face interaction that is used to strategically plan and coordinate not just my mundane day-to-day activities, but also bigger events and get-togethers with friends. Here are my general conclusions.

First, it is clear that although my everyday life is heavily bound up with technology, it is primarily used as a tool to organize in-person contact— various forms of which are strategically employed to fit the audience and specific, unique needs. For instance, Facebook invites or private thread messages are used to address a greater number of people and for events that will usually take place in the relatively far future. I use the mobile phone when I need to talk to an individual to discuss a matter that may take quite some time, while text messages are sent to individuals for short notifications or questions that do not require a lengthy discussion.

Second, mobile and online technologies are often used simultaneously, sometimes even while engaging in face-to-face interactions. I often find myself having multiple conversations with different people over a variety of media at the same time. Thus, there is a great deal of multitasking involved when it comes to planning and coordinating with friends, juggling the various forms of technology in ways that are best suited to the need.

Third, most events are planned many days if not even weeks in advance. However, at the onset of the planning process, they are often general agreements to meet with details continually negotiated and refined as the date of the event draws nearer. A look at my communication journal demonstrates this, as parties often begin with a Facebook invite or message with no location or time specified, but simply a vague description of the event. It is only days or even hours prior to the activity that my friends and I begin discussing what time we should meet, where we should meet, and how we should get there. More than that, we often use our mobile phones in the hours and minutes leading up to an event, calling and texting one another as a means of sending updates of our location and distance from the agreed upon meeting place.

Fourth, my friends and I rely heavily on the internet to help with our decision-making process. For instance, my friend Sarah and I actively use online generated reviews to determine potential venues to host her birthday party. I also use websites to check movie show times and listings. Mobile applications also play a role with location-based services providing various tips for users. These technologies provide us with invaluable information that assists us in the planning stages.

Finally, it is evident that the use of technology does not only occur prior to the event but also during and after. I often tweet about what I am doing as I am doing it, occasionally even tweeting photos live from the event. Mobile and online technologies are also used after the event with friends posting photos on Facebook, writing on each other's wall, or sending a text as a follow-up to our activities.

Mobile and online technologies are a constant presence in my life. They are strategic and most necessary tools when it comes to planning get-togethers with friends. Despite its pervasiveness or quite possibly because of it, this intricate interplay of mobile, online, and face-to-face interaction is one that I rarely ever consciously think about.

III   How to Operate in a Networked World, Now and in the Future

10 Thriving as a Networked Individual

The underlining theme of this book is that it is a networked world, and that being networked is not so scary. Rather, it provides opportunities for people to thrive if they know how to maneuver in it. Arguably, the emerging divide in this world is not the "digital divide" but the "network divide." Technology continues to spread through populations, so the emerging need is for people to learn how to cultivate their networks—and to get out from the cocoon of their bounded groups.

Those who want to thrive in the network operating system need insight into its realities and need to practice how to function effectively in this changed world. People and institutions exist now in information and communication ecologies that are strikingly different from the ones that existed just a generation ago. People's relationships remain strong—but they are networked. To be sure, neighbors and neighborhoods still exist, but they occupy a smaller portion of people's lives. It is hard to borrow a cup of sugar from a Facebook friend 1,000 miles away, but it has become easier to socialize, get advice, and exchange emotional support at any distance. Where commentators had been afraid that the internet would wither in-person ties, it is clear that they enhance and extend them. It is not an either in-person OR online dichotomy; it is an in-person AND the internet AND mobile contact comprehensiveness. They all intertwine in the ecology of the relationship.

These, then, are the hallmarks of the new operating system in which networked individuals function. The volume of information has grown and the pace is accelerating. The variety of information sources has boomed; the velocity of information is accelerating, especially at the level of personal news and niche communities. The places and times in which people encounter information, media, and each other have expanded to anywhere at any time as long as there are mobile connections. People's attention to communicating and gathering information has become simultaneously more fractured as sources multiply *and* more focused as their searches become more specific. Information itself has become networked and more densely packed, making people's experiences with it more immersive and participatory. The capacity of people to find relevant, abundant information on topics has radically improved in the era of instant online search. The ability of people to tell their own stories and share their own ideas has substantially broadened in the era of social media. Their capacity to rate and rank the value of information (and profit from the collected ratings/rankings of others) has grown by leaps. And their ability to build, tap into, and learn from personal networks has bounded upward with the rise of social networking sites and other applications.

These changes in the network operating system have affected individual behavior and attitudes. Among other things, people now expect to find information on almost every subject quickly. They expect that they are more findable and reachable at many more times and places than in the past—and they assume others are equally as likely to be accessible. They have reallocated the way they use their time and attention. They pack more information and communications exchanges into their days and they are interruptible in their activities more often. Their sense of place, distance, and presence with others is transformed as they participate in more encounters that feature "absent presence" or "present absence." Their sense of self transforms from a hard unitary shell to a reconfigurable amoeba with situationally changing pseudopods. Their sense of personal efficacy grows as they practice the art of seeking and gaining social, emotional, and economic support using new technologies. Those activities also highlight the extra effort that networking requires.

This is an operating system that confers social and economic advantages to those who behave effectively as networked individuals, blending significant personal encounters and new media as they solve problems and build social support.

Linda Evans Becomes a Networked Individual

One telling example is the story of how Linda Evans (her maiden name, but not the name she currently uses) has developed into a networked individual. Linda's life abruptly changed in January 1995 when her husband of twenty-three years told her he wanted a divorce. She had been a stay-at-home mother for years and had only recently returned to a manageable work schedule as a sign-language interpreter at her children's elementary school in a Texas suburb. As the divorce process began, the forty-two-year-old found herself financially strapped, and she slipped into a serious depression as she tried to think about a future with her daughters, ages fifteen and thirteen, and son, age eight.

Over the next decade, she rebuilt her life with the help of many allies. First, she decided to seek a semblance of normalcy after she saw her ex-husband striding purposefully into a new life with an annoying lack of scarring. "Time to return to the world," she told herself as she joined a class of newly divorced and separated people at her church. All of them were strangers to her before their encounters in class, but they soon joined a core group of caring friends to give Linda succor. She was also comforted by several inspiring books and video tapes recommended by friends and provided by the church.

Linda could tell she was beginning to mend when soon after the divorce was settled in 1997, she took her ten-year-old son to a two-week vacation at Disney World in Orlando. For the first time striding out on her own, she researched the trip through printed material and friends, made the reservations by phone, planned the driving itinerary via maps (strictly on back roads so they could enjoy the countryside), met her budget, handled all the typical trip glitches with aplomb, and got safely back home. But, she had not gotten seriously involved with the internet yet.

The church class for divorcees became such an important part of her life that Linda stayed on as a layperson helper in the class after she felt she had recovered. Lucky thing. One of the first new entrants to the class after she assumed the modest leadership role was freshly divorced John. He was pleasant, but also as distressed as Linda had been at the start of the divorce. Yet, as the class went on its regular round of activities, Linda and he began to discover common interests.

By the time her relationship with John ripened, Linda was a regular internet user and that helped her relationship with him build. John worked at home as an online support specialist for an electronics firm. On many days, he and Linda would spend a decent share of time emailing each other. They would start stories and played a little game of add on to each other's "mystery of the week." For example, Linda wrote:

It was a lovely spring morning and Susan was thinking of all the things she had to do that day. As she sipped her tea, she was drawn to the array of colors in her flower garden and decided to cut some blossoms to add a perfumed fragrance to her kitchen. Grabbing up her garden shears, she strolled out to the back yard, thinking about which blooms she would cut first. Stepping carefully around some bulbs which were just beginning to emerge from the soil, she let out a scream as her eyes came to rest upon . . .

Then John would append a few sentences and end with ellipses. Linda would pick up the story and they would go back and forth until one of them would close it out. It all took place on email. They only lived six miles apart, but she found that he was often more engaging online than in person. He was wittier and just "different" in his digital persona in ways she quite liked.

As her friendship with John grew, Linda turned for insight, support, and guidance to people in her life who were often different from those who had helped her during her own recovery. She got help as she pondered the romantic stage of their relationship from her teenage daughters, her pastor, and other leaders in the local Southern Baptist community who ran seminars and conferences for singles. "They were not the hurting divorced crowd, they were the right people for the next stage of a relationship," Linda says. "I grabbed at all the things and all the people I thought would help me, each step along the way." She settled all her questions with the support of this group. John found a similar group of allies to coach him through the many steps toward commitment. They married in March 2001.

As Linda emerged from her emotional chrysalis, she also wanted to improve her job skills. Her associate's degree in sign-language interpretation did not produce enough income, so she returned for seven semesters at a small, local university to get her bachelor's degree in business. While she was taking those courses, she also garnered a full-time job with a nearby firm that offered sign-language translation services. However, she did not feel at home in the commercial world. After getting her bachelor's degree she began to talk to a favorite professor and some former teaching colleagues to get advice about the steps she would need to take to work in education. One teacher on Linda's high school campus, Barb, was a particularly good sounding board for Linda. Barb was aware of Linda's personality, interests, and strengths, and pointed her to some occupational assessment tools and a couple of education-theory books. "Barb used her knowledge of my experiences and my interest in educational theory to help me tie things together," Linda notes. "This helped me decide teaching was for me."

In mid-2003, Linda enrolled in a program to get a master's degree in instructional technology from a large state university not far away. She had done some basic research about the program using the internet, but did not feel fully confident about the place until she'd driven the forty miles to the campus and quizzed an administrator in person. "I didn't feel quite comfortable with it until I had actually sat down with somebody and physically checked out the campus," Linda says. By this point, Linda had become an avid user of the internet. The coursework for the degree was given entirely online. She eventually decided to enroll that fall after another couple of rounds of internet research and consultation with the segment of her network who were especially tuned in to the world of educational careers. "I can't tell you where my reading stopped and my friends' advice began," Linda explains. "I'd read something and then huddle with people who had been down this road. That would trigger new online searches. Back and forth and forth and back—websites and people and books and pamphlets and then websites again."

It all worked out as Linda graduated with her master's degree two years later in 2005, a lot more internet-savvy. Soon after graduation, she enrolled in a PhD program in education at an out-of-state college. The program was run entirely online. Her decision in this case was aided less by human interaction than by resources on the internet. A friendly student named Marci inspired Linda by sending her emails with assurances she could succeed and brainstorming with Linda about her options. Linda consulted other acquaintances who already had PhDs to get information about how difficult the whole process was and what their own experiences had been like. Still, the vast majority of her digging for background information was done through online searches and queries.

By this time, Linda had gotten a broadband internet service at home. She decided to pay about 50 percent more for her internet connection in mid-2002 because she was tired of the login process of her dial-up connection, tired of waiting for web pages to load, and anxious to do more browsing for everything from recipes to textbook bargains. "By then, I wanted to do everything online," Linda says. She has stayed enrolled. She also holds down a full-time job helping her public school district figure out how to use new technologies to facilitate online learning, instruction, and greater teacher productivity in K–12 classrooms.

In addition to reconstructing her emotional and professional life, Linda also activated another part of her social network to gain mastery of her financial affairs. The two-year divorce process left her with more than $60,000 to invest and some child-support revenue. No one in her immediate circle of good friends seemed to be as plugged into the commercial and investment world as her friend Susan,[1] who ran a firm where Linda worked. When she sought advice from Susan about how to handle this nest egg, the older woman put her in touch with her investment advisor, who, in turn, gave Linda the name of a new junior associate at his investment firm.

That turned out to be a decent-enough relationship for the basics of investments, but Linda profited more from an even more casual relationship. Linda reached out to Troy, a financial advisor and the son of Cynthia, a local bookkeeper and tax consultant who had previously helped her and her husband John with financial basics like tax preparation. Linda and John transferred their financial affairs to Troy's care. He, in turn, opened up the world of online financial advice to them through his email newsletters that were filled with links and references to online resources for investors. On their own, Linda and John also have become omnivorous browsers of financial websites and discussion groups. They go to Troy's regular client gatherings armed with web printouts and questions about financial instruments, the tax implications of various investment strategies, and tips on the performance of various stocks. They and Troy have helped the nest egg grow several-fold.

Linda's mobile phone is now connected to the internet and has been adding to what has become her networked life. She says the phone allows her to stay in touch with more people, more often, and under more circumstances than would have been possible if she had been tethered to landlines. She often multitasks with multiple devices. For example, she was sending emails to individuals who were helping her make her choice about continuing her education on the same day she was receiving other emails from people in her church support group for the newly divorced. She made calls to plan her next meeting on finances within minutes of taking calls from the faculty who were grading her term papers. "All this technology makes it easier for me to take care of lots of things quickly," Linda says. "It's a juggling act with all the things I need to do, but I don't know how I'd be able to work with so many people on so many different subjects if I didn't have this technology."

When Linda purchased her first mobile phone in 1994, she quickly discovered its many virtues. Whenever she had a few idle minutes she would find herself thinking, "Who do I need to call?" She had never thought about acting so spontaneously in placing calls in the past. The mobile phone has also helped her ease the tension of awkward episodes: If she is running late for an engagement, she calls and makes apologies. When she herself is waiting for others, she uses the phone to do quick outreach to those to whom she "owes" calls or those who might feel good about a "how are you doing?" call. She says:

I use my cell for everything that I can, I just got a BB [BlackBerry] and am now learning how to do even more—for example, right now I am researching how to use it as a modem for my laptop! I send photos and videos back and forth between my children, text with them all the time, and use my cell for GPS capabilities. I also use it for quick research like checking movie times, answering questions, and so on.

Perhaps most interesting is that the mobile phone has brought Linda a modest sense of personal liberation—matched by a sense of loss of control when her children got their own mobile phones. The liberation came when she was glad not to have to wait her turn for access to the family's landline telephone. She also cherished her new power to place calls whenever she needed. She was comfortable when her children got their own phones because it made it much easier to check on their where-abouts and stay in contact with them throughout the day. Still, when they were free to conduct a lot of their phone conversations—and text sessions—outside the household's "public" zone, Linda was less thrilled: "I liked it more when I had an idea of who was on the other end of the phone, and I could hear a little of what was being said in my child's half of the conversation."

In addition to using the internet and her mobile phone to draw support, Linda has also reached out to others with new technologies. For years, she has run her own support group for people with myasthenia gravis, a chronic autoimmune neuromuscular disease characterized by varying degrees of weakness of the muscles. She was diagnosed with this disease while in the middle of her divorce and created the group using online outreach that focused on the type of disease she was diagnosed as having. Eventually, the diagnosis changed, but she stayed active in the original group and formed another one for those who had the same type of disease that her second diagnosis revealed. Linda has run an email "listserv" discussion for each group and a website for both for about fourteen years. The groups also have accounts on Flickr, the photo-sharing site, in order to document the disease and to celebrate the relationships that have been formed in these communities.

"This is a new kind of way to offer people hope and support," Linda says. "There is nothing quite like sharing a common place with people who are just like you." She regularly sends emails and makes phone calls to those who are anxious for two kinds of help: those who want aid in knowing more about the disease and those who want tech support from a savvy friend. "Since many of the folks who have this disease are invalids, I set up a recipe club, joke club, and the like to provide some email communication outside the more serious-focused listserv," she recalls "I wish I could say I have done more, but I can't. I have been in school for five straight years and it leaves little time." Looking back—and forward—Linda says:

The internet was the place I started, the place that gave me a reality check on things I heard, and it was the place that was available all hours of the day and night to help me. I love surfing the web—it is my entertainment (looking around, not games) and I pay all my bills online, use online banking, etc. If it can be done on the internet to make my life simpler, I want to learn how to do it. I even found my electricity provider, bought my cell phone, and send my kids and grandkids e-cards, stuff from Amazon, etc. I recently had laser eye surgery and scoped out YouTube videos on what was entailed!

How People Can Thrive as Networked Individuals

Like Peter and Trudy Johnson-Lenz in the opening chapter of this book, Linda Evans is a networked individual. There are several layers to Linda's social network and, as she puts it, she "activated" those different layers as her needs arose. She says she has eight extremely close family members, friends, and other allies. They are the kind of people with whom she regularly discusses important matters and on whom she can depend when she needs the most important kind of help. She also has eleven friends and other buddies that she feels somewhat close to. They are more than just casual acquaintances. From time to time they give her help and advice, but they are not as close as those in that extremely close group. And then she has scores of other people whom she knows and may be available and supportive when she reaches out. Remember, average Americans now have an estimated six hundred people in their networks.

Linda's story of how she used the various strands of her social network to meet her needs is typical of those who now navigate their personal networks to get the aid and comfort they need. There was very little overlap among the various groups who assisted her in different aspects of her life-rebuilding effort. The people who were most instrumental in helping her make her educational and career-development choices did not know the people to whom she turned when her emotional and social life was troubled. And people in those groups had no direct point of connection with the people and online resources that were important to Linda as she wrestled with questions about her finances.

Linda's story demonstrates how the social shift toward networked individuals changes the rules of the game—the operating system—for social, economic, and personal success. Abundant evidence shows that good, strong social networks of all kinds have important benefits.

Moreover, those with relatively big and diverse networks, including many weak-tie associates, gain special advantages. They get information, support, and advice from more—and more diversified—sources. They are freer to move between different networks, when one becomes too controlling or does not supply what they need. They have the capacity to construct their own groups and negotiate the terms of their engagement with others.

Without consciously following the "rules" of the new social operating system, Linda was practicing the precepts of those who generally do better in a world of networked individuals:

Invest in existing relationships via the Golden Rule so that help will be there when needed: Some old friends were the ones who checked in with Linda regularly right after her separation, commiserated during long phone calls when she felt poorly, and were sounding boards during the legal process leading to divorce. Other long-time pals served as coaches and technical support when she needed to master computers and the internet. And still others helped out with household logistics such as carpools for her children. "My best friends walked with me every step of the way," she says. Surely that was because Linda had given similar aid and comfort to them when they had times of need.

Use ICTs enthusiastically and nimbly: Linda developed from having little involvement with information and communication technologies (ICTs) to becoming an enthusiastic active user of a variety of internet and mobile technologies. Beyond appreciation of technology and having the skills to use it, media-literate people are in better shape as networked individuals in their ability to find information, assess it, react to it, and even remix it with their own creative spin on it. With this sort of media literacy, people can manage their networks better. Not surprisingly, those who have broadband internet connections and are serious internet and mobile users tend to have bigger and more diverse networks and have contact with a wider array of partners. They do not sacrifice the quality of their relations for the greater quantity of their ties. New digital tools give them more ability to function in larger networks.

This is not a blanket endorsement that urges everyone to be early adopters of new ICTs. Rather, it is to note that the evidence suggests that people do well as networkers if they are not timid about technology. Those in the Pew Internet tech-user typology who we classified as "motivated by mobile connections" were also the most likely to use the internet for health-related, civic-related, and news-related material and report good outcomes from their information searches. In other Pew Internet findings, more active technology users had bigger and more diverse networks, and more socially useful contact on a regular basis.

Use technology to develop your access to a wider audience that can share your interests: Look at how Linda's myasthenia gravis groups evolved as they became known via search engines and friends telling friends. The wisdom of crowds can now be tapped in ways that were impossible in the pre-internet era. The web is full of examples of people acting altruistically or seeking help from strangers and getting it. The "audience" layer of people's networks often stands ready to respond when a request for help comes along. Peter and Trudy Johnson-Lenz received help coping with her brain injury from many people they had been scarcely been in touch with for a while.

In a modest, unscientific sampling of 294 tech-savvy people who had agreed to take online surveys, Pew Internet found that roughly two-thirds had done something altruistic online on behalf of strangers or just to give a distant connection a helping hand. Many had responded on listservs and other discussion forums to strangers who had technology problems. Many others had given consumer tips on products and services. One respondent said he was happy to act as an informal trip adviser just for the fun of helping people have good travel experiences. Another performed language translation activities for non-English-speaking inquirers. Another acted as an advice giver and listener to anonymous victims of domestic violence. Another was a group leader on Griefnet and regularly interacted with newcomers to the site seeking to be consoled. Another who was an amateur genealogical expert after years of practicing it as a hobby online said she regularly assisted others get started on their own family trees and provided links to relevant resources. Another helped connect people seeking to adopt pets to those trying to get rid of pets on Craigslist. Another had created an online support group for people suffering from lupus and dealt with queries from strangers every day. Another did the same for those recovering from weight-loss surgery. Another provided free editing assistance for wannabe writers. Another described a prolonged online session where he comforted someone who had received a bogus death threat and extortion attempt. A photography enthusiast regularly posted free advice columns offering tips to amateur photographers. Another acted as an aggregator of links and advice to parents of gifted children. Another took it upon herself to become a patient advocate for strangers seeking help from the Veterans Benefit Administration. Another ran an online support group for families with soldiers in Iraq and Afghanistan. And on and on and on.

Stay active and nimble: Know how to scan the different segments of your social network so you can access the people who are the most suitable to provide the needed information and support. Linda used a variety of contacts as she was making decisions about her educational options—different ones during periods when she was pondering new options. "When I first went back to school, I ran it by a few girlfriends, but it wasn't a hard decision," she recalls. "I had to do it and there was an easy option. When I was going for the higher degrees I felt it was best to talk to professionals and experts—and do a lot more research on my own. My effort was more serious because the stakes were higher and I was less certain at the beginning what the best choice was."

Do not count on a single, tightly connected group of strong ties to help: Linda pulled together one-to-one encounters or ad hoc groupings on an as-needed basis. "I often had very specific needs and each new need was different enough from the last one that I needed another person to help me," she said. Groups in the networked age are often too weak, small, specialized, and uncoordinated to hang together over the course of an extended problem-solving period. They often lack access to a shifting diversified set of resources. One-to-one relationships and partial networks are the ones that usually are most effective and efficient.

Develop meaningful new ties as you go along and be especially alert to reaching into new social circles that serve your purposes: Linda made new friends in the church group of divorced people: her teacher buddy Barb who linked her to information relevant to entering the master's program, her classmate Marci who affirmed her decision to seek a PhD, and her tax preparer's son Troy who became her financial mentor. None knew the others or traveled in the same circles. They had varying interests and expertise. All were appropriate choices for advice and support when Linda was making decisions. Asked if she was consciously trying to diversify her social network in these choices, Linda demurred: "It would be more accurate to say that I had a sense of who had the most knowledge to help me or whose insights into me and my personality seemed accurate."

Develop larger and more diverse networks: Personal networks can now run to more than a thousand people, if you count the most remote but still meaningful consequential strangers. Look at all the ties that Linda was gathering up in little more than a decade, with many more ties to come and few to lose. No doubt she will rely on Facebook or similar software to keep in some contact with friends from past years and life experiences.

Although bigger is not always better, those with diverse, broad-ranging networks are often in better social shape and have a greater capacity to solve problems than those who have smaller networks. Quantity *does* equal quality. Not only do larger networks provide more overall sociability, support, information, and connections to the rest of the world, but preliminary research shows that each tie in a large network is likely to be more supportive than those in smaller networks.[2] A culture and network of support breeds more support. And, those with many functional "weak ties" can find support and solve problems more adeptly than those who are deeply embedded in a small, tight social network.

In the digital era, networking behavior turns out to be an efficient way to collect and verify information. Recent research has shown that connections between clusters of friends—or websites—are efficient structures for acquiring information. Jon Kleinberg's HITS algorithm (Hyperlink-Induced Topic Search) notes that websites should be considered important if they are both linked to by many others and also themselves link to many others. In other words, such sites connect to a wider-ranging array of sources, and in doing so illustrate the verb meaning of "networking." Most clusters contain superconnectors—people linked to large numbers of others in multiple social milieus—and these connectors rapidly diffuse information. At the same time, talking to friends within clusters allows people to assess and validate the information they have received.[3]

Act "transitively": Look beyond your friends to become aware of the people in their networks who could provide access to new worlds. That is what Linda did when she asked her boss Susan for help finding someone who could help her manage her divorce nest egg. That was the same impulse that prompted Linda to ask a similar question of Cynthia, the tax preparer. "I didn't know any good financial advisers on my own," recalls Linda, "so I had to get tips from the people who seemed more financially-savvy than I am. I figured there would be someone in their lives who'd be a good fit for me."

Act as autonomous agents to cultivate your personal networks: In the good old days of strong kinship systems and densely knit villages, people could sit back and let things happen to them. They'd wander to the pub and find friendship, or go to their mother-in-law's house every Sunday for dinner. They would know there would be support when trouble struck. With dispersed, sparsely knit person-to-person networks, those days are no longer here. Witness how Linda had to reach out and actively get help and advice. She even had to go to her church group for help rather than waiting for them to come to her.

This is the era of free agents and the ethic of personal agency. Social advantages and privileges accrue to those who prospect for network ties the way effective sales agents prospect for clients. The individuals primed to take advantage of this are the ones who are motivated to reach out to others, share their stories and support, and then invite conversation, feedback, and reciprocal gestures. The internet and mobile phones vastly expand the capacity of people to do the outreach and nurturing of friendships that are part of prospecting.

Monitor and manage your reputation—your personal brand: At the first level, individuals need to monitor the information about them because there is likely to be much more available in social media such as Facebook, blogs, listservs, photograph sites such as Flickr, video sites such as YouTube, and online discussion forums.

Information persists once it has been created and communicated online. As information scientist danah boyd notes, "What you say sticks around. This is . . . not so great when everything you've ever said has gone down on your permanent record. . . . [Moreover,] you can copy and paste a conversation from one medium to another, adding to the persistent nature of it. This is great for being able to share information, but it is also at the crux of rumor-spreading."[4]

In effect, there are limitless possible ways to recreate and reshare and recommunicate material after it has been created digitally. Even worse for those bent on anonymity, boyd points out that people and information are searchable even if you segment your network: "With social media, it's quite easy to track someone down or to find someone as a result of searching for content. . . . This is great in some circumstances, but when trying to avoid those who hold power over you, it may be less than ideal."

The increased prevalence of self-monitoring and observation of others creates a dynamic environment where people promote themselves or shroud themselves depending on their intended audience and circumstances. In several surveys about reputation management, Pew Internet found that 57 percent of American internet users had searched for material about themselves online.[5] Some 33 percent of internet users worried about how much information is available about them, and 8 percent had asked other users to remove information about them.

Among social networking site users, more than three-quarters had taken steps to limit the amount of information that they shared about themselves. For instance 65 percent of adult social networking users have changed the privacy settings on their profile to limit what they share with others online; 63 percent had "unfriended" contacts in their network—deleting people from their friends list—and 52 percent had kept some people from seeing certain updates; 44 percent had deleted comments that others have made on their profile, and 37 percent had removed their name from photos that were tagged to identify them. At the same time, many reported benefits from sharing information online: 40 percent of internet users had been contacted by someone from their past who had found them online, which is double the number who had reported that in a 2006 survey. Another 48 percent said they found it easier and more meaningful to meet new people because it was easy to learn things about them online. Overall, the message from the surveys was that when people consider the trade-offs of disclosure vs. privacy in the digital era, the majority of them see advantages in disclosure and the prospect of being findable. They take modest steps to limit information about themselves, but most are not deeply worried about disclosures they think are reasonable and confer benefits.

This is a reality for all networked individuals, but it is especially important for those whose livelihoods depend on the outside world knowing who they are and what their reputation is. In the 2009 Pew Internet survey, 12 percent of the respondents said their job required them to promote themselves or market their name. That population segment will surely grow and the portion of them monitoring their online identities will likely rise. Fully 84 percent of the people with those jobs had checked for information about themselves in 2009; 73 percent were social networking site users and 29 percent were bloggers. There will also likely be broader awareness of the online tools that are available to those who want to check up on their virtual selves. Those tools include email alerts and syndicated news feeds when an individual's name is mentioned on news or other prominent sites and buzz-monitoring tools for the rankings, ratings, and social media sites where that person's name might pop up.

The incentive to monitor and control information will become more pronounced as more information about people finds its way into databases and online venues. As Daniel Solove argues, "once information about us finds its way into the minds of others, we can't control what they think about it. Our ability to exercise control consists of being able to limit the circulation of information about us."[6] Solove points to the key conflict for networked individuals in reputation management: Too much information disclosure harms individuals' privacy and freedom to act; too little informationly inhibits people's ability to promote themselves and, more importantly, to build trust. Too little information also stifles free speech and the free flow of information.

Segment your identity: Linda's networked self operates in multiple networks: her church group, her divorce group, her teaching-support group, her graduate school friends, and two different groups focused on myasthenia gravis. She does not hide them from each other, but there is not much overlap.

Over time, more people will take at least a few steps to manage their identities by segmenting pieces of themselves—in effect, embracing a networked self in which different parts of themselves are on display to different audiences in their networks. It is not that they will have separate selves for different segments of their networks, or for online versus offline interactions. Rather than different personas, people's selves are networked: There is a core, but different aspects of that self get emphasized in different social situations.

These are difficult trade-offs for networked individuals to calculate. They know, though, that people can be especially hurt by mismanaging their identities in one of several ways. For example, if an individual does not disclose her needs, talents, and achievements she will miss opportunities to gain help or advance. Or, if she discloses too much inappropriate material about her life she may likewise find herself denied opportunities. And if she doe not monitor what is known or said about her online, she cannot know where her reputations falls short of reality or where it could be bolstered. As boyd points out, "social media scales things in new ways. Conversations that were intended for just a friend or two might spiral out of control and scale to the . . . whole world." That is why segmenting networks and watching boundaries make even more sense online.

Develop the knack of functioning effectively in different contexts and "collapsed contexts": For Linda, like all of us, the act of joining and belonging to multiple networks requires the development of contextual knowledge as each has different histories, norms, folklore structures, and dynamics. As boyd puts it: "Some behaviors are appropriate in one context but not another, in front of one audience but not others. Social media brings all of these contexts crashing into one another and it's often difficult to figure out what's appropriate, let alone what can be understood."

People must learn the ropes in these different milieus. The more gracefully they can do this, the quicker they can assume greater roles within multiple communities and networks.

This classic networking behavior—acting smartly in the situation in which you find yourself—has new complexity in the digital era. boyd argues that a central reality of life in the world of "networked publics" is that old social and group contexts collapsed as digital media emerged. She notes that in conventional social arrangements, people chose what to say, when to say it, how to frame it appropriately for the audience that is obviously listening, and how to set it in the most useful general context. "Some behaviors are appropriate in one context but not another, in front of one audience but not others," she argues.

Nowadays, a speech that is delivered to a professional audience can also be blogged in real time and posted quickly on YouTube, making it available to those outside the intended audience and viewable in bedrooms in the middle of the night or digested over morning coffee. As boyd maintains: "Social media brings all of these contexts crashing into one another and it's often difficult to figure out what's appropriate, let alone what can be understood." Effective networkers have to take this into account by expanding the milieus in which their activity can be understood, and explaining themselves in ways that multiple audiences can comprehend. Realize that intense scrutiny—even in unexpected situations—is a realistic possibility. As coauthor Rainie's Pew colleague Susannah Fox likes to point out, the new social networking environment means that people "should always be ready for their close-ups" nowadays, because that close-up moment could be viewed and reviewed and reviewed.

Build high levels of trust and social capital in each network segment: Linda has worked diligently at this as she has navigated her complex life. Social capital can be earned, amassed, to some extent banked, and often used. Note how Marlon Brando in *The Godfather* film built his capital: "Someday—and that day may never come—I'll call upon you to do a service for me. But until that day, accept this justice as a gift. . . ."[7]

A bedrock law of social networking is that people need to discover and interact with those who can provide resources. Although humans seem to be hardwired for reciprocity, social capital has its own rewards as it allows us to gain prestige with individuals or in networks, get things done, and enhance our sense of self. The essential point is that trust and reciprocity are primary currencies for networked individuals.[8]

Digital media and social networks provide new ways to offer and procure social capital online, and the basic value of social capital is growing because such networks are essential to people's success. Just as digital technology has changed the contours of identity and privacy for networked individuals, it has also added elements to the dynamics of trust building. One particularly challenging change is tied to what the technology community calls "transparency." Trust building at the personal level has always required evolving levels of disclosure, but now the audience for such disclosures is bigger and the expectation it will take place is also greater.

Another aspect of trust building that has intensified in the digital era relates to mistakes. Everybody makes them. But when they occur online or are captured digitally, people's mistakes can be more widely dissemi-nated and accessed. It is not clear yet whether social norms have become—or will become—more forgiving of prior indiscretions because of this or whether people will pay for their mistakes over a longer period of time because such mistakes are archived and findable by would-be friends, would-be employers, would-be romantic partners, would-be clients, and people from the past. The technology community's response is to encour-age people to admit their mistakes, correct them if possible, and seek forgiveness. This is part of why builders of digital media generally urge that people and institutions be transparent. Yet, many networked individuals will want to put up walls around parts of their lives and control who has access to what information, as well as find ways to prune the internet's digital memories.

Manage boundaries: As the power of formal, densely knit groups wanes in light of the buildup of networks, personal and community boundaries are less distinct. With digital technologies, more private information is potentially available to interested members of the public—and to govern-ment and organizational surveillance authorities. Networked individuals need to develop nuanced understandings of what to make public, which publics to make information available to, and how to intermix technolo-gies of privacy with those of public narrowcasting. People used to do that more or less routinely in real life, as they encountered the sights, sounds, smells, and the people of different social milieus. But online, all they face is a screen. As danah boyd puts it:

There's the blurring of public and private. These distinctions are normally structured around audience and context with certain places or conversations being "public" or "private." These distinctions are much harder to manage when you have to contend with the shifts in how the environment is organized. . . . Trying to keep social acts to one space online is futile, even though that is the norm in unmediated environments.

Other identity-shaping boundaries are also blurred or obliterated in the digital age. The edge between home and work (or school) becomes less distinct as people work at home outside "regular" work hours and do "leisure" activities while they are sitting at their desks on the clock at the office. Similarly, the border between education and entertainment is not as clear as it used to be in the era of "serious games" that have educational purposes at their core. Surely, the longstanding distinction between con-sumer and producer has eroded as people remix media, and as nonprofes-sionals broadcast their works on video-sharing websites and publish their thoughts on blogging sites.

Be aware of invisible audiences: In this book, we have argued that the converging internet and mobile environments facilitate creation of a distinctive new layer to social networks: the audience layer. Its inhabitants can be socially helpful when they are activated and motivated, as were some strangers when they learned of Trudy Johnson-Lenz's medical problems and donated money because friends of hers had put out a call for help. At the same time, lurkers may be less well-meaning. They may be stalkers or competitors who now can have easy access to much more personal and professional information about people than in the past. As boyd points out, there are also "visitors who access our content at a later date or in a different environment than where we first produced them. As a result, we are having to present ourselves and communicate without fully understanding the potential or actual audience."[9] The challenge in this for networked individuals is striking the right balance between disclosures that bring help and broadcasting too much information that invites trouble.

Manage time well; multitask strategically: People need to manage their attention more carefully than ever before. Effective networkers exploit this new digital environment more powerfully than those who get lost in their browsing or swamped by information inputs. Under the conditions of networked individualism, people need to work harder to stay on top of their own needs. They must spend more time maintaining ties in their networks and making sure that stores of social capital are replenished. Moreover, they know they are a part of others' networks and they have a heightened sense of obligation to meet the needs of those who consider them social ties. A new widely accepted etiquette for transitory networked relationships has not emerged to create acceptable social rules that allow people to more easily enter into or break off from networked relationships. But it is taking shape in the context of perpetual engagement and partial attention.

New Literacies for Networked Individuals

The networked individuals who thrive have a combination of talent, energy, altruism, social acuity, and tech-savviness that allows them to build big, diverse networks and tap into these networks when they have needs. They are mastering a set of new literacies to navigate the network operating system.[10]

They have *graphic literacy* that recognizes that more and more of life is experienced as communications and media on screens. They can interpret this material and feel some need to contribute to it. They know how to participate in digital conversation and creation. This literacy requires networking behavior that is often conducted graphically.

Networked individuals also have *navigation literacy*, a sense of internet geography that allows them to maneuver through multiple information channels and formats. They understand the change that has occurred as linear information formats such as print and broadcast media have given way to the nonlinear realities of hyperlinked, networked information. Not only do they know how to navigate, they also use their communications and contributions to help others navigate, often by recommending links in their digital communications or by creating their own posts to show others what they have learned.

Beyond that, they have *context and connections literacy* that helps them weave together the information and chatter that is flowing into their lives at a quickening pace. Even if the tidbits they gather are disaggregated from any larger context, they have the wherewithal—often with help from the network helpers—to puzzle through the material they collect. Networked individuals draw on their family, friends, and associates to make meaning of the things they encounter and the things that happen around them.

We suspect that networked individuals who are thriving also have *focus literacy*: the capacity to minimize the distractions of the digital cacophony and complete the work they need to do for their jobs and their personal enrichment, even as they multitask. The paradox is that even while connecting people to multiple social networks, digital technologies carry with them almost an insistence that people stay connected on their mobile phones or the internet. Yet, there are other times when people who are completing an urgent or personally satisfying project require a more individualized mastery of material and creation of knowledge. The most accomplished networked individuals are better at turning the solitude switch on so that they can focus.

Despite their focus, networked individuals also have *multitasking literacy:* the ability to do several things (almost) at once. With multiple inputs from family, friends, work, and institutions—and multiple in-person, internet, and mobile sources provide these inputs—thriving networkers such as Nelu Handa (in chapter 4) have gained the ability to attend to them all without fuss. Those who say it can't be done forget the reality of driving a car in a big city while routinely integrating steering, braking, checking gauges, scanning the surroundings, chatting with passengers, and listening to music.

One of the great challenges in this age of overabundance of information requires *skepticism literacy*. Internet veteran Howard Rheingold's new book about new literacies focuses on what he calls *"crap detection"*—the ability of individuals to evaluate what they encounter online. This is especially urgent for those who get an unrelenting stream of material from others in their social networks. The most successful networked individuals have the capacity to assess inputs from friends and media sources for their accuracy, authority, relevance, objectivity, and scope. Skepticism literacy is the ability to weed out the media and people who have outdated, biased, incomplete, and agenda-driven or just dead-wrong ideas to pass along.

Connected to this is the idea that networked individuals should have an *ethical literacy* as they network with others. Successful networked individuals build trust and value for their partners by being accurate and thoughtful with the information they create and pass along. When everyone can be a publisher and broadcaster, there are advantages that accrue to those who are found to be reliable and transparent about the information they share. In contrast, social penalties are conferred on those who cheat, misrepresent information, cut corners, exploit relationships, and are mysterious about the sources of their information.

In a sense, you could say that the basic argument of this chapter is that the most successful networked individuals have *networking literacy*. Like Linda Evans, they know how to move adroitly through their network operating system—personal, institutional, and digital—without getting locked into one world. They follow the Golden Rule. They scan their existing networks for the possibility of gaining introduction to new networks that can expand their reach and diversify their sources of information. They strike useful balances between being "on the grid" taking advantage of digital opportunities and being available to help others, and "off the grid" when they need time to rebalance and contemplate without interruption. In short, they are masters of the new network operating system's universe.

11 The Future of Networked Individualism

Large-scale forces in economics, politics, culture, and religion drive social change, and societies change as they embrace new information and communication technologies (ICTs).[1] Changes toward a network operating system have brought a host of unsettling things. For starters, new ICTs bring fundamental challenges to the societal role of experts and information gatekeepers. In response, they fight back, claiming that their institutional connections and credentials give them the authority that amateurs lack. Doctors have to deal with e-patients who get information from the internet, and journalists sniff at bloggers who riff on civic events. That is invariably followed by a populist backlash as people demand access to new tools to create and share information and new pathways to ever-larger stores of information. As a byproduct, the new tools also give new life to crackpot propagandists and scam artists.

Still, the march of social adjustments presses onward. New methods of accessing, cataloging, sorting, and searching information have burst on the scene as people try to cope with information overload. Wholly new institutions and organizational structures form as enterprises take advantage of the new technologies and try to sort through a mountain of new information. At the same time, fights over intellectual property break out as creators who existed in the old order are threatened by the way technology facilitates easier duplication and sharing of information. Subgroups based on personal identities and tastes multiply as people use technology to seek out and bond with others who have things in common with them: lifestyles, belief systems, medical conditions, political outlooks, hobbies, passions, phobias, enthusiasms for learning, life-stage similarities, common enemies, pop-culture favorites, professional aspirations, and on and on. New languages and ways of expression come into being—think of emoticons and cryptic abbreviations like POS ("parent over shoulder," which teens instant message to friends as a warning). New professions take shape, such as search engine optimizers who work to get their clients listed high in search engine results and privacy consultants who work to protect their clients' information from unauthorized disclosure. ICTs foster the reconfiguration of social norms and spaces as people adopt and fit them into their lives. What were once formal boundaries between public and private realms either become more permeable or break down completely.

The Triple Revolution—Social Network, Internet, and Mobile—has given rise to far-reaching consequences, though it is not yet clear what the outermost limits of the impact will be. What is evident is that networked individualism is tied to technological changes that are on the horizon. So, it is not too soon to project from current trends to predictions about the technosphere of the future, and to consider how upcoming developments will affect the pursuits of networked individuals. We start with an exploration of how ICTs will change, and then we explore how such changes could affect the lives of networked individuals.

The Enablers of Future Trends: The "Laws" of Digital Development

As the Triple Revolution moves on, what will it bring? The simplest—and perhaps the most accurate—answer is "We don't really know." We have nowhere near the hubris of Isaac Asimov's science-fiction hero Hari Seldon, who tried to predict a thousand years ahead for the galactic empire.[2] Seldon got it wrong anyway. Would-be forecasters often forget that human societies coevolve with technology. Within two decades, the internet and mobile phones have evolved into universal extensions of people's networked individualism. As Norman Augustine's "Second Law of Socioscience (and engineering)" puts it: "For every scientific (or engineering) action, there is an equal and opposite social reaction."[3]

But why be only cautionary when we can have fun thinking about near-term possibilities that the continuing mutual feedback of technical developments and social situations provide? We begin by thinking about some emerging technological changes: increased computing capacity; miniaturization; density of graphical displays; digital storage capacity, file compression, and faster connections. They are the basis for the technological trends we discuss in the next section, and the social trends we anticipate—and in one scenario, fear.

One of the most useful and formal exercises in futurism in recent years was the 2006–2007 work of the Metaverse Roadmap project. It was driven by the Acceleration Studies Foundation of Mountain View, California, and originally conceived of as a brief for the future of the World Wide Web as it became three dimensional. Once the leaders of the effort began to hear from several dozen thinkers, their own views branched out in other directions. The leaders, John Smart, Jamais Cascio, and Jerry Paffendorf, had started their inquiries with the notion of a "Metaverse" that was first conceived by the influential science fiction writer Neal Stephenson in his 1982 classic, *Snow Crash*.[4] To Stevenson, the Metaverse was an immersive, virtual space with 3D technologies. Yet, the Metaverse Roadmap thinkers went beyond seeing the Metaverse as a virtual domain. They saw it as the "convergence of 1) virtually enhanced physical reality and 2) physically persistent virtual space. It is a fusion of both, while allowing users to experience it as either."[5] In other words, it is the connection of the physical and virtual worlds. Although we do not foresee people living mostly in virtual space, the technological directions suggested by the Metaverse Roadmap provide guides for how networked individualism may proceed.

The "laws" we summarize here do not really describe universal physical or social phenomena. Rather, Such laws are forecasts based on what has happened and what the laws' creators expect to happen. Such laws are especially popular in the high-tech industry, and with its pundits, who are keen to capitalize on trends. Taken together, the laws identify the probability that pushing more bits through wires and over airwaves will get easier and cheaper. This will further enable networked individuals to share greater amounts of information and interact with various parts of their networks more quickly and affordably by using information-dense media in an increasingly portable manner. Most importantly, physical and digital phenomena will converge as networks go beyond linking networked individuals and ICTs to connecting them with objects.

Computing capacity—the price-to-performance ratio of computing hardware—may continue to double every eighteen to twenty-four months. The timeframe of every five years of continued development has been estimated to provide as much as ten times the computing capacity, and as much as one thousand times every fifteen years. This has come to be widely known as Moore's Law,[6] which is neither a real physical nor social law, but an empirical generalization of past trends that have proven to be true until now.

The *miniaturization* of technology will arguably have an even more powerful impact because of the efficiencies and computing power it creates in miniscule objects. It has become a cliché that we carry in our smartphones more computing power than the first manned space flights did.

The *density of graphical displays*—the capacity to represent the world visually—doubles every two years in what is known as "Nishimura's Law": The display size available for the same cost doubles every 3.6 years.[7] These developments make representations of people and processes more realistic and dynamic by transforming the internet from a series of static snapshots to more interactive and detailed graphic representations. Enabled by this, networking becomes more fluid and engaging.

Computer *storage capacity* has increased to terabytes: "Kryder's Law" asserts that digital storage capacities have doubled every twenty-three months since 1956.[8] At the same time, improved *file compression* could allow another ten times' worth of digital material to be packed into the space now occupied by today's typical MP3 files. In addition, there is currently a seismic change from inherently fragile hard disks to smaller, faster, and more reliable solid-state flash drives. These changes in usable file size would allow higher resolution for more detailed—and hence more realistic—images. They would also allow larger files to be routinely and quickly exchanged. "In combination with the storage growth trends, today's 60 GB, 15,000 song iPod will be capable of storing 2.4 million songs in 2014," the authors of the Metaverse Roadmap calculate.[9]

Other laws point to the doubling of *available bandwidth* within the wired environment every two years or so ("Gilder's Law"), *internet connection speeds* increasing by fifty percent every year ("Nielsen's Law"), and the number of *possible wireless communications* doubling every two and a half years ("Cooper's Law").[10] This enhanced speed coupled with the *greater reliability* of the internet and mobile networks makes it more feasible to move personal data from one's own computer and tablet to the internet cloud. Security remains an issue, not only from criminal elements but also because cloud-based storage is more easily subject to government and organizational surveillance and subpoena.[11]

Convergence of the Physical with the Digital

Remember Tom Cruise in the film *Minority Report,* when he waved his hands to access information from computers, and when the shop fronts blazed with personalized signs as he walked past?[12] This is a dramatic change from the current situation. Until now, humans have been external users of rather impersonal phone and computer networks that did not know much about them other than their phone numbers and login identifications. By contrast, we are at the start of an environment in which people are intimately intertwined with digital networks that follow them everywhere they go, with all of their individualized settings and information.

In coming years relatively ordinary objects—as well as computers and phones—will become ubiquitously networked with each other, and networked individuals will be able to augment their information through direct contact with databases and objects that have become smarter and more communicative. Increased computing power may make user involvement with virtual worlds more immersive and compelling, although experiences to date suggest that people are more apt to use computer networks that integrate with real life rather than becoming totally immersed in virtual worlds—virtual game players excepted.

Ubiquitous computing, sometimes called "the internet of things" (or "everyware"), describes human–computer interaction that goes beyond personal computing to an environment of objects processing information and networking with each other and humans.[13] Objects would share information: appliances, utility grids, clothing and jewelry, cars, books, household and workplace furnishings, as well as buildings and landscapes. They would learn additional information and preferred methods of use by gathering data about people who are in their environment. For example, cars could tell each other not to be in the same spot at the same time, and bicycles could tell car doors not to open when the bikes pass by. More grandly, nuclear arms could control themselves by broadcasting to the world when people start to activate them. Through such systems, people might engage with multiple computational systems without even knowing they are doing so. Although people do that now when they drive—aided by the car's geographical positioning system, antiskid control, and anticollision device—in the future, such networks would be even more connected and encompassing.

Smart objects are emerging, although they are far from ubiquitous. In addition to general-purpose computers and smartphones, there will be specialized smart appliances. Already, Microsoft's Surface is a table that recognizes, responds to, and networks with objects that are placed on it. At present, it can only recognize other electronic devices. For example, smartphones placed on it can show scheduled events in calendars and simultaneously display route maps to these events. Walls can be smart too. Painters and designers could work collaboratively and remotely as they draw on their networked surfaces as if sharing a single canvas, drafting table, or white board.[14]

Interactive "blogjects" would have memories and network with both other objects and people.[15] For example, a networked camera would have its own picture-sharing account as well as context-aware capabilities making it trackable by other networked objects and traceable via Google and GPS. The camera could develop a sense of its surroundings and tasks (such as surveillance), by knowing what other blogjects are around it and communicating with them.

On a grander scale, an intelligent travel system would rely on sensors in streets and cars to provide realistic current information to each driver about preferred routes and anticipated travel times. Sensors would communicate reports of accidents, alert other drivers, and dispatch help. Ultimately, cars might drive themselves, relying on their own computational power and ubiquitous computing in their environments,[16] freeing drivers to safely use their mobile devices.

This world of ubiquitous computing is one in which both aspects of *networked individualism* will be enhanced. *Individualism* will be heightened as people can seek, scan, sift, sort through, and make sense of more and more information on their own. Their technology and their smart agents (or bots) will increase their capacity to navigate the world. They will also have the ability to participate as individual agents in the world, leaving information and insights in their digital wakes. This material will be helpful to them, because it will teach their objects and technologies about their interests and personal data such as calendars, preferred vendors, friends, and communities. In short, improvements in technology will give people more capability to be on their own and to act and network when they choose.

At the same time, the *networked* side of networked individualism could be enhanced by such technological changes. The coming changes in technology will allow people to locate and join forces with others who have sought the same material or shared similar paths of experience and exploration. People will be better able to find experts who know more than they do and to locate communities—both formal organizations and informal tribes—that can help them solve problems and make decisions. They will also be able to reach out to others whose disclosures about themselves make them seem simpatico and emotionally helpful to become connected citizens with shared cultural, economic, social, or political interests.[17] Similarly, networked individuals who disclose what they know and what they have been through will find that others reach out to them as part of their networking activity. The fluidity and value of personally assembled networks is likely to develop in this environment and the effort required to create those networks should decrease. In sum, the value and appeal of the behaviors tied to networked individuals will increase.

Active agents and the semantic web: To accomplish ubiquitous computing, it would be useful to have a widely accessible computer language—such as the *semantic web* championed by World Wide Web inventor Tim Berners-Lee.[18] This system could enable virtual assistants to conduct tasks on behalf of people. Such networked agents would access the web more intelligently and systematically by finding relevant information and acting on it. They would compile and disseminate information, such as getting directions and aggregating and passing on news stories that are useful and pertinent for their users. As agents of networked individuals, they would also be able to schedule appointments in concert with other agents, resolving conflicts and presenting alternative plans, routes, and locales. They would also share personal information with appropriate network members based on the privacy preferences learned through their past uses. To some extent, agents could eliminate the need for people to comb through lists of search results, as the agent would understand how pieces of information relate to one another and would search for whole concepts and their linked notions instead of disconnected keywords. This would help alleviate information overload by filtering out unwanted information and communication. It would create more awareness among continuously updated network members but also increase social control.[19]

Many observers of technology believe that even if the particulars of the semantic web vision do not come to pass, there will be vast and unstoppable movement on the internet toward technology-driven understanding of information that will make the web smarter and more capable of helping people coordinate their lives, find the data they want, and locate the people and organizations that can help them get their needs met. Data mining, social network analysis tools, social-computing studies, and user-generated folksonomies (collaborative creation of tags and catalogues) will make the web easier to navigate and allow information now scattered in various places to be pulled together in meaningful ways for networked individuals. More layers of information will be added to things—for instance, as location-aware applications become widely used. Search engines will also have better algorithms to help locate high-quality information. More digital material that is currently difficult to parse will be much more searchable: especially videos, images, and material generated in foreign languages.

Interfaces will diversify. People's voices, and perhaps even their thoughts, will trigger interactions with machines and other humans. Interfaces manipulated by hands and body movements will become common. Screens that display data of all kinds will not have to be carried around, as active walls and tabletops become display areas.

Augmented reality (AR) systems and interfaces process and layer networked information on top of people's everyday perceptions of the world. They help people view the physical environment with computer-generated information enhancing what is being observed through images, videos, audio files, or text. As the Metaverse Roadmap authors see it: "The augmented reality scenario offers a world in which every item within view has a potential information shadow, a history and presence accessible via standard interfaces." Two widely-seen examples are the yellow first-down lines that are inserted on American TV screens during football games and the digital ads and logos that are displayed during televised athletic contents worldwide on fields or walls.

More often, augmented reality interacts with an individual in a symbiotic relationship. For example, the growing use of smartphones with built-in cameras has pushed the beginnings of handheld AR to the forefront. There are already dozens of AR apps that, among other things, allow people to find bars and restaurants near them, get directions to them and call up reviews, menu details, and discount coupons. Other apps allow people to load relevant Wikipedia articles into their phones as they gaze at objects, find WiFi hotspots in the storefronts they are passing, turn their smartphones into spyglasses through added compass and GPS information, or decipher constellations using digital star charts of the night sky layered over images taken by digital cameras. Other apps create digital graffiti by "painting" a building in ways that people with similar apps can see. Digital "painting" can also help find cars in parking lots by tagging their location and leading drivers back to them by putting arrows over video images of the parking lot.

As mobile devices become smaller and more aware of where they are located, and as objects become more intelligent and communicative, people could have greater dialogues with the world around them that are integrated with additional material drawn from the internet. Indeed, this is what the Brazilian security officials want to do at the 2014 World Cup. They hope to equip police officers with "Robocop" eyeglasses that will scan 400 facial images per second, acquire 46,000 biometric data points for each face, and send them to a central computer database storing up to thirteen million faces. Police officers expect to receive information back rapidly about the person at whom they are looking. If there is a match to a suspect, an alerting signal will appear on a screen in their eyeglasses, informing the police officers of the situation.[20] In less fraught circumstances, people might expect to receive visual briefings in their eyeglasses of the Wikipedia entry of nearby objects (and celebrities) they are viewing or Facebook profiles of the acquaintances they encounter. Perhaps, they would have their text messages streamed to a section of their eyeglasses. Utilizing such technology, different people may experience the same physical locations in quite different ways.[21]

Augmented reality offers a world in which many objects could have digital material appended to them, accessible to anyone by observing them and following the links they provide. The physical and the digital would become more mutually informed. Empty parking spots could show you where they are; restaurants could show you their menu offerings' ingredients—helpful for food allergies—and previous patrons could recommend their favorite dishes. Perhaps all patrons would be encouraged to rate and comment on the restaurant—much like Yelp does today—with ratings summaries interactively readable on the restaurant's outdoor sign and website. Objects that could change state (be turned on or off, change appearance, etc.) could be interactively controlled wirelessly. For instance, when entering a theater or meeting, mobile phones would identify it as a "silent zone," automatically switch to silent mode, and then go back to regular mode when leaving.

Mirror worlds are another form of physical and digital convergence to augmented depictions of environments. Mirror worlds augment representations of physical spaces, such as maps. They might involve geographic information systems (GIS) for mapping, modeling, annotating, sensor-based inputs, and location-aware technology. Google Earth is the prototypical mirror world as it is a web-based digital map that adds cartographic and informational overlays, providing contours and alerts about important buildings. Later, Google Street View developed to show specific buildings on streets: a boon to people finding their way at the cost of the privacy of people near those buildings. While Street View requires elaborate corporate arrangements, ordinary people can use GPS to add descriptions of places, paths, and tagged photos.[22] In the future, people could use geo-aware tools to scan places to see if friends, friends of friends, or followers are in the vicinity and interested in a meal, a drink, a stroll in the park, or a shopping binge. This is a particular kind of "smart mobbing"—to use Howard Rheingold's phrase[23]—that will be enabled more readily in an environment that is aware of who and what is where. Thus, mirror worlds can make networked individuals more aware of each other's physical environments and aid in-person meetings to find a suitable location and travel to it.

Automated technology could convert digital pictures, videos, and sensor inputs into jointly manipulable 3D models: a tabletop Grand Canyon or a model of a friend's home.[24] Going further, data-rich geographic simulations could provide insight into complex global systems such as climate change, migration patterns, and the expansions of human settlement.[25]

Virtual worlds differ from augmented reality and mirror worlds because they are online "places" where all content is graphically simulated, rather than projecting information onto the real world. Virtual worlds are quite vivid and have been featured in movies such as *Avatar* and *Tron*. Millions of people enter virtual gaming environments to join in massively multiplayer online role-playing games (MMORPGs), such as World of Warcraft, where they can virtually lose themselves.[26] Besides the games themselves—which involve bonding and movement among social networks as well as competition—people can try on different identities and roles. As science fiction author Neal Stephenson puts it in *Snow Crash:* "Your avatar can look any way you want it to. . . . If you're ugly, you can make your avatar beautiful. If you've just gotten out of bed, your avatar can still be wearing beautiful clothes and professionally applied makeup."[27]

Second Life is the most active virtual world that is not a game. People virtually dress in costumes, interact, and sometimes buy land and products.[28] Although many user activities involve sexual fantasies, Second Life can have serious uses. For example, coauthor Wellman participated in a grant review session in Second Life that was organized by the U.S. National Science Foundation. (He dressed as a mashup of Agent Smith from *The Matrix*, with Robocop wrap-around sunglasses and TV warrior princess Xena's lace-up high boots.) By working together on Second Life for two days, the review panelists saved themselves a three-day stay in Washington, D.C. Yet, interpersonal dynamics were muted, and panelists were free to do other things concurrently without being observed. Moreover, Wellman—and other participants new to Second Life—had difficulty using the avatar and navigating the NSF's walled (password-protected) island. Perhaps these difficulties—along with Second Life's lack of compellingness as compared to games—helps to explain why most of its twenty-three million registered users are not active. As registration is free, many may have just been sightseeing or sampling. Only about fifty thousand have recently been participating actively.[29]

While virtual games and Second Life allow people to get away from it all in the real world, we suspect that it is more likely that virtual reality—rather than fully immersive worlds—will become incorporated into everyday life. Although we doubt that societies will ever become like the one in Asimov's science fiction thriller *The Naked Sun*, where no one ever meets in person,[30] we expect that virtual drop-ins will become as routinely convenient for person-to-person contact as video phone calls have become today—another enhancement of "the next best thing to being there," as the old AT&T long-distance phone commercial used to say.

Convergence of Social, Internet, and Mobile Networks

This book has shown the rapid convergence of mobile and desktop internet networks. The following technological innovations can potentially further enhance these tendencies.

Convergence of information and communication: As social networking applications gain users and allow them to connect with and embed information-rich media, they become portals: points of access to information all over the web. Increasingly, social media sites are used for communicating information. For example, when Facebook users click on the "like" button of other websites, links to these websites are posted on their various news feeds. This simultaneously informs others in their Facebook network and informs the sites' operators about their users' interests.[31] There is already a mad rush in Silicon Valley to create products to embed this kind of social interplay in most kinds of information and media encounters, and it will likely accelerate in the future.

Networked selves become osmotic. Our notion of a networked self in this book has been amoeboid: going beyond the conceptions of a holistic self or a multiple self to seeing a core "me" that creates "pseudopods" as it responds to different persons, networks, and situations. But as information and communication become more available, we suggest a conception of an osmotic self that is able to absorb new information about people and situations even before encountering them directly. If in the past we actually needed to interact with almost all people to learn personal details about them, the online aggregation and synthesis of information in the future will provide a head start before interaction.

"Lifelogging" represents a convergence of past experiences, current life, and future possibilities. It is a combination of hardware and software that allows experiences to be captured, stored, integrated, remembered, and disseminated to others. It does this by automatically recording and sharing

people's daily lives, including the places they visit, interactions they have, the media they use, and the objects they interact with—be they magazines, meals, or automobiles. Lifelogging technologies offer two primary functions: First, they serve as a kind of video recorder for a person's life, recording the sights and sounds an individual encounters throughout the day, and making them more retrievable at later dates. Second, they enable collaborative sharing and aggregation of life experiences.[32] Lifelogging thus presents possibilities for the convergence of the past and present, and for personal memories and interactions with family, friends, and coworkers.

To give some examples of lifelogging: a small number of people, such as computer engineer Steve Mann and social scientist Theresa Senft, have recorded and streamed their lives on "vlogs" via headcams, wall-mounted cameras, or sensors.[33] Senft investigated camgirls: "women who broadcast themselves over the web for the general public while trying to cultivate a measure of celebrity in the process."[34] She showed how these networked creators are sharing and distributing their personal information, sometimes promoting themselves as micro-celebrities via the internet.[35]

There are more limited lifelogging situations. Nike and Apple have a partnership that uploads jogging statistics from Nike runners' shoes to Apple iPhones that can be shared with friends around the world. Devices such as Fitbit track people's exercise and sleep rhythms and upload them to their online communities who can provide monitoring and support.[36] Likewise, cameras in police cars can transmit a continuous video stream. This can both protect the officers from false charges and document criminal behavior. While lifeloggers are currently narrowcasting pictures, videos, opinions, and status updates to friends, in the future more people may use social media to share more information more widely.

In many ways, Facebook, Google, and mobile phone service providers are lifelogging as they record who people are connected to, what their interests are, what they talk about, and where they go. "We know your normal behavior and who is likely to be your friend," asserted a Facebook engineer at a Georgia Tech security seminar on April 22, 2011.[37] Google matches its ads to Gmail content and uses earlier messages to guess who people want to chat with even if the message is in Tagalog.[38] Apple's iPhone captures and stores all the places it (and its user) have been for at least one year—suitable for retrieval for lawsuits or by law-enforcement agencies.[39] Facebook's Timeline makes visible to members of their personal networks the historical record of people's posts, milestones, pictures, and websites they have liked.

More elaborate lifelogging may come to be embraced. A prototype life-logging exercise has been going on since 1998 in computer scientist Gordon Bell's "MyLifeBits" effort at Microsoft. Bell has built on Vannevar Bush's 1945 idea of creating a memex. But where Bush focused on external documents (see chapter 9, "Networked Information"), Bell focuses on himself and his interactions. He tries to capture and archive every bit of his life, wearing a camera capturing everything he sees and keeping records of:

- all the videos from his camera
- emails he has exchanged
- phone calls he has exchanged
- documents he has read
- DVD films he has viewed
- MP3 music files he has played
- the places he has been as captured by his mobile phone's GPS system
- web pages he has scanned
- pictures he has taken and observed
- regular recordings of his vital signs captured by sensors he wears

As Vannevar Bush suggested about the memex, this is "an enlarged intimate supplement"[40] to a person's memory. Now, Gordon Bell can search much of his own memory. He claims that his lifelog will provide the memory and data for better personal health, productivity, and interpersonal relationships. Bell argues (as does his research partner Jim Gemmell) that lifelogging is going to transform how people think about the meaning of their lives because the self-knowledge that comes from this would be revelatory. A deeply searchable past, accessible through a never-blinking memory changes people's notions of their identity and their role with others. Bell postulates that people's devices will automatically communicate with each other regarding privacy issues, recording permissions and so forth through artificial intelligence programs:

Total Recall will come about within the context of networks within networks, interconnecting everything from in-body networks to home networks to global networks and finally to networks that include satellites and space vehicles. Dust-sized sensors will automatically form wireless networks and connect to everything that can be sensed. In-body implants will communicate with each other to form a "body-area" network. The body network will connect to the car network when you're driving. The garden network will connect to the home network. The car and home network will connect to the worldwide internet. This vast network of networks will host huge farms of servers with millions of processers and petabytes of storage space. . . . From the microscopic to the heavens, all will be sensed, networked and stored. This is not a forty-year-out wild guess. This is a decade-out sure bet. And I don't lose many bets.[41]

One future step would be to develop techniques to combine Bell's lifelog with augmented reality so that people might see, for example, a visual clip of their last interaction with the people they are now meeting: a boon to salespeople, alumni get-togethers, and those with faltering memories.[42]

Despite the practical and sentimental benefits of having a cumulative, multimedia, and searchable lifelog, there are the inevitable costs of surveillance. Any such log could be obtained legitimately by the authorities and possibly by malefactors. Few lives can exist unblemished with that kind of in-depth scrutiny, and it would not be easy to erase embarrassing moments. Moreover, identities might be forged.

Privacy: As the lifelogging discussion suggests, the loss of privacy is an important trade-off for the benefits of the Internet and Mobile Revolutions. Networked individuals have to balance the recording and sharing of personal information with the potential risks that this information may be used against them. This is amplified in a "total recall" environment where digital footprints that can be traced back and reveal detailed patterns about people's tastes, preferences, and intimate thoughts.[43] Are we reverting to the total visibility of the preindustrial villages discussed in chapter 5?

How do internet users minimize the risk of leaving these digital footprints? It is likely that a new equilibrium will be established in the legal, commercial, and personal spheres. At the legal level, there will be pressure on legislators to pass new laws and courts to apply new standards about how personal information can and cannot be used by organizations and government agencies. At the commercial level, it is also likely that firms will see market opportunities in helping people find more personal ways to decide what aspects of their digital lives should be hidden from others. And at the personal level it is likely that people will develop new norms around how to treat the information they can discover about others—and how much importance to assign to facts about people's pasts or tastes.[44] Ultimately, the use of ICTs is inherently a trade-off between convenience and privacy. Going forward, both government agencies and service providers will play a role in developing new privacy standards and people may have some discretion about their threshold for potential risks.[45]

The assertion of national sovereignty online: One critical uncertainty about the extent of convergence between data worlds and real worlds will be the role that countries play. In a countertrend to convergence, they can put barriers on the free communication and information sharing of the global village. In principle, like the telephone network the internet connects everywhere. Indeed, governments and organizations have to go to some trouble to build private networks to keep their communications off the internet. Only a few countries—such as North Korea and Myanmar —block or severely restrict the internet, although others channel most communication through government internet service providers with surveillance capabilities.[46]

Yet, decades after the manifestation of the internet as a global network, some governments are asserting their online sovereignty. If "information wants to be free," as the saying goes,[47] not every government agrees to let it be free. The very success of the internet has given rise to the trend toward barriers. The internet is too important for governments to ignore. When the Egyptian authorities tried to shut it down within that country in January 2011 to quell a revolt, the chaos and loss of commercial functionality stopped this attempted disconnection in a few days.[48]

If governments do not want to stop the march of internet and mobile networks, they may surveill and regulate them. Hence some countries have blocked their citizens from accessing particular websites and some online services. More than a dozen countries block internet content for political, social, and security reasons.[49] China is one example where Facebook and Twitter are among the sites blocked under its Golden Shield policy and even tightened restrictions since the Arab spring.[50] And the U.S. government has the capability of looking at the huge amount of traffic through its main internet routing points.[51] Surveillance, even without blocking, can have a chilling effect on networked individuals' information access and communication. As the internet becomes more central in everyday life, there are likely to be further attempts by governments to establish their online sovereignty.[52]

The New Logic of Personal Transparency and Connectedness

We discussed in chapter 10 that a moral imperative of successful networking is to follow the Golden Rule. Similarly, the acquisition of social capital depends on the expenditure of social capital. There are other social realities that networked individuals face in the emerging technological environment. One potent imperative of social networking is for actively sharing information and creations. People cannot build networks without describing who they are, what talents or skills they possess, what they know, and what their needs are. There are also some pressures toward deliberate, considered disclosure in social media when people cannot fall back on close, long-term friends who perpetually stand ready to help them.

Besides the imperative to share, there is a push to be connected. People cannot easily ask for help from their networks without using digital tools and they cannot be available to help others if they are off the grid. The social requirement of the age of networked individuals is to be connected and findable. It is a precondition to successful networking and network building. It is also a reality that is anathema to privacy and solitude.

Already, online social networking allows people to activate their networks quickly and to generate immediate data inputs and responses from the crowd that can be as authoritative and useful for information as deep encyclopedic searches. In the Metaverse, interactions with people will not just be conscious. Machines and objects will play a role in social networking activities. They will amass vast amounts of information about people's behavior, transactions, and interactions and then try to anticipate their desires and shape their activities. Once a person's preferences are embedded in hardware and software, there will be digitally enhanced, automatic social networking. Networked individuals will be actively connecting with other people and things without much thought or exertion after they have checked off their preferences with the technology. The richness of their social connections should grow. The depth of their understanding of the people they know and the environment around them should increase dramatically. Their ability to tap into smart crowds and vast databases of information will expand—and, with that, their capacity to tap into greater levels of knowledge.

Alas, the cost of more transparency and connectedness will be additional transparency and connectedness, whether people like it or not. There will be ways in the emerging environment that people will find harder to hide or disconnect. When objects become networked, then people will be transmitting information as their appliances, their clothes, their goods, and their gadgets give digital readings that are assigned meaning by other people and even machines. Much of this disclosure will occur without nearly as much awareness as people now give to personal disclosures. As people get evermore defined by "learning" machines as they make choices in the media they consume, the friends they have, the things they purchase, the places they go, and the things they disclose, it becomes a societal danger that the machines will start surrounding them with an echo chamber of similar people and similar ideas. This "filter bubble," as Eli Pariser calls it, might make it harder for people to encounter other people and information that does not fit their "profiles" and that will balkanize the culture, diminish serendipitous encounters with fresh and diverse kinds of people and media.[53]

With all these trends rolling along into the future, there is still reason to be uncertain about how the environment of networked individuals will evolve. We offer two different scenarios that seem credible.

Scenario 1: Even More Networked Individuals

Waking up in a networked future, his digital agent's soft voice slowly grows into Harry Sanchez's hearing range. It's been monitoring his sleep rhythms and cross-referencing them with data from his ongoing brain scans to see when it's most appropriate to wake him. After stretching and rubbing the sleep from his eyes, Harry suddenly and happily recalls yesterday's purchase. He found a collaborative coupon on the web the other day for a deal on a new pair of augmented reality (AR) contact lenses and the haptic feedback implant everyone's been raving about.

The implantation was a simple and quick outpatient procedure that reminded him more of getting his ears pierced than of surgery. It was not the real Doctor McCoy in person, but a projection whose robot mimicked his every move as the doctor mimed the procedure. It was not as though Harry could really tell, however, since his AR glasses had "skinned" (covered) the robot with the doctor's virtual image. In this way, the doctor efficiently treats dozens of patients a day projecting in from his home.

Now that he is awake, Harry eagerly slips on his new AR contact lenses for the first time. Up until yesterday he had to use an old-fashioned pair of eyeglasses with a projected display on them. The contacts instantly network with his microcomputer, smartphone, and the internet. His personalized augmented overlay appears in his field of vision: the time and date, the weather and air quality, a few applications he left open from the previous night minimized into his peripheral vision, a faintly blinking icon notifying him of some messages he missed overnight, an icon notifying him of information updates on news stories aggregated for him by his agent, and an InterFace lifelog update showing what his friends did last night, cross-referenced with the media they consumed and the tagged conversations they had. He sees a call for participating in a political smart mob in the virtual world, but he tells his agent to disregard it.

His agent also warns him about his health. He hasn't been sleeping well, as his late-night virtual meetings with people in China have taken a toll on his system. Yet, he's happy to not have to fly there ever since they've been able to collaborate long-distance by using the Cavecat productivity system with active walls and tables holding spreadsheets, texts, drawings, and videos. Harry and his Chinese team members edited together, with each person using different colors for legibility. It was great to see how the webcams showed everybody's faces while emphasizing who was speaking.

As Harry settles in at the kitchen table, the surface notices he's put down his morning cup of coffee, but his agent tells the table to wait five minutes before it displays the morning news on the active wall facing him. When Harry is online, it always feels like high noon and happy hour combined, and he's not quite awake enough for that yet. It's another time zone effect as the majority of people from any one part of the world are always most actively online during midday work hours and for their evening entertainment. The news comes as manipulable augmented reality overlays of Harry's social network with pictures of each network member blinking when she or he posts updates such as messages, videos, or lifelog entries. The new haptic implant gives him a sensory understanding of the news: He can feel the continuing battle in Kabul, experiencing its sounds and vibrations as if he were at the scene. And it now feels as if the computer icons of his various applications have weight and texture.

Having not found any urgent messages, Harry's agent organizes his correspondence by topic and relevance. Noticing a conversation he had that he does not want many network members see, Harry has his agent make the information private across his entire InterFace network. His agent makes suggestions based on Harry's past experiences regarding who should see what, but Harry doublechecks these with his graphical network overlay to confirm who gets what. His agent also sends out a quick update to his entire network, letting them know his plans for the day. Unless his contacts are on special alert, Harry's calendar update doesn't interrupt them because their agents automatically direct his updates into their own daily calendars.

Harry is distracted by a knocking sound. His agent informs him that his best friend Neal is projecting in for their regular weekend virtual breakfast. Though they only live fifty kilometers apart, this is a nice way for them to check in on one another and spend some time together. Harry hasn't shaved, and so he puts on his shiny-face skin before he opens the virtual door. He uses his new haptic chip to get the sensation of shaking his friend's hand. It's a little strange at first since there's nothing actually present to shake, but his nervous system responds as though he had reached out and touched someone.

Harry and Neal chat about how everyone who was at the pub's avatar party last night has shared recordings of the evening with friends. Their agents have already automatically tagged these recordings with relevant information about the people and location. Avatar parties have become popular these days. Everyone dresses like their favorite game character; some even come looking like one another. It can be a lot of fun role playing like this, and the collected and tagged videos are highly amusing as people's voices, looks, and even smells can be altered in the virtual world.

After visualizing and flipping through these tags for mentions of his name, Harry updates the conversation file with some witty things he thought of after the fact, and his agent forwards the updates to the relevant people. He also tells his agent to delete information about last night's embarrassing ice-cube escapade at the avatar party, and to ask his friends to delete their versions. His agent also notifies him of music they listened to, especially the New Jersey Symphony's classic *O was für eine Nacht.* His agent tells his personal server to download this concerto from the universal library, put it on his Memories playlist, and play it. Perceiving the change in sound, his friend Neal turns his attention back to Harry, phasing out his other communications by uploading a temporary away message he recorded in the past.

Harry's agent softly chimes in just as he's saying goodbye to Neal, reminding him that he has to meet his sister Merril today. She has changed the schedule on him since she has to first drop her kids off at daycare. Noticing the change in times, his agent uses his preset preferences to work out the details by corresponding with her agent. Harry had originally wanted to meet Merril at their regular place downtown, but noticing a scheduling conflict due to travel time, their agents present both of them with other, more local options. They work with their agents to narrow the agents' search parameters and settle on a place down the road. It's local and the tables there get automatically reserved.

Never having been there, Harry wants to make sure the restaurant is right for them. He had fallen into this trap once before when his new agent hadn't learned his preferences yet and ended up sending him to a local theme restaurant where the servers had skinned themselves with former celebrities: It was uncomfortable to have Arnold Schwarzenegger (skinned as *Demolition Man)* serve him food. Harry projects himself into the restaurant's virtual space and finds that it keeps a good online presence, with a nice menu, list of ingredients, health report, and real-time webcam view. Wanting to talk privately with Merril, Harry checks the integrated RFID/ GPS personal locator software to make sure no friends are nearby. He's comfortable with their privacy, for most people leave their RFID trackers on as they've been doing since before kindergarten.

As Harry gets ready for the day, his agent presents him with a few clothing options. He decides to wear the new trousers suggested by his girlfriend, but calls up another app to make sure his sister would also approve. Harry's girlfriend had tagged the info to the trousers while doing some virtual window shopping and had a pair in his size set aside after asking his belt how big his waist was.

Not wanting to be late, Harry has his agent arrange a car for him through a collaborative consumption app that recognizes his high trust score. He rarely uses a car, as his fridge automatically schedules grocery deliveries. Slipping his microcomputer into his pocket, Harry goes to the car, has his agent set the restaurant's coordinates, and leans back to check his messages as the car pulls out.

Scenario 2: A Walled and Surveilled World

As Will Li rouses himself from sleep, he walks over to "his" computer to see what he's missed overnight. The computer isn't really his. He owns rights to its usage but isn't allowed to change its hardware or software or else he'd void his warranty or break the law. His computer is really only an access point as all his data are in the cloud, yet another thing that's owned—with all the data in it—by a big corporation. Before Will can reach for the cloud, the system completes its mandatory scan of his computer for viruses and copyright infringement.

The price of media access has spawned its own subculture of media pirates. They usually meet in person, sharing miniature portable terabyte flash drives back and forth packed with music, TV shows, movies, ebooks, and more. The pirates often get their "warez" from people who collected old computers from trash heaps, recycling centers, and garage sales. They've even developed a code language to arrange meet-ups, but Will hardly keeps up with the ever-evolving lingo: He would much rather rely on his best friend Spider to help him when they get together in person at Callahan's wine bar.

Leaning over his morning coffee, Will dreams of how nice it would be to have a personal agent, but he's heard most are double agents that also report back to the authorities and sell information to corporations. And he doesn't like the way FaceWall is collecting all the information on him whenever he uses it. He also can't afford to hire the technician it would require to help him set up the devices and access all the fragmented networks of media sites, search engines, and social applications online. Each has a tricky "right to information" form to sign. So he's reduced his online presence to a minimum, trying to limit himself to good old-fashioned emails and avoid social media.

However, Will needs to use FaceWall today to find something. He's forced to wait thirty seconds to let the mandatory ad play. It has his picture in it. CoffeeCo must have bought a recent photo that tagged him on a friend's wall. Will notices that his system slows down as the massive data file from the advertisement clogs up his bandwidth, but since the corporations pay more to guarantee themselves fast access, he endures the wait. It's almost ironic to see a return to the days of loading screens since the amount of available bandwidth has only increased, but all that bandwidth is either auctioned at sky-high prices or owned by a few companies.

Finally finding the photo, Will learns he cannot delete it because CoffeeCo now owns it. Perhaps he should make sure no one ever uploads anything about him again, although that would be difficult. Most people seem to put up with these situations because they want to keep going online. Will assumes that from now on he'll get peppered with ads geared to the tastes that FaceWall has observed online—both for him and for all those other forty-year-olds who became unemployed when countries set up their own walled-off internets, claiming morality and national security demanded it.

Giving the situation further thought, Will starts to browse his friends' profiles, and finds that his sister is earning extra money by selling her personal information to FaceWall, including links to his profile. Maybe that's how CoffeeCo found his photo. He'll ask her to never do it again. You can never be quite sure of who's informing on you, only in this case it's not only the state but data-aggregating organizations. Will remembers from history class how in the 1960s, FBI Director J. Edgar Hoover had used his dossiers on the Kennedys to keep power. Now, FaceWall has even more comprehensive dossiers on everyone.

Doing what he knows he shouldn't, Will reaches for a doughnut. Maybe he can sneak one without his insurance company's sensors registering it. At least Will made the right decision by paying extra for their privacy clause. Otherwise, his health data might have just been sold off to the highest bidder at an info auction. But, since he's not able to see the information himself, he can't be sure.

Will and Spider prefer to meet in person. There is less chance for any number of things happening. They remember how someone pretending to be an insurance representative once duped Spider online to steal private information. The latest scam is reverse-identity theft. The thieves pose as old friends, using detailed avatars whose digital image and voice have been reconstructed from public profiles. Too bad the government killed the trusted identities program.[54]

Will calls his sister Lorelei to meet for coffee. They avoid the local CofeeCo franchise because its low-cost internet access makes it always busy even though people must sign away their privacy rights to use it. Will and Lorelei decide instead to go to a local restaurant. Lorelei hopes the guards will let her through the city gate this time: She's brought her travel pass up to date.

Abruptly, Will's computer comes to a stop as the local net's virus filter kicks in. Someone must have accidently shared a virus with him, and the system has gone haywire. His gated community monitors his network continuously and has a strict quarantine procedure that cuts infected computers off from the net until their "owners" can have the manufacturers clean them. He shuts off his computer, grabs his phone and travel pass, and goes out past the security scanner.

After a wait, Lorelei pulls up giggling about the whole-body security scan at the gate. "Hope they got a better picture this time." She's also worried that the guards may have found the incriminating photo of her online. She's already lost one job because of it, even though it was taken without her permission and out of context. They head off for their meal, but arrive just in time to see the last open table become occupied.

The Possible Futures of Networked Individuals

Although present technologies are still far from being capable of implementing either scenario in its entirety, each represents a potential evolution from current trajectories. The first scenario assumes a move toward more networked individualism based on continued technological progress and trust in computer and human networks—including the withering of boundaries. The second scenario assumes more boundaries, more costs, more corporate concentration, and more surveillance. At present, the Western world is trending in the direction of the first scenario, but we would be naïve to think that the second scenario could not happen here.

We close the book by highlighting a selected set of premises that we have considered while envisioning these scenarios. At the same time, we take into account how the technological enablers discussed earlier can contribute to their realization.

Networked Individuals ++

This moment in ICT history marks the stage at which a fuller impact of digital technology will become evident.[55] The impact of technology unfolds in three stages. The first stage is *substitution* as new technology performs older technology's tasks more efficiently. The second stage is *enlargement* as new technology is used to increase the volume and complexity of tasks that old technology used to perform. The final state is *reconfiguration* as new technology fundamentally changes the nature of the things it was created to address.[56]

In not too many years, technology will have afforded changes in people's relationships with others and reconfigured information. For networked individuals, that will mean a number of things. The transformations wrought by the Triple Revolution that we identified at the beginning of this book will intensify and become more widespread. People will be more connected as individuals and less embedded in groups. They will get their needs met more often in spatially dispersed and loosely knit networks, rather than relying on a small number of tightly knit connections. They are likely to test partial membership in more milieus and rely less on permanent membership in settled groups. The nature of their networks will change as they find more ways to connect with weak ties. That, in turn, will lead to more diversified social networks.

We can see how Harry in Scenario 1 relies on collaborative consumption when searching for new products, by using both a collaborative coupon and by sharing resources such as a car with his peers. The notion of collaborative consumption actually integrates Harry with consequential strangers who share his consumption needs. This process is enabled by the convergence of communication and information, which allows Harry to access valuable resources while effectively managing his membership in different groups.

We can further see in both scenarios how people's contributions to the online world will multiply and perhaps change character as they become more aware of how their postings are used, aggregated, and surveilled and of how tweets, texts, blogs, searches, and social media posts are captured, stored, analyzed, and reassembled into profiles that affect individuals' reputations. Their relationships will be built more around networks. In line with these notions, we can see in Scenario 1 how Harry aggregates social media as well as updating and sharing lifelogs. To learn what his friends did last night, he links the updates of their lifelogs to their social network information. Harry also uses an augmented pictograph to help him quickly visualize his network to assess who he does—and does not—want to know about his recent activities. By doing so, he can better guard his privacy and shape his online representation. By contrast, Will's sister Lorelei in Scenario 2 has had her reputation damaged by the circulation of an incriminating image.

Scenario 1 also shows how people's social networks will have a more vivid presence in their lives as they carry their networks with them by using mobile connections, content creation tools, and crowdsourced methods to help them draw upon helpers to solve problems, make decisions, and provide social support. In Scenario 1, Harry owns a computer, but he also relies on ubiquitous computing and objects. Harry especially relies on his agent to access content gathered by his network: Netsourcing from trusted others has replaced indiscriminate crowdsourcing. He consults these sources while making decisions about important matters, such as where to travel this summer. By sensing his interest, his agent can notify him about relevant information: comparing his interest with the available content generated through netsourcing. One example for this can be seen in Harry's exposure to new music that came from his friends' music preferences. By contrast, in Scenario 2, Will uses limited online sources: He owns neither hardware, nor software, nor agent, nor his own files. What he does use is utilized by others, and his behavior—like his sister's—is subject to government and organizational scrutiny. Their networks bind them, rather than liberate them.

Centrifugal and Centripetal Forces

Scenario 1 shows the Triple Revolution is a complex interplay of centrifugal and centripetal forces. Harry is working across an ocean with Chinese colleagues. His interactions with friends and his sister are dispersed. His life merges the corporeal and the digital. His agent-based connections with the internet give him comprehensive knowledge about his friends' whereabouts and plans for the day. This makes easier both the initiation and coordination of potential meetings.

Yet, the internet draws his network together centripetally as both the avatar party and Harry's virtual meeting with his physically distant best friend show. This centripetal process is enabled by the information generated and shared via the internet of things, linking digital calendars, RFID/GPS tags, and people. Technologies such as lifelogging, social aggregation tools, and smart tagging keep track of friends' past and present activities, as well as facilitating further interactions with them. Harry's inquiries about his virtual pub escapade can be seen as an example of this, where Harry was able to retrospectively shape his input through comments and delete the notorious ice-cube incident.

This centripetal social process also has the potential to enhance in-person interactions. After all, there will always be people who prefer to meet in person: Even Harry feels that it is a little strange to shake his friend's hand in an AR encounter. In Scenario 1, we can see how the network of things makes it easier for Harry to coordinate an in-person meeting with his sister, and how mirror and virtual worlds have been used for choosing a suitable location.

By contrast, Scenario 2 shows the future going back to the past: to a more thoroughgoing centripetal society that hearkens back to the prein-dustrial door-to-door towns described in chapter 5. The city is gated and national internets have become walled off because of cost and security fears. Social media also form gated virtual communities with quarantine policies that instantly wall-off infected computers from the net.

How could this second scenario happen? Science fiction has repeatedly shown a post-apocalyptic reversion to small, local, walled-off groups who fear outsiders and trust only their own tribe members whom they see every day. It is a continuing theme in movies since at least the 1930s, such as *Things to Come* (starring Ralph Richardson), *Mad Max 2: The Road Warrior* (Mel Gibson), the *Terminator* (Arnold Schwarzenegger), *George Romero's Land of the Dead* (Dennis Hopper), and *The Book of Eli* (Denzel Washington).[57] This reversion also appears in novels such as Walter Miller's *A Canticle for Leibowitz* (1960) through Doris Lessing's *Memoirs of a Survivor* (1974) to Margaret Atwood's *The Year of the Flood* (2009).[58]

Nor is the possibility only fictional. It has happened before. The open Greek cities gave way to walled Roman and medieval towns designed to protect against external attack and to control the residents.[59] To some extent, it is happening now, as street gangs work hard to control local territories and American suburbs erect gates to keep their terrors out. Thus it is possible, but hopefully not likely, that a bounded world of small, insular groups might reemerge.

Augmented Reality Augments, and Does Not Replace

As Scenario 1 shows, ubiquitous computing, augmented reality, mirror worlds, and virtual worlds have the potential to enhance the external physical world for networked individuals. The interpenetration of ICTs and people's activities will grow as the boundary between the digital and the corporeal becomes as transparent as Harry's AR lenses.

Scenario 1 shows how augmented reality and virtual worlds enable Harry to receive medical treatment by the remotely located doctor, a treatment that is actually performed by a robot that mimics every move made by the human doctor. The same scenario also illustrates location-aware systems and interfaces that process and layer networked information on top of people's everyday perceptions of the world. An example of this development is Harry's exploration of the restaurant by means of a mirror world, while at the same time he checks on his friends' physical proximity to the location by analyzing their RFID/GPS signals. Another important component in this scenario is that people may contribute material through tagging, blogging, commenting on, and mashing up virtual spaces and avatars. By contrast, in the second scenario, the surveillance risks for Will are too high for him to benefit from an agent.

Digital Divides

The net surfers in Scenario 1 appear to have transcended the digital divide, although they surely vary in how they use their tools and handle their agents. Scenario 2 shows an economic divide where costs have cut off many from much control over their digital lives. People curtail their use of the internet because of cost and fear; many become jaded and used to having their information marketed and surveilled.

Another digital divide debate appears in Scenario 2, with regard to the speed of connections and the cost of agents. Will has trouble paying for sufficient bandwidth to access media-rich websites, and he cannot take advantage of the capabilities that augmented reality, virtual worlds, and mirror worlds offer. He does not have special eyeglasses to see skins or manifestations of ubiquitous computing.

Privacy and Digital Shadows

Harry blithely erases the ice-cube incident in Scenario 1. In Scenario 2, governments and organizations constantly scrutinize online activities. As a result, Will has reduced his online presence to a minimum. He has also tried to avoid having anyone upload anything about him again—but that's impossible in a world of automatic photo tagging. Will also realizes that it is so easy to falsely burnish one's status—it's called "gilding the FaceWall"—that he has become skeptical of the validity of online content. He prefers meeting people in person.

Online information also plays a major role in determining the reputation of his sister Lorelei. She suffers from being on the wrong side of the law—or at least her internet-circulated picture is—a situation of continuing stress as she cannot erase her past. As it becomes possible for people to access stored references about someone or something, it is easy to conceive of circumstances where this will help people burnish their status and build social capital. It is also easy to conceive, just as in Lorelei's case, how this could become a world where, in the words of blogger and CUNY professor Jeff Jarvis, people live in a state of "mutually assured humiliation."[60]

The management of online privacy plays a role in both scenarios. The key questions concern control: Who controls the data, the software, and the hardware? This is more than a national matter, for as Scenario 1 shows, many connections are international. Ultimately the use of new technologies will remain a trade-off between convenience and privacy. Going forward, both government agencies and service providers will have to develop new, secure privacy standards and networked individuals will have to receive usable information to evaluate their own potential risks.

The Future Will Be What It Will Be

The future reality presented in our two scenarios is probably too tame and homogeneous for what the likely reality of networked individuals will be in the coming years. Even if one scenario largely comes true, the digital and social events will not appear coherently. There are visible elements of both of these scenarios in today's more advanced networkers—people such as the eager digital natives who networked with us to write this book—and those among their elders who have similar tastes. That would include people like Peter and Trudy Johnson-Lenz, the couple whose story about her fall and their networked aftermath began our story. But much is to come—and not at one time—and some potential developments will never be.

We envision at least four arenas for future issues:

• The internet itself—what it is and how its architecture functions: literally the plumbing and the code;
• The legislative and legal realms—where big policy battles about privacy and intellectual property will shape the future of who owns what; who shares what; who pays for what;
• The world of social norms and etiquette—where we figure out the social rules of engagement; and
• The commercial technology (apps and devices) realm—where people create the new tools that will help shape what is possible for networked individuals.

The Triple Revolution—in social networks, the internet, and mobile connectedness—will change but never end in the ongoing turn to a networked operating system. That means that the foreseeable future holds the prospect that individuals will be able to act more independently with greater power to shape their lives, if they choose to do so and if the circumstances will enable them to do so. Yet, the foreseeable future also contains the burden of knowing that people will have to work harder on their own to get their needs met. Tightly knit permanent groups will continue to be stable cores for some, and social networks will play greater roles in all human activities. The work of networked individuals is never quite done—and the satisfactions and uncertainties of netweaving are always available.

Notes

Chapter 1

1. Jeffrey Boase, Christine Ensslen, Tracy Kennedy, Stephen Perelgut, Beverly Wellman, and Xiaolin Zhuo gave good advice for this chapter.
2. See Kurtelling.com/forum.
3. Lisa Kimball's company is called Group Jazz and its website is located at http://www.groupjazz.com.
4. Jessica Lipnack heads NetAge, an organization focused on developing networked teams within organizations. See Jessica Lipnack, "NetAge Endless Knots," December 7, 2007, http://endlessknots.netage.com/endlessknots/2007/12/pt-paying-it-ba.html.
5. Lotsa Helping Hands says it is a free, private, web-based community for organizing friends, family, and colleagues—people's "circles of community"—during times of need. http://www.lotsahelpinghands.com.
6. INFJ (Introversion, iNtuition, Feeling, Judging) is one of sixteen personality types in the Myers-Briggs personality scale. See Isabelle Briggs Myers, *Gifts Differing: Understanding Personality Type*. [New York: Davies-Black, 1995].
7. At our request, Peter and Trudy Johnson-Lenz wrote us about their experiences on June 24, 2008. They added details in a series of interviews and email messages that took place between December 2008 and January 2011.
8. For example: Robert Putnam, *Bowling Alone* (New York: Simon & Schuster, 2000).
9. For example, Sherry Turkle, *Alone Together* (New York: Basic Books, 2011). See also the review in Jeffrey Boase and Barry Wellman, "Personal Relationships," in *Cambridge Handbook of Personal Relationships,* ed. Anita Vangelisti and Dan Perlman, 709–723 (Cambridge: Cambridge University Press, 2006).
10. Hua Wang and Barry Wellman, "Social Connectivity in America," *American Behavioral Scientist* 53, no. 8 (2010): 1148–1169.
11. Eric Schmidt is quoted by Derek Thompson, "Google's CEO: 'The Laws Are Written by Lobbyists,'" *The Atlantic,* October 1, 2010, http://www.theatlantic.com/technology/archive/2010/10/googles-ceo-the-laws-are-written-by-lobbyists/63908.

Chapter 2

1. Vincent Chua contributed to this chapter. Special thanks to Mohammad Haque, who did most of the data gathering and prepared the figures for the "Toward the Social Network Revolution" section of this chapter; Anatoliy Gruzd for constructing the Twitter network; and Valdis Krebs for constructing figure 2.16, which shows the links between book purchases before the 2008 U.S. presidential election. We appreciate the advice of Juan-Antonio Carrasco, Wenhong Chen, Rochelle Côté, David Gillen, Christine Ensslen, Mirna Ghazarian, Keith Hampton, Zack Hayat, Caroline Haythornthwaite, Bernie Hogan, Tracy Kennedy, Mohammad Keyhani, Alexandra Marin, William Michelson, Diana Mok, Adrienne Redd, Tom Smith, Mo Guang Ying, and Esther and Irving Zeitlin; as well as the assistance of Juan-Antonio Carrasco, Justin Couture, Zack Garcia, Anatoliy Gruzd, Gregory Jancelewicz, Shawna Korosi, Tom Snijders, Lilia Smale, David Toews, Mehdi Zabet, and Yu Janice Zhang.

2. Although the indicators we present in this section come from authoritative sources, scholars have debated their exact measurement. We avoid those debates as our goal here is to put the indicators together to suggest the turn toward a networked society. For a social history of America leading up to the trends we discuss here, see Claude Fischer, *Made in America* (Chicago: University of Chicago Press, 2010); for comparing U.S. values with those of other countries, see Wayne Baker, *America's Crisis of Values* (Princeton: Princeton University Press, 2005).

On affordances, see Donald Norman, "Affordance, Conventions, and Design," *Interactions* (May 1999): 38–44; Erin Bradner, "Understanding Groupware Adoption: The Social Affordances of Computer-Mediated Communication among Distributed Groups." Working Paper, Department of Information and Computer Science, University of California, Irvine, February 2000.

3. Figure 2.1: Automobile travel data from Patricia Hu and Timothy Reuscher, "Summary of Travel Trends: 2001 National Household Travel Survey," *USDOT Federal Highway Administration Report 2004*, http://nhts.ornl.gov/2001/pub/STT.pdf. See table 17: Availability of Household Vehicles. U.S. Department of Transportation Federal Highway Administration, Washington, DC. 1969 data do not include light trucks (pickups, vans). 1969, 1977, 1983, 1990, and 1995 surveys: the Nationwide Personal Transportation Survey. 2001 survey: the National Household Transportation Survey. Automobile reliability data from the J. D. Power Vehicle Dependability Study is only available since 2000 when there were an average of 4.45 problems for each three-year-old car. In 2010 the average was 1.55: a two-thirds (65 percent) decrease in problems per car from the start of the decade. 2010 data are from the National Highway Traffic Safety Administration, "Early Estimate of Motor Vehicle Traffic Fatalities in 2010," http://www-nrd.nhtsa.dot.gov/Pubs/811451.pdf.

4. Figure 2.2: Per-capita airline boardings data from Research and Innovative Technology and Administration Bureau of Transportation Statistics (RITA), Historical Air Traffic, Annual 1954–2007, http://www.bts.gov/programs/airline_information/

air_carrier_traffic_statistics/airtraffic/annual/1954_1980 and http://www.bts.gov/
programs/airline_information/air_carrier_traffic_statistics/airtraffic/annual/1981
_present.html. The traffic data presented are reported to the Department of Transpor-
tation on BTS Form 41 by Large Certificated Air Carriers and includes these carrier
groups: Majors, Nationals, Large Regionals, and Medium Regionals. Traffic statistics
for Small Certificated Air Carriers and Commuter Air Carriers are not included. En-
planements were then divided by the estimated U.S. population based on the U.S.
Census Bureau's population figures, available at http://www.census.gov/compendia/
statab/2010/tables/10s0002.xls. Note that the 1954 figure of 0.22 *airplane boarders*
per capita is misleading, as a few flew probably often and most never flew. This is a
problem of interpreting from statistics when only averages are available. Joyce Dargay
and Mark Hanly give similar decline data for the United Kingdom for both leisure
and business flights: "The Determinants of the Demand for International Air Travel
to and from the UK," Universities' Transport Studies Group Conference, Edinburgh,
November 2001. http://www2.cege.ucl.ac.uk/cts/tsu/papers/UTSGAIR2002.pdf.

5. Figures 2.3 and 2.4: Telephone data from the U.S. Census Bureau, "Table
1110: Telephone Systems Summary," *Statistical Abstract of the United States: 2010.*
http://www.census.gov/compendia/statab/2010. See also Claude. Fischer, *America
Calling* (Berkeley: University of California Press, 1992); U.S. Federal Communications Commission, 2008, Industry Analysis and Technology Division, Wireline
Competition Bureau, "Trends in Telephone Services," http://hraunfoss.fcc.gov/
edocs_public/attachmatch/DOC-284932A1.pdf.

6. Stephen J. Blumburg and Julian V. Luke, "Wireless Substitution: Early Release of
Estimates from the National Health Interview Survey, July-December 2010." Centers
for Disease Control and Prevention, June 2011, http://www.cdc.gov/nchs/data/nhis/
earlyrelease/wireless201106.pdf.

7. Figure 2.5: Personal computer data from International Telecommunication Union
(ITU) World telecommunication/ICT indicators database 2008 ed. Geneva. ITU. For
review of change in nature of cities, see Zachary Neal, "From Central Places to Network Bases," *City & Community* 10, no. 1 (2011): 49–75.

8. Figure 2.6: Data on international wars from "Number of State-Based Armed-
Conflicts by Type, 1946–2006," *Human Security Brief 2007* (Vancouver: Human
Security Report Project, 2007). Available as table 3.3 at http://www.hsrgroup.org/
human-security-reports/2007/graphs-and-tables.aspx.

9. Figure 2.7: Data on international trade obtained via Wolfram Alpha data archive
and analysis program (http://www.wolframalpha.com).

10. Figure 2.8: Photomontage of fruit signs by Gregory Jancelewicz and Barry Wellman, Fiesta Farms Supermarket, Toronto May 2010. For a broad account of globalization, see Nobel prizewinner Joseph Stiglitz's book *Globalization and Its Discontents*
(New York: Norton, 2002).

11. Statistics Canada time-use data for non-sleep activities at home for ages 18–65,
from the General Social Survey Cycle 7 (1992) and Cycle 19 (2005). Thanks to William Michelson for guiding this book's authors through this. The data on meals

eaten alone are from Martin Turcotte, "Time Spent with Family during a Typical Workday, 1986 to 2005," *Canadian Social Trends*, November 21, 2008, http://www .statcan.gc.ca/pub/11-008-x/11-008-x2006007-eng.htm.

12. Robert Putnam, *Bowling Alone* (New York: Simon & Schuster, 2000); Sidney Tarrow, *Power in Movement*, updated and revised ed. (Cambridge: Cambridge University Press, 2011); Beth Kanter and Allison Fine, *The Networked Nonprofit* (New York: Jossey-Bass, 2010); U.S. Bureau of Labor Statistics, updated and revised from "Families and Work in Transition in 12 Countries, 1980–2001," *Monthly Labor Review,* September 2003 plus unpublished data, http://www.census.gov/compendia/statab/ 2008/tables/08s1304.pdf.

13. Table 2.1: Data from the Pew Forum on Religion and Public Life/U.S. Religious Landscape Survey, June 2008, http://religions.pewforum.org/reports. See also Michael Hout and Claude Fischer, "Unchurched Believers." Working Paper, Survey Research Center, University of California, Berkeley, December 2009; Peggy Levitt, "Redefining the Boundaries of Belonging," *Sociology of Religion* 65, no. 1 (2004): 1018; Heidi Campbell, *Exploring Religious Community Online* (New York: Peter Lang, 2005); Robert Putnam and David Campbell, *American Grace* (New York: Simon & Schuster, 2010).

14. Figure 2.10: U.S. Census Bureau, "Table 1095 Utilization of Selected Media," Statistical Abstract of the United States, 2010. Based on Television Bureau of Advertising, "Trends in Television" annual report, http://www.census.gov/compendia/ statab/2010/tables/10s1095.xls; TVB Research Central, "Media Trends Track: TV Basics TV Sets per Household," http://www.tvb.org/media/file/TV_Basics.pdf; http:// www.tvb.org.central/MediaTrendsTrack/tvbasics/07_5_TV_Per_HH.asp.

15. Figure 2.11: Percentage of Creative Class source, courtesy of Kevin Stolarick, Martin Prosperity Institute, University of Toronto. The "Creative Class" and its "Super-Creative" subset come from Richard Florida, *The Rise of the Creative Class* (New York: Basic Books, 2002). Florida defines Creatives as "people whose economic function is to create new ideas, new technology, and/or new creative content" (p. 8) and Super-Creatives as "the thought leadership of modern society . . . [who produce] new forms or designs that are readily transferable and widely useful" (p. 69). We do not need to get into the controversies about Florida's data to agree that many more workers are now using computers to work with texts, numbers, and drawings.

16. Figure 2.12: In sociological terms, greater tolerance and less discrimination is a shift from "ascriptive" to "achieved" attitudes and behavior. It is documented by Molly Andolina, and Jeremy Mayer, "Demographic Shifts and Racial Attitudes," *Social Science Journal* 40 (2003): 19–31. The intermarriage attitude data come from James A. Davis and Tom Smith, General Social Surveys, 1972–2008 [machine-readable data file] (Chicago: National Opinion Research Center), using all valid responses (excluding "Don't Know," etc.) to the General Social Survey's question: "Do you think there should be laws against marriages between African-Americans and whites?" The percentage of respondents is based on valid responses (yes or no) to the question. The sample consisted of non-institutionalized English-speaking residents aged eighteen years or older and living in the United States between 1972

and 1977; only non-black respondents were asked this question. Starting in 1978, all respondents were asked this question. This difference is indicated by the break in the figure. See also John Iceland, Daniel Weinberg, and Erika Steinmetz, *Racial and Ethnic Residential Segregation in the United States: 1980–2002*, U.S. Census Bureau, Series CENSR-3 (Washington, DC: U.S. Government Printing Office, 2002), p. 60. The actual intermarriage data come from Jeffrey Passel, Wendy Wang, and Paul Taylor, "Marrying Out," Pew Research Center, June 4, 2010, http://pewsocialtrends .org/2010/06/04/marrying-out.

17. The GSS, funded by the U.S. National Science Foundation, is the preeminent large-scale national sociological survey. See http://www.norc.org/GSS+Website.

18. Will Herbert's 1955 book is *Protestant, Catholic, Jew* (Garden City, NY: Doubleday). Contemporary trends toward lower religious affiliation are documented in Mark Chaves's book *American Religion* (Princeton, NJ: Princeton University Press, 2011). The "love that dare not speak its name" is from Lord Alfred Douglas, "Two Poems" (1894), and was crucial evidence in the "gross indecency" trial of his lover Oscar Wilde.

19. Data from the Employee Benefit Research Institute, http://www.ebri.org. See also Steven Greenhouse, "Pensions on the Move," *New York Times*, March 1, 2011, B1, B4.

20. Our argument resonates in parts with thoughts expressed in Albert-Laszlo Barabasi, *Linked* (New York: Plume, 2003); Yochai Benkler, *The Wealth of Networks* (New Haven, CT: Yale University Press, 2006); Manuel Castells, *The Rise of the Network Society*, 2nd ed. (Oxford: Blackwell, 2000); Nicholas A. Christakis and James H. Fowler, *Connected* (Boston: Little, Brown, 2009); Dalton Conley, *Elsewhere USA* (New York: Pantheon, 2009); Charles Kadushin, *Making Connections* (New York: Oxford University Press, 2011); Duncan Watts, *Six Degrees* (New York: Norton, 2003); David Weinberger, *Small Pieces Loosely Joined* (New York: Perseus, 2002); Barry Wellman, "Structural Analysis: From Method and Metaphor to Theory and Substance," in *Social Structures: A Network Approach*, ed. Barry Wellman and S. D. Berkowitz, 19–61 (Cambridge: Cambridge University Press, 1988); Barry Wellman, "From Little Boxes to Loosely-Bounded Networks," in *Sociology for the Twenty-First Century*, ed. Janet Abu-Lughod, 94–114 (Chicago: University of Chicago Press, 1999); Bernie Hogan, "The Networked Individual: A Profile of Barry Wellman," *The Semiotician* 14 (2009): 5.

21. Linton Freeman, Sue Freeman, and Alaina Michaelson, "How Humans See Social Groups," *Journal of Quantitative Anthropology* 1 (1989): 229–238.

22. Émile Durkheim, *The Division of Labor in Society*, (New York: Macmillan, 1933 [1893]).

23. Norman Shulman, "Urban Social Networks," Ph.D. dissertation, Department of Sociology, University of Toronto, 1972. Barry Wellman, Renita Yuk-Lin Wong, David Tindall, and Nancy Nazer, "A Decade of Network Change," *Social Networks* 19, no. 1 (1997): 27–51.

24. The Buzz story is nicely summarized in Barbara Ortutay, "Google Tweaks Buzz Social Hub after Privacy Woes," Associated Press, February 12, 2010; and Riva

Richmond, "What You Need to Know about Google Buzz," New York Times, February 18, 2010, http://gadgetwise.blogs.nytimes.com/2010/02/17/what-you-need-to-know -about-google-buzz.

25. Harriet Jacobs [writing as "Fugitivus"], "Fuck You, Google," February 11, 2010, http://www.fugitivus.net/2010/02/11/fuck-you-google.

26. Eric Goldman, "Google Settles Buzz User Privacy Litigation," *Technology and Marketing Law Blog,* September 9, 2010, http://blog.ericgoldman.org/ archives/2010/09/googles_buzz_us.htm; Richard Wray, "Google Boss Says 'Nobody Was Harmed' by Buzz Debacle," *Guardian,* February 17, 2010, http://www.guardian .co.uk/technology/2010/feb/17/google-buzz-schmidt. Schmidt's no privacy comment and Harbour's rejoinder are in Emily Steel, "Google Buzz Exemplifies Privacy Problems, FTC Commissioner Says" *Wall Street Journal Blogs,* March 17, 2010; http://blogs.wsj.com/digits/2010/03/17/google-buzz-exemplifies-privacy-problems -ftc-commissioner-says; "FTC Charges Deceptive Privacy Practices in Google's Rollout of Its Buzz Social Network," FTC news release, http://ftc.gov/opa/2011/03/ google.shtm.

27. "FTC Charges Deceptive Privacy Practices," see note 26.

28. Ian Hacking, *Why Does Language Matter in Philosophy?* (Cambridge: Cambridge University Press, 1975); M[ark] E. J. Newnan, "Complex Systems: A Survey," *American Journal of Physics* 79 (2011): 800–810, arXiv:112.1440v1.

29. James Bagrow, Sune Lehmann, and Yong-Yeol Ahn, "Robustness and Modular Structure in Networks," *arXiv* (archive) 1102–5085v1, February 24, 2011.

30. Richard Leowontin, "The Corpse in the Elevator," *New York Review of Books,* January 20, 1983, p. 36.

31. For the argument that society is nothing more than the sum of two-person relationships, see sociologist George Homans, *Social Behavior: Its Elementary Forms* (New York: Harcourt Brace Jovanovich, 1961).

32. Michael Adams, *Fire and Ice* (Toronto: Penguin Canada, 2003); Wayne Baker, *America's Crisis of Values* (Princeton, NJ: Princeton University Press, 2004).

33. Craig Kinsley, personal communication to Sinye Tang and Barry Wellman, April 17, 2009. Kelly Lambert and Craig Kinsley, *Clinical Neuroscience* (New York: Worth, 2004).

34. For a modern take on old boy networks, see Parag Khanna, "Davos: Congress of the New Middle Ages," *Ideas Market* blog, *Wall Street Journal,* January 24, 2011. http://blogs.wsj.com/ideas-market/2011/01/24/davos-congress-of-the-new-middle -ages. The classic book on interlocking corporate directorships is Beth Mintz and Michael Schwartz, *The Power Structure of American Business* (Chicago: University of Chicago Press, 1985). A more recent update is Mark Mizruchi, "What Do Interlocks Do?" *Annual Review of Sociology* 22 (2003): 271–298. For replacement of board members, see R. Jack Richardson, "Directorship Interlocks and Corporate Profitability," *Administrative Science Quarterly* 32 (1987): 367–386. For European interlocks, see Eelke Heemskerk, "The Social Field of the European Corporate Elite," *Global Networks* 22, no. 4 (2011): 1–21.

35. "Text of Tiger Woods' Statement" as transcribed by *ASAP Sports*, Associated Press, February 19 2010, http://www.cbsnews.com/stories/2010/02/19/sportsline/main6223331.shtml.

36. Beverly Wellman, "Lay Referral Networks." *Sociology of Health Care* 12 (1995): 213–238.

37. Membership in the major professional organization devoted to social network analysis has doubled since 2000. The International Network for Social Network Analysis, founded by coauthor Wellman in 1976–1977, now has more than 1,300 members (http://www.insna.org). For a summary of the social network perspective see Alexandra Marin and Barry Wellman, "Social Network Analysis: An Introduction," in the *Handbook of Social Network Analysis,* ed. Peter Carrington and John Scott, 11–25 (Thousand Oaks, CA: Sage, 2011). The basic guide to social network analysis is Stanley Wasserman and Katharine Faust, *Social Network Analysis* (Cambridge: Cambridge University Press, 1994).

38. Google NGram source includes lower-case only. Smoothing = 2. http://ngrams.googlelabs.com/graph?content=social+networks,social+network&year_start=1950&year_end=2008&corpus=0&smoothing=2.

39. Barry Wellman and Kenneth Frank, "Network Capital in a Multi-Level World: Getting Support in Personal Communities," in *Social Capital*, ed. Nan Lin, Karen Cook, and Ronald Burt, 233–273 (Chicago: Aldine DeGruyter, 2001).

40. Nicholas Christakis and James Fowler, *Connected* (New York: Little Brown, 2009).

41. J. Niels Rosenquist, James Fowler, and Nicholas Christakis, "Social Network Determinants of Depression," *Molecular Psychiatry* 16 (March 16, 2011): 273–281, http://jhfowler.ucsd.edu/social_network_determinants_of_depression.pdf.

42. J. Niels Rosenquist, Joanne Murabito, James H. Fowler, and Nicholas A. Christakis, "The Spread of Alcohol Consumption Behavior in a Large Social Network," *Annals of Internal Medicine* 152, no. 7 (2010): 426–433.

43. Miller McPherson, Lynn Smith-Lovin, and James M. Cook, "Birds of a Feather: Homophily in Social Networks," *Annual Review of Sociology* 27 (2001): 415–444; Sinal Aral, Lev Muchnik, and Arun Sundarajan, "Distinguishing Influence-Based Contagion from Homophily-Driven Diffusion in Dynamic Networks," *PNAS* [Proceedings of the National Academy of Science] 106, no. 51: 21544–21549; Damon Centola, "The Spread of Behavior in an Online Social Network Experiment," *Science* 329 (September 3, 2010): 1194–1197.

44. James Fowler, Jaime Settle, and Nicholas Christakis, "Correlated Genotypes in Friendship Networks," *PNAS: Proceedings of the National Academy of Science*, January 18, 2011, http://www.phas.org/content/108/5/1993.full; on network gregariousness, see also Barry Wellman and Milena Gulia, "A Network Is More Than the Sum of Its Ties," in *Networks in the Global Village*, ed. Barry Wellman, 83–118 (Boulder, CO: Westview Press, 1999). For critiques of Christakis and colleagues' work, see Ethan Cohen-Cole and Jason Fletcher, "Detecting Implausible Social Network Effects in Acne, Height, and Headaches: Longitudinal Analysis," *British Medical Journal* 337 (2008), http://www.bmj.com/content/337/bmj.a2533.full;

Ethan Cohen-Cole and Jason Fletcher, "Is Obesity Contagious?" *Journal of Health Economics* 27, no. 5 (2008): 1382–1387, http://www.sciencedirect.com/science/article/pii/S0167629608000362; Russell Lyons, "The Spread of Evidence-Poor Medicine via Flawed Social-Network Analysis," *Statistics, Politics, and Policy* 2, no. 1 (2011), http://arxiv.org/abs/1007.2876.

45. Bernie Hogan, "The Networked Individual," *Semiotix,* January 2009, http://www.semioticon.com/semiotix/semiotix14/sem-14-05.html. See also Linton Freeman, *The Development of Social Network Analysis* (Vancouver: Empirical Press, 2004).

46. Simmel's life and work appears in a number of English-language compilations, including Georg Simmel, *The Sociology of Georg Simmel*, trans. and ed. Kurt Wolff (Glencoe, IL: Free Press, 1950); Georg Simmel, *Conflict and the Web of Group Affiliations,* ed. Kurt Wolff (Glencoe, IL: Free Press, 1955); Georg Simmel, *On Individuality and Social Forms*, trans. and ed. Donald Levine (Chicago: University of Chicago Press, 1971); Ferdinand Tönnies, *Community and Organization [Gemeinschaft und Gesellschaft]* (London: Routledge & Kegan Paul [1887] 1955).

47. Georg Simmel, "The Metropolis and Mental Life" (1903), translated and published in *The Sociology of Georg Simmel*, ed. Kurt Wolff (Glencoe, IL: Free Press, 1950), 409–424.

48. For Africa, see J. Clyde Mitchell, ed., *Social Networks in Urban Situations* (Manchester: Manchester University Press, 1969). Alvin Wolfe provides a complementary anthropologist's account in "Anthropologists View of Social Network Analysis and Data Mining," *Social Network Analysis and Mining* 1, no. 1 (2001): 3–19.

49. Elizabeth Bott, *Family and Social Network* (London: Tavistock, 1957).

50. Joe Feagin and Harlan Hahn, *Ghetto Revolt* (New York: Macmillan, 1973); for a general historical account at that time, see Charles Tilly, "Collective Violence in European Perspective," in *Violence in America*, ed. Hugh Graham and Ted Gurr, 4–45 (Washington, DC: U.S. Government Printing Office, 1969); There are many more recent studies, reviewed in Doug McAdam, Sidney Tarrow, and Charles Tilly, *Dynamics of Contention* (Cambridge: Cambridge University Press, 1973).

51. Herbert Gans, *The Urban Villagers* (New York: Free Press, 1962); Elliot Liebow, *Tally's Corner* (Boston: Little, Brown, 1967); Carol Stack, *All Our Kin* (New York: Harper & Row, 1974); Sudhir Venkatesh, *Gang Leader for a Day* (New York: Penguin, 2008).

52. Jeffrey Boase and Barry Wellman, "A Plague of Viruses," *Current Sociology* 49, no. 6 (2001): 39–55.

53. For an overall review, see Thomas Valente, *Network Models of the Diffusion of Innovations* (Cresskill, NJ: Hampton Press, 1995). For recent work on AIDS networks, see Susan Cotts Watkins and Ann Swidler, "Hearsay Ethnography," eScholarship series, March 2006, California Center for Population Research, UC Los Angeles, http://www.escholarship.org/uc/item/3bq2n770. For discussions on how insurgent networks are analyzed, see the work of Kathleen Carley, such as her 2009 paper "Dynamic Network Analysis for Counter-Terrorism Overview," http://citeseerx.ist.psu.edu/viewdoc/download?doi=10.1.1.137.9475&rep=rep1&type=pdf.

54. Everett Rogers, *Diffusion of Innovations,* 5th ed. (New York: Free Press, 2003). Rogers' bell curve is still useful, although further research has suggested more complex models. See the review in Raghurim Iyengar, Christophe Van den Buite, and Thomas Valente, "Opinion Leadership and Contagion in New Product Diffusion," *Marketing Science* 30, no. 2 (2011): 195–212.

55. Simon Cauchemez, Achuyt Bhattari, Tiffany Marchbanks, Ryan Fagan, Stephen Ostroff, Neil Ferguson, David Swerdlow, and the Pennsylvania H1N1 working group, "Role of Social Networks in Shaping Disease Transmission During a Community Outbreak of 2009 H1N1 Pandemic Influenza," *Publications of the National Academy of Sciences,* January 31, 2011, http://www.pnas.org/content/early/2011/01/28/1008895108.full.pdf+htm.

56. For identification of influential spreaders in the core and periphery, see Maksim Kitsak, Lazaros Gallos, Shlomo Havlin, Fredrik Liljeros, Lev Muchnik, H. Eugene Stanley, and Hernán Makse, "Identification of Influential Spreaders in Complex Networks," *Nature Physics* 6 (2010): 888–893.

57. Duncan Watts, *Everything is Obvious* (New York: Crown Business, 2011).

58. Alberto László Barabasi *Linked* (Cambridge, MA: Perseus, 2002). See also Joseph Kong, Nima Sarshar, and Vwani Roychowdhur, "Experience versus Talent Shapes the Structure of the Web," *PNAS* [Proceedings of the National Academy of Sciences] 105, no. 37 (2008): 13724–13729.

59. Robert K. Merton, "The Matthew Effect in Science," *Science* 159 (January 5, 1968): 56–63.

60. Barry Wellman, "Structural Analysis: From Method and Metaphor to Theory and Substance," in *Social Structures: A Network Approach*, ed. Barry Wellman and S. D. Berkowitz, 19–61 (Cambridge: Cambridge University Press, 1988).

61. David Easley and Jon Kleinberg, *Networks, Crowds and Markets* (Cambridge: Cambridge University Press, 2010); Andrei Broder, Ravi Kumar, Farzin Maghoul, Prabhakar Raghavan, Sridhar Rajagopalan, Raymie Stata, Andrew Tomkins, and Janet Wiener, "Graph Structure in the Web," *Proceedings of the 9th International World-Wide Web Conference* (2000), pp. 309–320. See also Watts, *Six Degrees,* note 20.

62. Mark Granovetter, "The Strength of Weak Ties: A Network Theory Revisited," *Sociological Theory* 1 (1983): 201–233.

63. Ronald Burt, *Neighbor Networks* (New York: Oxford University Press, 2009).

64. Ronald Burt, "Structural Holes in Virtual Worlds," University of Chicago, June 2011, http://faculty.chicagobooth.edu/ronald.burt/research/files/SHVW.pdf.

65. Robert K. Merton, "Patterns of Influence: Cosmopolitans and Locals," in *Social Theory and Social Structure*, ed. R. Merton, 387–420 (Glencoe, IL: Free Press, 1957).

66. Nathan Eagle, Michael Macy, and Rob Claxton, "Network Diversity and Economic Development," *Science* 328 (May 21, 2010): 1029–1031.

67. Burton Pasternak and Janet Salaff, *Cowboys and Cultivators* (Boulder, CO: Westview, 1993).

68. Twitter analyses come from Alex Cheng, "Six Degrees of Separation, Twitter Style," April 2010, http://www.sysomos.com/insidetwitter/sixdegrees; Dierderik van

Liere, "How Far Does a Tweet Travel?" Modeling Social Media conference, Toronto, June 2010; Yuri Takhteyev, Anatoliy Gruzd, and Barry Wellman, "Geography of Twitter Networks." NetLab working paper, 2010, http://homes.chass.utoronto.ca/~wellman; Gruzd and Wellman's network data is from August 2010, and was prepared by Anatoliy Gruzd. It is discussed more in Anatoliy Gruzd, Barry Wellman, and Yuri Takhteyev, "Imagining Twitter as an Imagined Community. *American Behavioral Scientist* 55, no. 10: 1294–1318.

69. Scott Feld, "The Focused Organization of Social Ties," *American Journal of Sociology* 86 (1981): 1015–1035. More than forty-five years ago, network analysis pioneer Harrison White called the combination of shared characteristics and network membership a "catnet" for the intersection of categories (people with similar attributes) and networks: "Notes on the Constituents of Social Structure," working paper (Harvard University: Department of Social Relations, 1965). For more recent work, see Mason Porter, Jukka-Pella Onnela, and Peter Mucha, "Communities in Networks," *Notices of the AMS* 56, no. 9 (2009): 1082–1097, 1164–1166.

70. Ronald Breiger, "The Duality of Persons and Groups," *Social Forces* 53 (1974): 181–190. A community example is in John Hipp and Andrew Perrin, "Nested Loyalties," *Urban Studies* 43, no. 13 (2006): 2503–2523.

71. Figure 2.16 was constructed from the standard Amazon book purchasing note: "People who bought ThisBook also bought ThatBook and ThatOtherBook." "To reduce noise, the data were cleaned slightly with k cores to remove weak associations, with k=2. All books that were not connected or had only one link with another book were removed. However no bridges were removed: the clusters are truly disconnected." Valdis Krebs, personal communication, April 24, 2010.

72. Stanley Milgram, "The Small-World Problem," *Psychology Today* 1 (1967): 62–67. See also Watts, *Six Degrees*, note 18. However, Judith Kleinfeld has noted that Milgram's original experiment had flaws. See her "The Small World Problem," *Society* 39, no. 2 (2002): 61–66.

73. Anatoliy Gruzd did the retweeting analysis. For the airline-like connectivity of Twitter, see Yuri Takhteyev, Anatoliy Gruzd, and Barry Wellman, "Geography of Twitter Networks," *Social Networks* 34 (2012): 73–81.

74. Molière, *The Bourgeois Gentleman [Le Bourgeois Gentilhomme]*, trans. Timothy Mooney, 1670, act 2, scene 4, http://moliere-in-english.com/bourgeois.html (about the adaptation).

75. Benoit Mandelbrot, *The Fractal Geometry of Nature* (San Francisco: Freeman, 1983) p. 1.

76. Robert K. Merton, "Patterns of Influence: Cosmopolitans and Locals," in *Social Theory and Social Structure,* ed. R. Merton, 387–420 (Glencoe, IL: Free Press, 1957); Mark Granovetter, "The Strength of Weak Ties: A Network Theory Revisited," *Sociological Theory* 1 (1983): 201–233.

77. Personal communication from Hogan to Wellman, June 10, 2010. See also Hogan's doctoral dissertation, "Networking in Everyday Life," Department of Sociology, University of Toronto, 2008.

Chapter 3

1. We appreciate the advice, assistance, and contributions to this chapter of Christine Ensslen and Mirna Ghazarian (especially) as well as Mo Guang Ying and Barbara Barbosa Neves.

2. For an engaging account of how the first personal computers came to be, see John Markoff, *What the Dormouse Said* (New York: Viking, 2005).

3. Ray Tomlinson has written about the first email, "The First Network Email," at http://openmap.bbn.com/~tomlinso/ray/firstemailframe.html.

4. Survey by Southern New England Telephone, conducted by Louis Harris & Associates, September 1 to September 11, 1983 and based on telephone interviews with a national adult sample of 1,256. Polling done by Louis Harris & Associates with data from The Roper Center for Public Opinion Research, University of Connecticut. See: http://webapps.ropercenter.uconn.edu/CFIDE/cf/action/catalog/abstract.cfm?label=&keyword=USHARRIS1983-2033&fromDate=&toDate=&organization=Any&type=&keywordOptions=1&start=1&id=&exclude=&excludeOptions=1&topic=Any&sortBy=DESC&archno=USHARRIS1983-2033&abstract=abstract&x=32&y=9.

5. Peter Lewis, "The Executive Computer: A Growing Internet Is Trying to Take Care of Business," *New York Times*, December 12, 1993, http://www.nytimes.com/1993/12/12/business/the-executive-computer-a-growing-internet-is-trying-to-take-care-of-business.html.

6. The creators of HTML thus perniciously inflected on billions of websites (and users) the prefix "http" for "hypertext transfer protocol."

7. For a more detailed history of the internet, see Katie Hafner and Matthew Lyon, *Where Wizards Stay Up Late* (New York: Simon & Schuster, 1996). See also Henry Edward Hardy, "A Short History of the Net," in *The Internet World Guide to Multimedia on the Internet*, ed. Gary Welz, (Westport, CT: Mecklermedia, 1995).

8. Kevin Werbach, "Digital Tornado," OPP Working Paper No. 29, March 1997, http://www.fcc.gov/Bureaus/OPP/working_papers/oppwp29.pdf.

9. Gordon Moore, "Cramming More Components onto Integrated Circuits," *Electronics* 38, no. 8 (April 19, 1965), ftp://download.intel.com/museum/Moores_Law/Articles-Press_Releases/Gordon_Moore_1965_Article.pdf.

10. "Moore's Law: 40th Anniversary," http://www.intel.com/pressroom/kits/events/moores_law_40th/index.htm?iid=tech_mooreslaw+body_presskit.

11. "Google Brings High-speed Broadband Network to Kan," Associated Press, March 30, 2011, http://abcnews.go.com/US/wireStory?id=13256735.

12. "10 Gigabit Ethernet," Wikipedia, http://en.wikipedia.org/wiki/10_Gigabit_Ethernet.

13. Jason Oxman, "The FCC and the Unregulation of the Internet," OPP Working Paper No. 31, July 1999, http://transition.fcc.gov/Bureaus/OPP/working_papers/oppwp31.pdf, p 14. See also Jacob Nielsen's articulation of "Nielsen's Law of Internet Bandwidth" at http://www.useit.com/alertbox/980405.html.

14. Wellman invented the term "network of networks" well before the advent of the internet. See Paul Craven and Barry Wellman, "The Network City," *Sociological Inquiry* 43 (1973): 57–88. For the possible balkanization of the Internet, see Nancy Scola, "When the Internet Nearly Fractured, and How It Could Happen Again," *The Atlantic,* February 24, 2011, http://www.theatlantic.com/technology/archive/2011/02/when-the-internet-nearly-fractured-and-how-it-could-happen-again/71662.

15. Chip Walker, "Kryder's Law," *Scientific American*, July 2005, http://www.sciam.com/article.cfm?id=kryders-law.

16. "Falling Through the Net II: New Data on the Digital Divide," National Telecommunication and Information Administration, July 1998, http://www.ntia.doc.gov/ntiahome/fttn99/part2.html.

17. Kristen Purcell, "Search and Email Still Top the List of the Most Popular Online Activities," Pew Internet & American Life Project, August 2011, http://pewinternet.org/Reports/2011/Search-and-email.aspx.

18. Joshua Meyrowitz, personal communication. See also: Claude Fischer, *America Calling* (Berkeley: University of California Press, 1992), p. 40.

19. Affordances quotation (p. 228) and description is from Donald Norman, *Living with Complexity* (Cambridge, MA: MIT Press, 2011). For a discussion of how affordances affect social interaction, see: Erin Bradner, Wendy Kellogg, and Thomas Erickson, "Social Affordances of BABBLE," paper presented at the European Computer Supported Cooperative Work Conference, Copenhagen, November 1998.

20. John Markoff, "A Fight to Win the Future: Computers vs. Humans," *New York Times,* February 14, 2011, http://www.nytimes.com/2011/02/15/science/15essay.html. As Markoff shows in his 2005 book *What the Dormouse Said* (New York: Penguin), the artificial intelligence debate goes back to the earliest days of human-centered computing.

21. Sheila Htoo is now a community planner, but was a University of Toronto student. Her story is used with her permission.

22. Lee Rainie, Amanda Lenhart, Susannah Fox, Tom Spooner, and John Horrigan, "Tracking Online Life," Pew Internet & American Life Project, May 10, 2000, http://pewinternet.org/Reports/2000/Tracking-Online-Life.aspx.

23. Rainie, Lenhart, Fox, Spooner, and Horrigan, "Tracking Online Life," see note 22.

24. Rainie, Lenhart, Fox, Spooner, and Horrigan, "Tracking Online Life," see note 22.

25. Pew Internet & American Life Project report, "Getting Serious Online," March 3, 2002, http://www.pewinternet.org/~/media/Files/Reports/2002/PIP_Getting_Serious_Online3ng.pdf.pdf.

26. Leonard Witt's blog is at http://pjnet.org.

27. John Horrigan, "The Broadband Difference," Pew Internet & American Life Project, June 23, 2002, http://www.pewinternet.org/Reports/2002/The-Broadband-Difference-How-online-behavior-changes-with-highspeed-Internet-connections.aspx.

28. Kathleen Moore, "71% of Online Adults Now Use Video-Sharing Sites," Pew Internet & American Life Project, July 26, 2011, http://pewinternet.org/Reports/2011/Video-sharing-sites.aspx.

29. In Pew Internet surveys this means internet users said in a phone interview they had used the internet "yesterday"—or the day before they were reached in the survey.

30. Leigh Estabrook, Evans Witt, and Lee Rainie, "Information Searches That Solve Problems," Pew Internet & American Life Project. December 30, 2007, http://www.pewinternet.org/Reports/2007/Information-Searches-That-Solve-Problems.aspx.

31. Tom Rosenstiel, Amy Mitchell, Lee Rainie, and Kristen Purcell. "The Local News Ecology," Pew Internet & American Life Project, September, 2011, http://pewinternet.org.

32. Statistics on the composition of the American online population and the general activities Americans pursue on the internet are updated several times a year after national surveys by the Pew Internet & American Life Project. The latest findings from Pew Internet are available at http://pewinternet.org/Static-Pages/Trend-Data.aspx. For information on rural Internet use by physically isolated northern Canadians, see Jessica Collins and Barry Wellman, "Small Town in the Internet Society," *American Behavioral Scientist* 53, no. 9 (2010): 1344–1366.

33. A general reading on digital differences in Pew Internet & American Life Project data can be found on the "Who's Online" page at http://pewinternet.org/Static-Pages/Trend-Data-%28Adults%29/Whos-Online.aspx. In addition, these Pew Internet reports have focused on why some Americans do not use the internet: Aaron Smith, "Home Broadband 2010," August 11, 2010, http://pewinternet.org/Reports/2010/Home-Broadband-2010.aspx; John Horrigan, "Home Broadband Adoption in Rural America," February 26, 2006, http://pewinternet.org/Reports/2006/Home-Broadband-Adoption-in-Rural-America.aspx; and Amanda Lenhart, "Who's Not Online," September 21, 2000, http://pewinternet.org/Reports/2000/Whos-Not-Online.aspx.

34. For digital skills, see various papers by Eszter Hargittai, including "Differences in Actual and Perceived Online Skills" (with Steven Shafer), *Social Science Quarterly* 87, no. 2 (2006): 432–448; "Digital Inequality" (with Amanda Hinnant), *Communication Research* 35, no. 5 (2006): 602–621; "The Digital Reproduction of Inequality," in *Social Stratification,* ed. David Grusky, 936–944 (Boulder, CO: Westview, 2008, 3rd ed); Philip Howard, Laura Busch, and Penelope Sheets, "Comparing Digital Divides: Internet Access and Social Inequality in Canada and the United States," *Canadian Journal of Communication* 35, no. 1 (2010), http://www.cjc-online.ca/index.php/journal/article/view/2192; Keith Hampton, "Internet Use and the Concentration of Disadvantage," *American Behavioral Scientist* 53, no. 8 (2010): 1111–1132; Howard Rheingold, *Net Smarts* (Cambridge, MA: MIT Press, 2012).

35. Manuel Castells's most thorough explanation appears in his book *The Internet Galaxy* (Oxford: Oxford University Press, 2001). See esp. pp. 36–63; quotation from pp. 46–47. See also Manuel Castells, *The Rise of the Network Society,* 2nd ed (Oxford: Blackwell, 2000).

36. Scott Cleland, "Google's Self-Serving 'Innovation without Permission,'" *The Precursor Blog,* November 1, 2006, http://www.precursorblog.com/node/212. See also Castells, note 35, pp. 46–47. A good description of early hacker culture is John Markoff, *What the Dormouse Said* (New York: Viking, 2005).

37. For a good example, see Howard Rheingold, *The Virtual Community,* 2nd ed. (Cambridge, MA: MIT Press, 2000).

38. Castells, note 35, pp. 52–55.

39. See Bernard Girard, *The Google Way* (San Francisco: No Starch, 2009); Doug Coupland, *Microserfs* (New York: HarperCollins, 1995).

40. William Dutton, "Through the Network (of Networks)—the Fifth Estate." Lecture, Examination Schools, University of Oxford, October 15, 2007, http:// people.oii.ox.ac.uk/dutton/wp-content/uploads/2007/10/5th-estate-lecture-text .pdf. The idea of the Fifth Estate is debated and explored on Dutton's blog: http:// people.oii.ox.ac.uk/dutton.

41. Mary Madden and Aaron Smith, "Reputation Management and Social Media," Pew Internet & American Life Project, May 26, 2010, http://pewinternet.org/Reports/ 2010/Reputation-Management.aspx. Susannah Fox, Mary Madden, Aaron Smith, and Jessica Vitak, "Digital Footprints," Pew Internet & American Life Project, December 16, 2007, http://www.pewinternet.org/Reports/2007/Digital-Footprints.aspx.

Chapter 4

1. Thanks to Anna Brady, Loi Sessions Goulet, Bernie Hogan, Lee Humphreys, and Rhonda McEwen for their advice and assistance.

2. Pamela Rabe, "Surviving Mumbai," *The Passionate Eye,* CBC News TV Documentary, November 29, 2009.

3. Nate Orenstam, "Doctor Cellphone: Can You Hear Me Now?" *Valley of the Geeks,* December 20, 2007, http://www.valleyofthegeeks.com/Features/Cooper.html. Also, coauthor Rainie's interview with Martin Cooper, April 29, 2011.

4. Rich Ling, *The Mobile Connection* (San Francisco: Morgan Kaufmann, 2004), pp 6–11. See also AT&T, "1946: First Mobile Telephone Call," http://www.corp.att.com/ attlabs/reputation/timeline/46mobile.html.

5. Wikipedia, "History of Mobile Phones," Wikimedia Foundation, Inc, March 19, 2010, http://en.wikipedia.org/wiki/History_of_mobile_phones.

6. For the latest U.S. statistics on wireless subscribers, see "CTIA Semi-Annual Wireless Industry Survey," http://www.ctia.org/media/industry_info/index.cfm/ AID/10316.

7. Centers for Disease Control and Prevention data through 2010 reported in "Wireless Substitution: Early Release of Estimates from the National Health Interview Survey, July-December 2010," http://www.cdc.gov/nchs/data/nhis/earlyrelease/ wireless201106.pdf. Future releases can be found by doing a search on "wireless substitution" at this site for the National Health Interview Survey: http://www.cdc.gov/ nchs/nhis.htm.

8. Hotspex, "Survey Reveals Less than 10 Per Cent of Canadians State They Can Live without Their Mobile Device," *Exchange,* September 10, 2010, http://www.exchange magazine.com/morningpost/2009/week37/Thursday/091023.htm#anchor.

9. Kristen Purcell, "Half of Adult Cell Phone Owners Have Apps on Their Phones," Pew Internet & American Life Project, November 2, 2001, http://pewinternet.org/ Reports/2011/Apps-update.aspx.

10. James Katz and Mark Aakhus, eds., *Perpetual Contact* (Cambridge: Cambridge University Press, 2002); see also Mary Chayko, *Portable Communities* (Albany: SUNY Press, 2008).

11. John Horrigan, "The Mobile Difference," Pew Internet & American Life Project, March 25, 2009, http://www.pewinternet.org/Reports/2009/5-The-Mobile -Difference--Typology.aspx.

12. International Telecommunication Union statistics are available at http:// www.itu.int/ITU-D/ict/statistics/at_glance/KeyTelecom.html. See also Samuel Greengard, "Upwardly Mobile." *Communications of the ACM* 51, no. 12 (2009): 17–19.

13. Rich Ling and Scott Campbell, *The Reconstruction of Space and Time* (New Bruns-wick, NJ: Transaction, 2009); Kathleen Diga, "Mobile Cell Phones and Poverty Reduction," MS thesis, School of Development Studies, University of KwaZulu-Natal, 2007.

14. Nyaga Mbatia, Paul Palackal, Antony Dzorgbo, Dan Bright, Ricardo Duque, Marcus Antonious Ynalvez, and Wesley Shrum, "Mobile Telephony and Core Network Expansion in Kenya," Working Paper, Department of Sociology, Louisi-ana State University, 2009. See also Tom Standage, "Mobile Marvels," *The Econo-mist,* September 24, 2009, http://www.economist.com/specialreports/displayStory. cfm?story_id=14483896; James Katz, ed., *Handbook of Mobile Communication Studies* (Cambridge, MA: MIT Press, 2008). For information by country, see Richard Heeks, "Mobile Phone Penetration," *ICTs for Development* blog, November 30, 2009, http:// ict4dblog.wordpress.com.

15. "China's Mobile Phone Users Top 879m in February," China Daily, March 25, 2011, http://www.chinadaily.com.cn/bizchina/2011-03/25/content_12229631. htm.

16. Daniel Miller, "The Unpredictable Mobile Phone," *BT Technology Journal* 24, no. 3 (2006): 41–48.

17. Mark Milian, "Why Text Messages Are Limited to 160 Characters," *Los Angeles Times*, May 3, 2009, http://latimesblogs.latimes.com/technology/2009/05/invented -text-messaging.html.

18. Amanda Lenhart, Rich Ling, Scott Campbell, and Kristen Purcell, "Teens and Mobile Phones," Pew Internet & American Life Project, April 20, 2010, http://www .pewinternet.org/Reports/2010/Teens-and-Mobile-Phones.aspx.

19. Rhonda McEwen, "A World More Intimate: Exploring the Role of Mobile Phones in Maintaining and Extending Social Networks," PhD dissertation, Faculty of Infor-mation, University of Toronto, 2009. For accounts of microcoordination in Japanese

life, see Mazuko Ito, Misa Matsuda, and Okabe Daisuke, eds., *Portable, Personal, Pedestrian* (Cambridge, MA: MIT Press, 2005).

20. Tess Kalinowski, "TTC Driver Caught on Break Suspended," *Toronto Star,* February 4, 2005, http://www.thestar.com/news/gta/article/760507--ttc-driver-caught -on-break-suspended. See also "TTC Workers' Facebook Site Swamped with Complaints," *CBC News,* February 8, 2010, http://www.cbc.ca/news/canada/toronto/story/ 2010/02/08/ttc-facebook.html.

21. Debra Donston, "The Decade's Most Important Tech," *eWeek*, January 4, 2010, p. 10; Kakuko Miyata, Barry Wellman, and Jeffrey Boase, "The Wired—and Wireless —Japanese," in *Mobile Communication*, ed. Rich Ling and Per Pedersen, 427–449 (London: Springer, 2005).

22. Kristen Purcell, Roger Entner, and Nichole Henderson, "The Rise of Apps Culture," Pew Internet & American Life Project, September 14, 2010, http://www .pewinternet.org/Reports/2010/The-Rise-of-Apps-Culture.aspx.

23. Jeff Bertolucci, "10 Things Killed by the Smartphone," *PC World*, April 15, 2010, http://www.pcworld.com/article/225372/10_things_killed_by_the_smartphone .html.

24. Chris Anderson and Michael Wolff, "The Web Is Dead. Long Live the Internet," *Wired,* August 17, 2010, http://www.wired.com/magazine/2010/08/ff_webrip/all/1.

25. Pew Research Center's Internet & American Life Project. Unpublished data from survey conducted in May 2011.

26. See Pew Internet survey data available at http://pewinternet.org/Shared -Content/Data-Sets/2010/May-2010--Cell-Phones.aspx.

27. John Packzkowski, "Microsoft/Danger. Enough Said," *All Things D* blog, http:// digitaldaily.allthingsd.com/20091012/sidekick/; BBC, "'Iranian cyber army' hits Twitter," *BBC News,* December 18, 2009, http://news.bbc.co.uk/2/hi/technology/ 8420233.stm.

28. Manuel Castells, "Afterword," in *Handbook of Mobile Communication Studies*, ed. James Katz, 448–449 (Cambridge, MA: MIT Press, 2008).

29. Scott Campbell and Michael Kelley, "Mobile Phone Use in AA Networks," *Journal of Applied Communication Research* 34, no. 2 (2006): 191–208.

30. Aaron Smith, "Americans and Their Cell Phones," Pew Internet & American Life Project, August 2011, http://pewinternet.org/Reports/2011/Cell-Phones.aspx.

31. Cartoon by "Gregory," *New Yorker,* August 2, 2010, p. 56. See also Naomi Baron, *Always On* (New York: Oxford University Press, 2008).

32. Howard Rheingold, *Smart Mobs* (New York: Perseus, 2002); Ling, *The Mobile Connection*, see note 4.

33. Ling, *The Mobile Connection*, see note 4.

34. Rheingold, *Smart Mobs*, see note 32. Ling, *The Mobile Connection*, see note 4.

35. Ling, *The Mobile Connection*, see note 4. See also Ling and Campbell, *The Reconstruction of Space and Time*, see note 13.

36. Rich Ling, *The Mobile Connection,* see note 14; Eva Thulin and Bertil Vilhelmson, "Mobiles Everywhere," *Young* 15, no. 3 (2007): 235–253.

37. Frances Cairncross quotation is from the back cover of *The Death of Distance: How the Communications Revolution Is Changing Our* Lives, 2nd ed. (Boston: Harvard Business School Press, 2001).

38. Email to Barry Wellman, December 20, 2010, based on her tweet of December 19, 2010.

39. For home-based internet, see Diane Mok, Barry Wellman, and Juan-Antonio Carrasco, "Does Distance Matter in the Age of the Internet? *Urban Studies* 47, no. 13 (2010): 2747–2783.

40. Manuel Castells, *The Rise of the Network Society,* Oxford: Blackwell, 2000).

41. Scott Campbell and Yong Jin Park, "Social Implications of Mobile Telephony," *Sociology Compass* (January 2008): 371–387.

42. Kenneth Gergen, *The Saturated Self* (New York: Basic Books, 1991).

43. Mary Madden and Lee Rainie, "Adults and Cell Phone Distractions," Pew Internet & American Life Project, June 18, 2010, http://pewinternet.org/Reports/2010/Cell-Phone-Distractions.aspx.

44. Madden and Rainie, "Adults and Cell Phone Distractions," see note 43. Also "Sixth of Cell Phone Owners Have Bumped into Someone or Something while Using Their Handhelds," June 18, 2010, http://www.pewinternet.org/Reports/2010/Cell-Phone-Distractions/Major-Findings/5-Bumping-into-people-and-objects.aspx.

45. The studies, which may not be representative, are summarized by Michael Kesterson, "Social Studies," *Toronto Globe and Mail,* December 19, 2009: A22; Ira Hyman Jr., S. Matthew Boss, Breanne Wise, Kira McKenzie, and Jenna Caggiano, "Did You See the Unicycling Clown?" *Applied Cognitive Psychology* 23, no. 2 (2010), http://onlinelibrary.wiley.com/doi/10.1002/acp.1638/abstract.

46. Abigail van Buren, "Cell Phone Users Should Give It a Rest in the Ladies Room," *Dear Abby on uExpress,* October 24, 2008, http://www.uexpress.com/dearabby/?uc_full_date=20081024.

47. "A Funny Thing Happened on the Way to New York (Or: Pillsbury Associates, Brace Yourselves)," *Above the Law,* an insider website devoted to gossip about law firms and courts, February 2009, http://abovethelaw.com/2009/02/a-funny-thing-happened-on-the-way-to-new-yorkor-pillsbury-associates-brace-yourselves-.

48. Erving Goffman, *Behavior in Public Places* (New York: Free Press, 1963). See also Georg Simmel, "The Metropolis and Mental Life," in *The Sociology of Georg Simmel,* ed. Kurt Wolff, 409–424 (Glencoe, IL: Free Press, [1903] 1950); Ruth Rettie, "Mobile Phone Communication: Extending Goffman to Mediated Interaction," *Sociology* 43 (2009): 421–438.

49. Lynn Loucester, "At My Wedding Twittering [sic] and Facebooking at the Altar," *YouTube,* November 22, 2009, http://www.youtube.com/watch?v=VSkT5XykJzo.

50. John B. Horrigan, "Mobile Access to Data and Information," Pew Internet & American Life Project, March 2008, http://www.pewinternet.org/PPF/r/244/report_display.asp.

51. "Continuous partial attention" is a concept developed by internet analyst Linda Stone, http://lindastone.net.

Chapter 5

1. Sharanpreet Kelley, Rhonda McEwen, and Justine Yu contributed to the writing of this chapter. We thank Kay Axhausen, Kristen Berg, Jeffrey Boase, Juan-Antonio Carrasco, Julia Chae, Jessica Collins, Sabrina Cutaia, Claude Fischer, Mirna Ghazarian, Melissa Godbout, Keith Hampton, Bernie Hogan, Chang Lin, Julia Madej, Eric Miller, Mo Guang Ying, John Robinson, Telus, Hua Helen Wang, Erin Weinkauf, and Yu Janice Zhang for their advice and assistance.

2. Miller McPherson, Lynn Smith-Lovin, and Matthew Brashears, "Social Isolation in America," *American Sociological Review* 71 (2008): 353–375.

3. Sebastian Mallaby, "Why So Lonesome?" *Washington Post*, June 26, 2006, p. A21, http://www.washingtonpost.com/wp-dyn/content/article/2006/06/25/AR2006 062500566.html; Douglas Cornish, "Is Computer-Glow the New Hearth-Light?" *Toronto Globe & Mail*, October 13, 2006, p. A13, http://www.theglobeandmail.com/news/technology/is-computer-glow-the-new-hearth-light/article848997.

4. Leo Marx, *The Machine in the Garden* (New York: Oxford University Press, 1964); Claude Fischer, *America Calling* (Berkeley: University of California Press, 1992); Jane Jacobs, *The Death and Life of Great American Cities* (New York: Random House, 1961); Andres Duany, Elizabeth Plater-Zyberk, and Jeff Speck, *Suburban Nation* (New York: North Point, 2000); Kay Axhausen, "Geographies of Somewhere," *Urban Studies* 37, no. 10 (2000): 1849–1864.

5. David Riesman, Reuel Denney, and Nathan Glazer, *The Lonely Crowd* (New Haven, CT: Yale University Press, 1950); William Kornhauser, *The Politics of Mass Society* (New York: Free Press, 1959); Robert Nisbet, *The Quest for Community* (Oxford: Oxford University Press, 1953).

6. Maurice Stein, *The Eclipse of Community* (Princeton, NJ: Princeton University Press, 1960).

7. Jim Hightower is quoted in Robert Fox, "Newstrack," *Communications of the ACM*, 38, no. 8 (1995): 11–12.

8. The original article is Robert Kraut, Michael Patterson, Vicki Lundmark, Sara Kiesler, Tridas Mukhopadhyay, and William Scherlis, "Internet Paradox," *American Psychologist* 53, no. 9 (1998): 1017–1031. One front-page story about this article is Amy Harmon's "Sad, Lonely World Discovered in Cyberspace," *New York Times*, August 30, 1998. The follow-up study is Robert Kraut, Sara Kiesler, Bonka Boneva, Jonathon Cummings, Vicki Helgeson, and Anne Crawford, "Internet Paradox Revisited," *Journal of Social Issues* 58, no. 1 (2002): 49–74.

9. William Gibson, *Neuromancer* (New York: Ace, 1984).

10. Sherry Turkle, "Who Am We?" *Wired*, January 1996, http://www.wired.com/wired/archive/4.01/turkle.html; Sherry Turkle, *Alone Together* (New York: Basic Books, 2011).

11. Keith Hampton, Lauren Sessions, Eun Ja Her, and Lee Rainie, "Social Isolation and New Technology," Pew Internet & American Life Project, November 2009, http://www.pewinternet.org/Reports/2009/18--Social-Isolation-and-New-Technology.aspx.

12. Marshall McLuhan, *Understanding Media: The Extension of Man* (New York: McGraw-Hill, 1964). A good guide to McLuhan's thoughts is Phillip Marchand and Neil Postman, *Marshall McLuhan: The Medium and the Messenger*, rev. ed. (Cambridge MA: MIT Press, 1998).

13. Paul McCartney, "Eleanor Rigby," *Revolver* 1966 (Parlophone), London.

14. Not opening door to strangers: Stanley Milgram, "The Experience of Living in Cities." Science, 167 (1970), 1461–1468. John Perry Barlow, "Is There a There in Cyberspace?" *Utne Reader* 68 (March–April 1995): 50.

15. Pre-internet research is summarized in Barry Wellman, "Physical Place and Cyber Place," *International Journal of Urban and Regional Research* 25, no. 2 (2001): 227–252; Barry Wellman, "The Persistence and Transformation of Community: From Neighbourhood Groups to Social Networks," Report to the Law Commission of Canada, Ottawa, October 2001, http://homes.chass.utoronto.ca/~wellman/publications/lawcomm/lawcomm7.PDF.

16. Robert Putnam, *Bowling Alone*, (New York: Simon & Schuster, 2000), pp. 111–112. For a critique of his argument, see Robert Andersen, James Curtis, and Edward Grabb, "Trends in Civic Association Activity in Four Democracies," *American Sociological Review* 71 (2006): 376–400.

17. Barry Wellman, "The Community Question," *American Journal of Sociology* 84 (1979): 1201–1231; Barry Wellman and N. Scot Wortley, "Different Strokes from Different Folks," *American Journal of Sociology* 96, no. 3 (1990): 558–588; Diana Mok and Barry Wellman, "How Much Did Distance Matter before the Internet?" *Social Networks* 29, no. 3 (2007): 430–461; Gabriele Plickert, Rochelle Côté, and Barry Wellman, "It's Not Who You Know, It's How You Know Them," *Social Networks* 29, no. 3 (2007): 405–429; Claude Fischer, *To Dwell among Friends: Personal Networks in Town and City* (Chicago: University of Chicago Press, 1982) and *Still Connected* (New York: Russell Sage Foundation, 2011).

18. Wellman, "The Community Question," note 17; Barry Wellman, Peter Carrington, and Alan Hall, "Networks as Personal Communities," in *Social Structures,* ed. Barry Wellman and S. D. Berkowitz, 130–184 (Cambridge: Cambridge University Press, 1988); Wellman and Wortley, "Different Strokes," note 17; Barry Wellman and David Tindall, "How Telephone Networks Connect Social Networks," *Progress in Communication Science* 12 (1993): 63–94.

19. Lauren Sessions, "How Offline Gatherings Affect Online Communities," *Information, Communication & Society* 13, no. 3 (2010): 375–395.

20. "Limited liability" is the key concept in sociologist Scott Greer's *The Emerging City* (New York: Free Press, 1962), the first book to document Americans' participation in multiple glocalized communities.

21. Christena Nippert-Eng, *Islands of Privacy* (Chicago: University of Chicago Press, 2011); Nicole Ellison, Jessica Vitak, Charles Steinfeld, Rebecca Gray, and Cliff Lampe, "Negotiating Privacy Concerns and Social Capital Needs in a Social Media Environment," in *Privacy Online*, eds. Sabine Trepte and Leonard Reinecke (New York: Springer, 2011), pp. 19–32; danah boyd and Alice Marwick, "Social Steganography:

Privacy in Networked Publics," paper presented to International Communication Association, Boston, May 2011.

22. Wellman and Wortley, "Different Strokes," see note 17.

23. Constantine Sedikides and Steven Spencer, eds., *The Self* (New York: Psychology Press, 2007).

24. Sherry Turkle, *The Second Self* (New York: Simon & Schuster, 1984) and *Life on the Screen* (New York: Simon & Schuster, 1995).

25. See also Kenneth Gergen, *The Saturated Self* (New York: Basic Books, 1991), Gergen, "Self and Community in the New Floating Worlds," in *Mobile Democracy*, ed. Kristof Nyiri, 103–114 (Vienna: Passagen, 2002).

26. Jay David Bolter and Richard Grusin, *Remediation* (Cambridge, MA: MIT Press, 1999), quotation p. 232. Note that despite its title, Zizi Papacharissi's edited work *A Networked Self* (London: Routledge, 2010) is more about online social networks than about the self.

27. Amanda Lenhart, Sousan Arafeh, Aaron Smith, and Alexandra Macgill, "Writing, Technology and Teens," Pew Internet Project, April 24, 2008, http://www.pewinternet .org/Reports/2008/Writing-Technology-and-Teens.aspx.

28. Marshall McLuhan and Bruce Powers, *The Global Village* (Oxford: Oxford University Press, 1989).

29. Pope Benedict XVI, "Message for the World Communication Day," January 24, 2011, http://www.vatican.va/holy_father/benedict_xvi/messages/communications/ documents/hf_ben-xvi_mes_20110124_45th-world-communications-day_en.html.

30. Juan-Antonio Carrasco, Barry Wellman, and Eric Miller, "How Far—and with Whom—Do People Socialize?" *Transportation Research Record* 2076 (2008): 114–122; Diana Mok, Barry Wellman, and Juan Carrasco, "Does Distance Matter in the Age of the Internet?" *Urban* Studies 47, no. 13 (2010): 2747–2783; Steven Farber and Antonio Paez, "Running to Stay in Place," *Journal of Transport Geography* 19, no. 4 (2011): 782–793; Lauren Dugan, "Survey: Twitter Users Are More Active in Their Real Life Communities," July 15, 2010.

31. Hua Wang and Barry Wellman, "Social Connectivity in America," *American Behavioral Scientist* 53, no. 8 (2010): 1148–1169; Jeffrey Boase, John Horrigan, Barry Wellman, and Lee Rainie, "The Strength of Internet Ties," Pew Internet & American Life Project, 2006, http://www.pewinternet.org/Reports/2006/The-Strength-of -Internet-Ties.aspx.

32. Nancy Baym and Andrew Ledbetter, "Tunes that Bind?" *Information, Communication & Society* 12, no. 3 (2009): 408–427; Keith Hampton, Lauren Sessions Goulet, Lee Rainie, and Kristen Purcell, "Social Networking Sites and Our Lives," Pew Internet & American Life Project, June 2011, http://www.pewinternet.org/Reports/2011/ Technology-and-social-networks/Summary.aspx; Randy Lynn and James Witte, "Social Network Sites, Social Ties, and Social Capital," paper presented to American Sociological Association, Las Vegas, August, 2011; Keith Hampton, Chul-joo Lee, and Eun Ja Her, "How New Media Affords Network Diversity," *New Media & Society* 13 (2011); Keith Hampton, "Comparing Bonding and Bridging Ties for Democratic

Engagement," *Information, Communication & Society* 14, no. 4 (2011): 510–528; Jennifer Earl and Katrina Kimport, *Digitally Enabled Social Change* (Cambridge, MA: MIT Press, 2011); Stefan Bauernschuster, Oliver Falck, and Ludger Woessmann, "Surfing Alone? The Internet and Social Capital," CESifo Working Paper 3469, Ifo Institute, Munich, August, http://www.cesifo-group.de/portal/pls/portal/docs/1/1205909.PDF.

33. Personal email to coauthor Wellman, February 5, 2010.

34. Wang and Wellman, "Social Connectivity," see note 31.

35. Nardi's quotation is from http://darrouzet-nardi.net/bonnie/. Her research is reported in her work *My Life as a Night Elf Priest* (Ann Arbor: University of Michigan Press, 2010). See also William Bainbridge, *The Warcraft Civilization* (Cambridge, MA: MIT Press 2010); Eleni Stroulia, "Smart Services across the Real and Virtual Worlds," in *The Smart Internet,* ed. James Cordy, Mark Chignell, Joanna Ng, and Yelena Yesha, 178–196 (Berlin: Springer, 2010).

36. Keith Hampton and Barry Wellman, "Neighboring in Netville," *City and Community* 2, no. 3 (2003): 277–311.

37. Keith Hampton, "Neighborhoods in the Network Society," *Information, Communication and Society* 10, no. 5 (2007): 714–748.

38. Aaron Smith, "Neighbors Online," Pew Internet & American Life Project, Washington, DC, June 2010, http://www.pewinternet.org/Reports/2010/Neighbors-Online.aspx.

39. Robin Dunbar, email communication to coauthor Wellman, March 30, 2010.

40. Robin Dunbar, "How Many Friends Do You Need? Dunbar's Number," *BBC News*, February 4, 2010, http://news.bbc.co.uk/today/hi/today/newsid_8497000/8497541.stm. See also Dunbar, "The Social Brain Hypothesis," *Evolutionary Anthropology* 6, no. 5 (1998): 178–190; Robin Dunbar, Clive Gamble, and John Gowlett, eds., *Social Brain, Distributed Mind* (Oxford: Oxford University Press, 2010), especially Sam Roberts, "Constraints on Networks," pp. 115–134. See also Sam Roberts, Robin Dunbar, Thomas Pollet, and Toon Kuppens, "Exploring Variation in Active Network Size," *Social Networks* 31 (2009): 138–146; Alistair Sutcliffe, Robin Dunbar, Jens Binder, and Holly Arrow, "Relationships and the Social Brain," *British Journal of Psychology* 102, no. 4 (2011).

41. For an example of computer scientists and media coverage of Dunbar's number, see Chris Gourlay, "OMG: Brains Can't Handle All Our Facebook Friends," [London] *Sunday Times*, January 24, 2010, http://technology.timesonline.co.uk/tol/news/tech_and_web/the_web/article6999879.ece.

42. Putnam, *Bowling Alone*, see note 16.

43. John Cacioppo and William Patrick, *Loneliness* (New York: Norton, 2008).

44. Thomas Valente, *Social Networks and Health* (New York: Oxford University Press, 2010).

45. Keith Hampton and Barry Wellman, "The Not So Global Village of Netville," in *The Internet in Everyday Life,* ed. Barry Wellman and Caroline Haythornthwaite, 345–371 (Oxford: Blackwell, 2002).

46. Jeffrey Boase and Barry Wellman, "A Plague of Viruses," *Current Sociology* 49, no. 6 (2001): 39–55; Emanuel Rosen, *The Anatomy of Buzz Revisited* (New York: Broadway Business, 2009).

47. Mark Granovetter, "The Strength of Weak Ties: A Network Theory Revisited," *Sociological Theory* 1 (1983): 201–233.

48. Émile Durkheim, *The Division of Labor in Society* (New York: Macmillan, [1893] 1993).

49. Lijun Song, Joonmo Son, and Nan Lin, "Social Support," in *Sage Handbook of Social Network Analysis,* ed. John Scott and Peter Carrington (London: Sage, 2011), 116–128; Lijun Song, Joonmo Son, and Nan Lin, "Social Capital and Health," in *The New Companion to Medical Sociology,* ed. William Cockerham (Oxford: Wiley-Blackwell, 2010), 184–210; Lijun Song, "Social Capital and Psychological Distress," *Journal of Health and Social Behavior* (forthcoming); Sheldon Cohen, William Doyle, David Skoner, Bruce S. Rabin, and Jack Gwaltney, "Social Ties and Susceptibility to the Common Cold," *Journal of the American Medical Association* 277 (1997): 1940–1944; Nicholas Christakis and James Fowler, *Connected* (New York: Little, Brown, 2009).

50. Jeffrey Boase and Barry Wellman, "Personal Relationships: On and Off the Internet," in *Cambridge Handbook of Personal Relationships*, ed. Anita Vangelisti and Daniel Perlman, 709–723 (Cambridge: Cambridge University Press, 2006); Boase, Horrigan, Wellman, and Rainie, "The Strength of Internet Ties," see note 31.

51. Barry Wellman and Milena Gulia, "A Network Is More Than the Sum of Its Ties," in *Networks in the Global Village*, ed. Barry Wellman, 83–118 (Boulder CO: Westview, 1999).

52. H. Russell Bernard, Peter Killworth, Eugene Johnsen, Gene Shelley, and Christopher McCarty, "Estimating the Ripple Effect of a Disaster," *Connections* 24, no. 2 (2001): 18–22.

53. Tyler McCormick, Matthew Salganik, and Tian Zheng, "How Many People Do You Know? Efficiently Estimating Personal Network Size," *Journal of the American Statistical Association* 105, no. 489 (2010): 59–70; Tian Zheng, Matthew Salganik, and Andrew Gelman, "How Many People Do You Know in Prison?" *Journal of the American Statistical Association* 101, no. 474 (2006): 409–423. See also Thomas DiPrete, Andrew Gelman, Tyler McCormack, Julien Teitler, and Tian Zhang, "Segregation in Social Networks Based on Acquaintanceship and Trust," *American Journal of Sociology* 116, no. 4 (2011): 1234–1283.

54. Hampton, Sessions, Rainie, and Purcell. "Social Networking Sites and Our Lives," see note 32.

55. Melinda Blau and Karen Fingerman, *Consequential Strangers* (New York: Norton, 2009).

56. Jeremy Boissevain, *Friends of Friends* (Oxford: Blackwell, 1974).

57. Raymond Firth, Jane Hubert, and Anthony Forge, *Families and Their Relatives* (London: Routledge, 1969).

58. Boase, Horrigan, Wellman, and Rainie. "The Strength of Internet Ties," see note 31.

59. Melissa Godbout, Tracy Kennedy, Barry Wellman, and Yu Janice Zhang, "The Colors of Closeness," International Sunbelt Social Network Conference, St. Pete Beach, FL, February 2011.

60. Wellman, Carrington, and Hall, "Networks as Personal Communities," see note 18.

61. Barry Wellman, "The Place of Kinfolk in Personal Community Networks," *Marriage and Family Review* 15, no. 1/2 (1990): 195–228; Wellman and Wortley, "Different Strokes," see note 17; Jacob Habinek, John Levi Martin, and Benjamin Zablocki, "Long Term Persistence and Re-Formation of Close Personal Ties," Working Paper, University of Chicago Department of Sociology, April 2010.

62. Beverly Wellman and Barry Wellman, "Domestic Affairs and Network Relations," *Journal of Social and Personal Relationships* 9 (1992): 385–409.

63. Fischer, *To Dwell among Friends*, see note 17.

64. Rochelle Côté, Gabriele Plickert, and Barry Wellman, "Does the Golden Rule, Rule?" in *Contexts of Social Capital*, ed. Ray-May Hsung, Nan Lin and Ronald Breiger, 49–71 (London: Routledge, 2009).

65. "Oxford Word of the Year 2009: Unfriend," *OUPblog*, http://blog.oup.com/2009/11/unfriend; Chirstopher Sibona and Steven Walczak, "Unfriending on Facebook," *Proceedings of the 2011 Hawaii International Conference on System Science* (Washington, DC: IEEE Computer Society, 2011), http://dx.doi.org/10.1109/HICSS.2011.467.

66. Barry Wellman, Renita Wong, David Tindall, and Nancy Nazer, "A Decade of Network Change," *Social Networks* 19 (1997): 27–51; Niles Eldredge and Stephen Gould, "Punctuated Equilibria" in *Models in Paleobiology*, ed. Thomas Schopf, 82–115 (San Francisco: Freeman, Cooper, 1972).

67. While we discuss substantive matters here, see also the methodological discussions in Claude Fischer, "The 2004 GSS Finding of Shrunken Social Networks," *American Sociological Review* 74 (2009): 657–669; Miller McPherson, Lynn Smith-Lovin, and Matthew Brashears, "Models and Marginals," *American Sociological Review* 74 (2009): 670–681; Peter Marsden and Karen Campbell, "Measuring Tie Strength," *Social Forces* 63 (1984): 482–450.

68. Peter Bearman and Paolo Parigi, "Cloning Headless Frogs and Other Important Matters," *Social Forces* 83, no. 2 (2004): 535–557.

69. Fischer, *To Dwell among Friends*, note 17.

70. The "target" closeness question started with: "Here is the sheet where we will draw your social network. It will look something like this when it is done. [Show them the example sheet.] Start with the very close names. Put the people who know each other closer together, and put the people who you feel closest to nearest to you." The second set of closeness questions began this way: "In the survey, we asked you about people who are Very Close and Somewhat Close to you. VERY CLOSE: Discuss important matters with, or regularly keep in touch with, or there for you if you need help. This is our Name Template. On each of the little strips, you will be able to write down the names of people you know. Okay, now think of people who fit that 'Very Close'

description. Please write down all the names of the people you feel very close. Please do not include people who you live with." For further information, see the interview and survey instruments themselves: http://homes.chass.utoronto.ca/~wellman/cgi-bin/counter.php?url=http://chass.utoronto.ca/~wellman/publications/ConnectedLives/InterviewGuide.htm&f=InterviewGuidea&mode=1. For more information, see: Barry Wellman, Bernie Hogan, Kristen Berg, Jeffrey Boase, Juan-Antonio Carrasco, Rochelle Côté, Jennifer Kayahara, Tracy Kennedy, and Phuoc Tran, "Connected Lives: The Project," in *Networked Neighborhoods,* ed. Patrick Purcell (Guildford , UK: Springer, 2006), 157–211. The study only looked at relationships outside of the household and not with spouses or others living at home.

71. McPherson, Smith-Lovin, and Brashears, "Social Isolation," see note 2.

72. As does a recent Pew Internet survey reported in: Keith Hampton, Lauren Sessions, and Eun Ja Her, "Core Networks and New Media," *Information Communication and Society* 14, no. 1 (2011): 130–155.

73. Email to coauthor Wellman, October 2, 2009.

74. Barry Wellman, "Domestic Work, Paid Work and Net Work," in *Understanding Personal Relationships,* ed. Steve Duck and Daniel Perlman (London: Sage, 1985), 159–191.

75. Gwen Moore, "Structural Determinants of Men's and Women's Personal Networks," *American Sociological Review* 55 (1990): 726–735; Arlie Russell Hochschild, *The Time Bind* (New York: Metropolitan, 1997); Wellman, "Domestic Work," see note 74.

76. Eszter Hargittai and Yu-li Patrick Hsieh, "Predictors and Consequences of Differentiated Practices on Social Network Sites," *Information, Communication and Society* 13, no. 4 (2010): 515–536.

77. Jon Kleinberg, "The Human Texture of Information," World Question Center, 2010, http://edge.org/q2010/q10_9.html#kleinberg.

78. Mary Madden and Kathryn Zickuhr. "65% of Online Adults Use Social Networking Sites." Pew Internet & American Life Project survey. August 2011, http://pewinternet.org/Reports/2011/Social-Networking-Sites.aspx.

79. Hampton, Sessions, Rainie, and Purcell, "Social Networking Sites and Our Lifes," see note 32.

80. KABC News Los Angeles, "Nielsen: Google Most Visited Site, But More Time Spent on Facebook," April 14, 2011, http://abclocal.go.com/kabc/story?section=news/consumer&id=8072533.

81. David Kirkpatrick, *The Facebook Effect* (New York: Simon & Schuster, 2010), p. 199.

82. Jacob Habinek, John Levi Martin, and Benjamin Zablocki, "Long Term Persistence and Re-Formation of Close Personal Ties," Working Paper, Department of Sociology, University of Chicago, 2010; Bernie Hogan, Nai Li, and William Dutton, "A Global Shift in the Social Relationships of Networked Individuals," Working Paper, Oxford University, Oxford Internet Institute, February 14, 2010, http://papers.ssrn.com/sol3/papers.cfm?abstract_id=1763884.

83. For Facebook's growing international presence, see Miguel Helft, "Friending the World," *New York Times,* July 8 2010; Ben Parr, "Facebook Is the Web's Ultimate

Timesink," February 16, 2010, http://mashable.com/2010/02/16/facebook-nielsen -stats.

84. Scott Golder and Sarita Yardi, "Structural Predictors of Tie Formation in Twitter," *Proceedings of the Second IEEE International Conference on Social Computing*, August 20–22, 2010, Minneapolis, MN; Nicole Ellison, Cliff Lampe, Charles Steinfield, and Jessica Vitak, "With a Little Help from My Friends," in *A Networked Self*, ed. Zizi Papacharissi (London: Routledge, 2010), 124–145; danah boyd, "Why Youth (Heart) Social Network Sites," in *Youth, Identity and Digital Media*, ed. David Buckingham (Cambridge, MA: MIT Press, 2007), 119–142.

85. Leopoldina Fortunati, James Katz, and Raimonda Riccini, *Mediating the Human Body* (Mahwah, NJ: Erlbaum, 2003).

86. Hampton, Sessions, Her, and Rainie, "Social Isolation and New Technology," see note 11; John Robinson and Steven Marin, "Of Time and Television," *Annals of Political and Social Sciences* 625, no. 1 (2009): 74–86.

87. Nicole Ellison, Charles Steinfield, and Cliff Lampe, "The Benefits of Facebook 'Friends,'" *Journal of Computer Mediated Communication* 12, no. 4 (2007): article 1, http://onlinelibrary.wiley.com/doi/10.1111/j.1083-6101.2007.00367.x/pdf; Moira Burke, Cameron Marlow, and Thomas Lento, "Social Network Activity and Social Well-Being," paper presented to the *CHI conference* (International Conference on Human Factors in Computing Systems), Atlanta, April 2010. Michael Stern, Jessica Collins, and Barry Wellman, ed., "The Internet in Rural North American Life," *American Behavioral Scientist* 53, no. 9 (2010): 1344–1366.

88. Irina Shklovski, Sara Kiesler, and Robert Kraut, "The Internet and Social Interaction," in *Computers, Phones and the Internet*, ed. Robert Kraut, Malcolm Brynin, and Sara Kiesler (New York: Oxford University Press, 2006), 250–264; Hampton, Sessions, and Eun Ja, "Core Networks," see note 71; Wang and Wellman, "Social Connectivity," see note 31.

89. Keith Hampton, Chul-joo Lee, and Eun Ja Her, "How New Media Affords Network Diversity," *New Media & Society* 13, preprint doi: 10.1177/1461444810390342.

90. Stephen Rains and Valerie Young, "A Meta-Analysis of Research on Formal Computer-Mediated Support Groups," *Human Communication Research* 35 (2009): 309–336; Boase, Horrigan, and Wellman, "The Strength of Internet Ties," see note 31; Jeffrey Boase, "America Online and Offline," Ph.D. dissertation, Department of Sociology, University of Toronto, 2006; Lijun Song, Joonmo Son, and Nan Lin, "Social Support," in *Sage Handbook of Social Network Analysis*, ed. John Scott and Peter Carrington (London: Sage, 2011), 116–128; Kristen Berg, "Health Management in the Age of the Internet," PhD dissertation, Faculty of Social Work, University of Toronto, 2011.

91. Susannah Fox, "Peer to Peer Healthcare," Pew Internet & American Life Project, Feb. 28, 2011, http://www.pewinternet.org/Reports/2011/P2PHealthcare. aspx, p. 2.

92. Quoted by Claire Cain Miller, "Social Networks a Lifeline for the Chronically Ill," *New York Times*, March 24, 2010, http://www.nytimes.com/2010/03/25/ technology/25disable.html.

93. Patti Valkenburg and Jochen Peter, "Online Communication and Adolescent Well-Being," *Journal of Computer-Mediated Communication* 12, no. 4 (2007): article 2, http://jcmc.indiana.edu/vol12/issue4/valkenburg.html.

94. Jessica Vitak and Nicole Ellison, "There's a Network Out There That You Might as Well Tap," paper presented at the National Communication Association, San Francisco, November 2010.

95. Juan-Antonio Carrasco, Eric Miller, and Barry Wellman, "How Far and with Whom Do People Socialize?" *Transportation Research Record* 2076, (2008): 114–122.

Chapter 6

1. Tracy L. M. Kennedy coauthored this chapter. We appreciate the advice and assistance we received from Mohammad Haque, Arlie Hochschild, Maria Majerski, Melissa Milkie, and Yu Janice Zhang.

2. Hillary Rodham Clinton, *It Takes A Village* (New York: Simon & Schuster, 1996), p. 14.

3. Martin Turcotte, "Time Spent with Family during a Typical Work Day, 1986 to 2006," Statistics Canada, *Canadian Social Trends*, April 13, 2010; Melissa Milkie, Marybeth Mattingly, Kei Nomaguchi, Suzanne Bianchi, and John Robinson, "The Time Squeeze," *Journal of Marriage and Family* 66, no. 3 (2004): 739–761.

4. This is further described in Tracy L. M. Kennedy, "The Household Internet," PhD dissertation, Department of Sociology, University of Toronto, 2010.

5. Throughout this chapter we use the term "partner" instead of "married" to include those North Americans cohabiting without being formally married. (In Canada, they would often be called "living common-law.") We use "spouse" to refer to a cohabiting partner. Where our sources only report about formally married people, we use "married" instead of "partner."

6. Tom Edwards, "As Baby Boomers Age, Fewer Families Have Children under 18 at Home," U.S. Census Bureau, U.S. Department of Commerce, February 25, 2009, http://www.census.gov/newsroom/releases/archives/families_households/cb09-29 .html.

7. Figure 6.1: Household size data from Gretchen Livingstone and D'Vera Cohn, "Childlessness Up among All Women; Down among Women with Advanced Degrees," Pew Social & Demographic Trends, Pew Research Center, http://pewso cialtrends.org/pubs/758/rising-share-women-have-no-children-childlessness; Statistics Canada, Census 1971, Publication number 93–715, Ottawa; Statistics Canada, Census 1981: Profile of Census Subdivisions, Catalogue no. SDP81A10, Ottawa; Statistics Canada, Census 1981–1991: Profile of Census Subdivisions, Statistics Canada Catalogue no. 95F0072XCB, Ottawa; Statistics Canada, Census 1996: Profile of Federal Electoral District, Catalogue no. 95F0180XDB, Ottawa, 1998); Statistics Canada, Census 2001: Profile of Federal Electoral District, Catalogue no. 95F0495X-CB2001007, Ottawa, 2003; Statistics Canada, Census 2006: Profile of Federal Electoral District, Statistics Canada Catalogue no. 94–581-X2006007, Ottawa, 2008; U.S.

Census Bureau, Current Population Survey, March and Annual Social and Economic Supplements, 2008 and earlier, Table FM-3: http://www.census.gov/population/ www/socdemo/hh-fam.html (1960/1, 1970/1, etc., refers to data available for 1960 in the United States and for 1961 in Canada, up to 2005 for United States and 2006 for Canada).

8. Later marriage data are from D'Vera Cohn, "The States of Marriage and Divorce: Lots of Ex's Live in Texas," Pew Social & Demographic Trends. Pew Research Center, http://pewsocialtrends.org/2009/10/15/the-states-of-marriage-and-divorce/; Richard Fry and D'Vera Cohn, "Women, Men and the New Economics of Marriage," Pew Social & Demographic Trends, Pew Research Center, http://pewsocialtrends.org/ pubs/750/new-economics-of-marriage; Jason Fields, "America's Families and Living Arrangements: 2003," U.S. Census Bureau, U.S. Department of Commerce, Economics and Statistics, November 2004, http://www.census.gov/prod/2004pubs/p20-553 .pdf.

9. Divorce rate data from "Divorces," *The Daily*, March 9, 2005, http://www.statcan .gc.ca/daily-quotidien/050309/dq050309b-eng.htm; R. D. Fraser, "Section B: Vital Statistics and Health," Series B75-81: Number of marriages and rate, average age at marriages, number of divorces and rate, net family formation, Canada, 1921 to 1947, *Historical Statistics of Canada*, Statistics Canada Catalogue no. 11–516-X Ottawa, http://www.statcan.gc.ca/pub/11-516-x/pdf/5500093-eng.pdf [1960–1970 source]; Statistics Canada, "Population Estimates," *Annual Demographic Statistics 2005*, Catalogue no. 91–213-XPB, Ottawa, 2006, http://www.chass.utoronto.ca./datalib/code books/dsp/ads2000.htm [1975–2000 population estimates]; Statistics Canada, *Table 053–0002—Vital statistics, divorces, annual (number)* (table), http://www40.statcan .gc.ca/l01/cst01/famil02-eng.htm; Statistics Canada, *Table 101–6501—Divorces and crude divorce rates, Canada, provinces and territories, annual* (table), http://estat.statcan .gc.ca/cgi-win/cnsmcgi.exe?Lang=E&EST-Fi=EStat/English/CII_1-eng.htm[2005 source]; U.S. Census Bureau, Statistical Abstract of the United States: 2009, based on U.S. National Center for Health Statistics, *Vital Statistics of the United States*, and *National Vital Statistics Report (NSVR): 1960–2006*, http://www.census.gov/ prod/2008pubs/09statab/vitstat.pdf. U.S. divorce rate data include reported annulments, but exclude rates from 2000 for California, Colorado, Indiana, and Louisiana, and from 2005 for California, Georgia, Hawaii, Indiana, Louisiana, and Minnesota.

Household stability data from Richard Fry and D'Vera Cohn, "Women, Men and the New Economics of Marriage," Pew Social & Demographic Trends, see note 8.

10. Doug Anderson and Radha Subramanyam, "The New Digital American Family," Nielsen Report, April 2011, New York. http://www.mobimatter.com/wp-content/ uploads/2011/04/New-Digital-American-Family.pdf.

11. Elizabeth Abbott, *A History of Marriage* (Toronto: Penguin, 2010); Andrew Cherlin, *The Marriage Go-Round* (New York: Random House, 2009). For a more alarmed view, see Paul Amato, Alan Booth, David Johnson, and Stacy Rogers, *Alone Together* (Cambridge, MA: Harvard University Press, 2007).

12. Dual-earner household data from Frank T. Denton, "Section D: The Labour Force," Series D8-85: Work force, by industrial category and sex, census years 1911 to 1971 (gainfully occupied 1911 to 1941, labour force 1951 to 1971), *Historical Statistics of Canada,* Statistics Canada Catalogue no. 11–516-X, Ottawa, section D, p. 12, http://www.statcan.gc.ca/pub/11-516-x/pdf/5500094-eng.pdf; Statistics Canada, "Population Estimates and Projections: Citizenship and Immigration 1921–1971," report 91–512 (1973): pp. 50 and 60; Statistics Canada, "Labour Force Historical Review, 2008," labor force estimates by detailed age groups, sex, Canada, province, annual averages, Statistics Canada Labour Statistics Division, Ottawa; Communication Canada, Depository Services Program, STC cat. 71F0004XCB, 2009, http://datalib.chass.utoronto.ca/inventory/3000/3833.htm; U.S. Census Bureau, Statistical Abstract of the United States: 2008, based on U.S. Bureau of Labor Statistics, Bulletin 2307 and Employment and Earnings, monthly, January 2008 issue: "Table 569. Civilian Population—Employment Status by Sex, Race, and Ethnicity," http://www.census.gov/prod/2007pubs/08abstract/labor.pdf. Authors' calculations for Canadian percentages based on number of employed individuals divided over population estimates.

Time use data from Liana C. Sayer, "More Work for Mothers?" in *Time Competition,* ed. Tanja van der Lippe and Pascalla Petters, pp. 41–56 (Northampton, MA: Elgar, 2007); Liana C. Sayer, "Gender, Time and Inequality: Trends in Women's and Men's Paid Work, Unpaid Work and Free Time," *Social Forces* 84, no. 1 (2005): 285–303.

13. Statistics Canada, Table 355-0001: Restaurant, caterer, and tavern statistics, monthly dollars, CANSIM last updated November 2007, http://www5.statcan.gc.ca/cansim/pick-choisir?lang=eng&searchTypeByValue=1&id=3550001; U.S. Census Bureau, "County Business Patterns," 1998–2007 (Table 1238—Service Related Industries).

14. Francine Blau, Marianne Ferber, and Anne Winkler, *Economics of Women, Men, and Work,* 5th ed. (Englewood Cliffs, NJ: Prentice Hall, 2006); U.S. Bureau of Labor Statistics, "Employment Characteristics of Families Summary," United States Department of Labor, Bureau of Labor Statistics, http://www.bls.gov/news.release/famee.nr0.htm.

15. Arlene Moscovitch, "Electronic Media and the Family," Ottawa: Vanier Institute of the Family, 1998; Martin Turcotte, "Time Spent with Family during a Typical Work Day, 1986 to 2006," *Canadian Social Trends* (Ottawa: Statistics Canada, 2008), http://www.statcan.gc.ca/pub/11-008-x/2006007/9574-eng.htm; Jerry Jacobs and Kathleen Gerson, "Overworked Individuals or Overworked Families?" *Work and Occupations* 28, no. 1 (2001): 40–63.

16. Liana Sayer, "Gender, Time and Inequality: Trends in Women's and Men's Paid Work, Unpaid Work and Free Time," *Social Forces* 84, no. 1 (2005): 285–303; Ben Veenhof, "The Internet: Is It Changing the Way Canadians Spend Their Time?" Statistics Canada, Science, Innovation and Electronic Information Division, 2006. Shira Offer and Barbara Schneider, "The Gender Gap in Time-Use Patterns," *American Sociological Review* 20 (2011): 809–833.

17. Glenn Stalker, "Change in the Social and Environmental Context of Canadian Leisure Time, 1986–1998." Presented to the International Association for Time Use Research, Halifax, Nova Scotia, November 2005; William Michelson, *From Sun to Sun* (New York: Oxford University Press, 2005); Heather Menzies, *No Time* (Vancouver: Douglas & McIntyre, 2005).

18. Barry Wellman and Beverly Wellman, "Domestic Affairs and Network Relations," *Journal of Social and Personal Relationships* 9 (1992): 385–409; Barry Wellman, "Domestic Work, Paid Work and Net Work," in *Understanding Personal Relationships,* ed. Steve Duck and Daniel Perlman (London: Sage, 1985), 159–191.

19. Leslie Haddon, "Explaining ICT Consumption," in *Consuming Technologies,* ed. Roger Silverstone and Eric Hirsch (Routledge, London), 82–96; Roger Silverstone and Leslie Haddon, "Design and the Domestication of Information and Communication Technologies," in *Communication by Design,* ed. Roger Silverstone and Robin Mansell (Oxford: Oxford University Press, 1996), 44–74. Environics Analytics and Delvinia, "New Database Reveals Social Media Habits Tied to Canadian Lifestyles," March 21, 2011, http://www.environicsanalytics.ca/media_room.aspx?tab =news&item=2011Mar21_Database.

20. Survey data from Pew Internet & American Life Project, August 2011, unpublished.

21. Stephen Blumberg and Julian Luke, "Wireless Substitution: Early Release of Estimates From the National Health Interview Survey, July-December 2010," Centers for Disease Control and Prevention, http://www.cdc.gov/nchs/data/nhis/earlyrelease/ wireless201106.pdf; Statistics Canada, "Household Internet Use Survey," *The Daily,* July 8, 2004, http://www.statcan.gc.ca/daily-quotidien/040708/dq040708a-eng .htm; Eric Newburger, "Home Computers and Internet Use in the United States: August 2000," *Current Population Reports,* U.S. Census Bureau, September 2001, http://www.census.gov/prod/2001pubs/p23-207.pdf.

22. Annenberg Center for the Digital Future, "Family Time Decreasing with Internet Use," University of Southern California, Los Angeles, 2009, http://uscnews.usc.edu/ digital_media/family_time_decreases_with_internet_use.html; *USA Today,* "Family Eroding in U.S. as Internet Use Soars," June 15 2009; Garey Ramey and Valerie Ramey, "The Rug Rat Race," National Bureau of Economic Research, Washington, DC, August 2009.

23. Melissa Milkie, Marybeth Mattingly, Kei Nomaguchi, Suzanne Bianchi, and John Robinson, "The Time Squeeze," *Journal of Marriage and Family* 66, no. 3 (2004): 739–761; Marybeth Mattingly and Suzanne Bianchi, "Gender Differences in the Quantity and Quality of Free Time," *Social Forces* 81, no. 3 (2003): 999–1030.

24. Pew Research Center, "62%—Crowded Tables This Thanksgiving," March 25, 2011, http://pewresearch.org/databank/dailynumber/?NumberID=1137.

25. Raelene Wilding, "'Virtual' Intimacies," *Global Networks* 6, no. 2 (2006): 125–142; Michael Stern and Chris Messer, "How Family Members Stay in Touch," *Marriage & Family Review* 45 (2009): 654–676.

26. Telus, "Technology Usage and Attitudes of the Consumer," August 2009, unpublished report.

27. Katie Hafner, "If the Kitchen's Warm, It May be the PC," *New York Times*, December 11, 2003, http://www.nytimes.com/2003/12/11/technology/if-the-kitchen-s-warm-it-may-be-the-pc.html.

28. Maria Lianos-Carbone, "Kids Who Love Technology: My Kid is Tech Savvy," *Amother World* blog, http://amotherworld.com/main/parenting/tech-savvy-kids, September 28, 2009.

29. Quoted in Hilary Stout, "Toddlers' Favorite Toy: The iPhone," *New York Times*, October 15, 2010, http://www.nytimes.com/2010/10/17/fashion/17TODDLERS. html; see also Avivia Lucas Gutnick, Michael Robb, Lori Takeuchi, and Jennifer Kotler, "Always Connected," New York: Joan Ganz Cooney Center, 201.

30. Telus, "Technology Usage," see note 26.

31. Amanda Lenhart, Kristen Purcell, Aaron Smith, and Kathryn Zickuhr, "Social Media & Young Adults," Pew Internet & American Life Project, February 3, 2010, http://www.pewinternet.org/Reports/2010/Social-Media-and-Young-Adults/Part-2/1-Cell-phones.aspx. See also the detailed ethnography of teens' online life in Mizuko Ito, Sonja Baumer, Matteo Bittanti, danah boyd, Rachel Cody, Becky Herr-Stephenson, Heather Horst, Patricia Lange, Dilan Mahendran, Katynka Martinez, C. J. Pascoe, Dan Perkel, Laura Robinson, Christo Sims, and Lisa Tripp, *Hanging Out, Messing Around, and Geeking Out* (Cambridge, MA: MIT Press, 2009).

32. Jill Dimond, Erika Shehan Poole, and Sarita Yardi, "The Effects of Life Disruptions on Home Technology Routines," Proceedings of the Group '10 Conference (New York: ACM, 2010), 85–88, http://dl.acm.org/citation.cfm?doid=1880071.1880085.

33. Tracy Kennedy and Barry Wellman, "The Networked Household," *Information, Communication & Society* 10, no. 5 (2007): 644–669.

34. Caroline Haythornthwaite and Barry Wellman, "Work, Friendship, and Media Use for Information Exchange in a Networked Organization," *Journal of the American Society for Information Science* 49 (1988): 1101–1014.

35. Margaret Nelson, *Parenting Out of Control* (New York: New York University Press, 2010).

36. Sarita Yardi and Amy Bruckman, "Social and Technical Challenges in Parenting Teens' Social Media Use," Presented to the *CHI 2011* conference, May 7–12, 2011, Vancouver, http://www.cc.gatech.edu/~yardi/pubs/Yardi_ParentsTechnology11.pdf.

37. Caroline Haythornthwaite and Richard Andrew, *E-Learning Theory and Practice* (London: Sage, 2011).

38. Hua Wang and Arvind Singhal, "Entertainment Education through Digital Games," in *Serious Games*, ed. Ute Ritterfeld, Michael Cody, and Peter Vorderer (London: Routledge, 2009), 271–291.

39. Sara Rimer, "Play with Your Food, Just Don't Text!" *The New York Times*, May 26 2009, http://www.nytimes.com/2009/05/27/dining/27text.html; Laura Holson, "Text Generation Gap: U R 2 Old (JK)," *The New York Times*, March 9, 2008, http://www.nytimes.com/2008/03/09/business/09cell.html; "Family Time Eroding in U.S. as Internet Use Soars," *USA Today*, June 15, 2009, http://www.usatoday.com/tech/news/2009-06-15-internet-family_N.htm.

40. Robert Paul Smith, *Where Did You Go? Out. What Did You Do? Nothing* (New York: Norton, 1957).

41. Sherry Turkle, *Alone Together* (New York: Basic Books, 2011).

42. Steve Collins, "There's No We in iPod," *Metro Toronto*, May 18, 2010, p. 21.

43. "No Texting @ Dinner! Parenting Digital Kids," *Associated Press*, October 1, 2009, http://today.msnbc.msn.com/id/33122598/ns/today-parenting_and_family.

44. Alexandra Macgill, "Parent and Teen Internet Use," Pew Internet & American Life Project, Oct. 24, 2007, http://pewinternet.org/Reports/2007/Parent-and-Teen -Internet-Use.aspx.

45. Margaret Nelson, *Parenting Out of Control* (New York: New York University Press, 2010); Robert Tokunaga, "Following You Home from School," *Computers in Human Behavior* 26 (2010): 277–287; Gustavo Mesch, "Parental Mediation, Online Activities, and Cyberbullying,." *Cyberpsychology and Behavior* 12, no. 4 (2009): 387–393.

46. Amanda Lenhart, "Social Networking Websites and Teens," Pew Internet & American Life Project, http://www.pewinternet.org/Reports/2007/Social-Networking -websites-and-Teens.aspx.

Chapter 7

1. Wenhong Chen coauthored this chapter. We appreciate the major assistance provided by Anna Brady and the advice and assistance provided by Julie Amoroso, Amy Bruckman, Joseph Cothrel, Maya Collum, Dimitrina Dimitrova, Richard Florida, Melissa Godbout, Jonathan Grudin, Mohammad Haque, Cliff Lampe, Zara Matheson, Stephen Perelgut, Anabel Quan-Haase, Kevin Stolarick, Charles Steinfield, and Justine Yu.

2. Aficionados of *Mad Men* will note the exception that proves the rule. One woman, "Peggy," does rise on merit from secretarial ranks to become a much-remarked-on account executive. See also William H. Whyte, Jr., *The Organization Man* (New York: Simon & Schuster, 1956).

3. Key statements about the nature of networked work and networked organizations include Manuel Castells, *The Rise of the Network Society*, 2nd ed. (Oxford: Blackwell, 2000); David Limerick and Bert Cunnington, *Managing the New Organisation*, 2nd ed. (London: Allen & Unwin, 2003); David Knoke, *Changing Organizations* (Boulder, CO: Westview Press, 2001); Andrew Hargadon, "Firms as Knowledge Brokers," *California Management Review* 40, no. 3 (1998): 209–227; Wayne Baker, *Networking Smart* (New York: McGraw-Hill, 1994); Rob Cross and Andrew Parker, *The Hidden Power of Networks* (Boston: Harvard Business School Press, 2004); Charles Heckscher and Paul Adler, eds., *The Firm as a Collaborative Community* (New York: Oxford University Press 2006); Peter Monge and Noshir Contractor, *Theories of Communication Networks* (New York: Oxford University Press 2003); Nitin Nohria and Robert Eccles, eds., *Networks and Organizations*, 2nd ed. (Boston: Harvard Business School Press, 1992); Andrew McAfee, *Enterprise 2.0.* (Boston: Harvard Business School Press, 2009); Rob Cross, Andrew Parker, and Lisa Sasson, eds., *Networks in the Knowledge*

Economy (New York: Oxford University Press, 2003); Katherine Chudoba, Eleanor Wynn, Mei Lu, and Mary Watson-Manheim, "How Virtual Are We?" *Information Systems Journal* 15, no. 4 (2005): 279–306; Manju Ahuja and Kathleen Carley, "Network Structure in Virtual Organizations," *Organization Science* 10, no. 6 (1999): 741–757; Michael Barrett, David Grant, and Nick Wailes, "ICT and Organizational Change," *Journal of Applied Behavioral Science* 42, no. 1 (2006): 16–22, see 17; Peter Monge and Noshir Contractor, *Theories of Communication Networks* (Oxford University Press, New York, 2003); Ronald Burt, *Structural Holes* (Cambridge, MA: Harvard University Press, 1992); and Lynn Wu, Ching-Yung Lin, Sinan Aral, and Erik Brynjolfsson, "Value of Social Network," presentation, Winter Information Systems Conference, Salt Lake City, Utah, February 2009, http://smallblue.research.ibm.com/projects/snvalue.

4. Richard Florida, *The Rise of the Creative Class* (New York: Basic Books, 2002), p. 8.

5. As far as we can tell, the concepts of "atom workers" and "bit workers" come from Nicholas Negroponte (former head of the MIT Media Lab) in the early 1990s. See Nicholas Negroponte, *Being Digital* (New York: Random House Vintage, 1995). See also Florida, *The Rise of the Creative Class*, see note 4.

6. Thomas Friedman, *The World Is Flat* (Vancouver: Douglas & McIntyre, 2007). See also John Friedmann, *The Prospect of Cities* (Minneapolis: University of Minnesota Press, 2002).

7. The quotation is from pp. 28–29 of Charles Heckscher and Paul Adler, "Towards Collaborative Community," in *The Firm as a Collaborative Community*, ed. Charles Heckscher and Paul Adler (New York: Oxford University Press, 2006), 11–105.

8. For a nice example of networks in both hierarchies and markets, see Thomas Gold, Doug Guthrie, and David Wank, eds., *Social Connections in China* (Cambridge, UK: Cambridge University Press, 2002).

9. On nineteenth-century ships see Allan Pred, *Urban Growth and the Circulation of Information* (Cambridge, MA: Harvard University Press, 1973).

10. For the history of computers see Paul Ceruzzi, *A History of Modern Computing*, 2nd ed. (Cambridge, MA: MIT Press, 2003).

11. Andrew McAfee, "IT Is Everywhere. Why?" *Harvard Business Review*, HBR Blog Network, April 29, 2010, http://blogs.hbr.org/hbr/mcafee/2010/04/it-is-everywhere-why.html; data on ICT use from Andrew McAfee, "Corporate America's Ongoing Love Affair with Geek Gear," *Harvard Business Review*, HBR Blog Network, April 16 2010, http://blogs.hbr.org/hbr/mcafee/2010/04/corporate-americas-ongoing-lov.html.

12. Lee Rainie, "Wired Workers: Who They Are, What They're Doing Online," Pew Internet & American Life Project, Current Population Survey (CPS) Reports, CPS September 2000, Detailed Tables—Computer and Internet Use in the United States, PPL-175, http://www.pewinternet.org/Reports/2000/Wired-Workers.aspx.

13. Data on hours of work at the computer per week from the University of Southern California's Center for the Digital Future's annual survey, headed by Jeff Cole and compiled by Andromeda Salvador.

14. Mary Madden and Sydney Jones, "Networked Workers," Pew Internet & American Life Project, September 24, 2008, http://www.pewinternet.org/Reports/2008/Networked-Workers.aspx.

15. Jacques Bughin and Michael Chui, "The Rise of the Networked Enterprise," *McKinsey Quarterly,* December 2010, http://www.mckinseyquarterly.com/Organization/Strategic_Organization/The_rise_of_the_networked_enterprise_Web_20_finds_its_payday_2716?pagenum=3.

16. Networked office design discussed by Malcolm Gladwell, "Designs for Working," in *Networks in the Knowledge Economy*, ed. Rob Cross, Andrew Parker, and Lisa Sasson, 180–189 (New York: Oxford University Press, 2003).

17. Benjamin Waber, Sinan Aral, Daniel Olguin Olguin, Lynn Wu, Erik Brynjolfsson, and Alex Pentland, "Sociometric Badges," March 17, 2011, http://ssrn.com/abstract=1789103.

18. Gary Olson and Judith Olson, "Mitigating the Effects of Distance on Collaborative Intellectual Work," *Economic Innovation and New Technologies* 12, no. 1 (2003): 27–42; Jim Suchan and Greg Hayzak, "The Communication Characteristics of Virtual Teams," *IEEE Transactions on Professional Communications* 44, no. 3 (2001): 174–186; Wu, Lin, Aral, and Brynjolfsson, "Value of Social Network," see note 3.

19. Dimitrina Dimitrova and Emmanuel Koku, "Managing Collaborative Research Networks," *International Journal of Virtual Communities and Social Networking* 20, 2 (2010): 1–22.

20. Neil Gandal, Charles King, and Marshall Van Alstyne, "The Social Network within a Management Recruiting Firm," *Review of Network Economics* 8, no. 4 (2009): 302–324. See also Ronald Burt, *Structural Holes* (Chicago: University of Chicago Press, 1992); Valdis Krebs, "Managing the 21st Century Organization," *IHRIM Journal* [International Association for Human Resources Information Management] 11, no. 4 (2007): 2–8; Arent Greve, Mario Benassi, and Arne Dag Sti, "Exploring the Contributions of Human and Social Capital to Productivity," *International Review of Sociology* 20, no. 1 (2010): 35–58.

21. Caroline Haythornthwaite and Barry Wellman, "Work, Friendship, and Media Use for Information Exchange in a Networked Organization," *Journal of the American Society for Information Science* 49, no. 12 (Oct. 1998): 1101–1114; Seungyoon Lee and Peter Monge, "The Consolidation of Multiplex Communication Networks in Organizational Communities," *Journal of Communication* 61, no. 4 (August 2011): 758–779.

22. Alexander Pentland quotation is from Sarah Underwood, "Making Sense of Real-Time Behavior," *Communications of the ACM* 53, no. 8 (2010): 17–18. See also Alexander Pentland, *Honest Signals* (Cambridge, MA: MIT Press, 2010).

23. Dimitrina Dimitrova, "Controlling Teleworkers," *New Technology, Work and Employment* 18, no. 3 (2010): 181–195.

24. Google's former CEO Schmidt's quotation is from Elizabeth Tenety and Andrea Useem, "On Leadership: Google CEO Eric Schmidt on Google Culture," *Washington Post*, May 5, 2009, http://www.washingtonpost.com/national/on-leadership

-google-ceo-eric-schmidt-on-workplace-culture/2011/07/15/gIQAajkPGI_video .html. The description of Google is from "The Google Culture," 2009, http://www .google.ca/intl/en/corporate/culture.html. See also Bernard Girard, *The Google Way* (San Francisco: No Starch Press, 2009).

25. Etienne Wenger, Richard McDermott, and William Snyder, *Cultivating Communities of Practice* (Boston: Harvard Business School Press, 2000). The quotation is from p. 142 in Arent Greve, "Social Networks and Creativity," in *Routledge Companion to Creativity,* ed. Tudor Rickards, Mark Runco, and Susan Moger (London: Routledge, 2008), 132–145. See also Arent Greve and Janet Salaff, "The Development of Corporate Social Capital in Complex Innovation Processes," *Research in the Sociology of Organizations* 18 (2001): 107–134.

26. Peter Monge and Noshir Contractor, *Theories of Communication Networks* (Oxford: Oxford University Press, 2003).

27. Ronald Burt, *Neighbor Networks* (Oxford: Oxford University Press, 2010).

28. Wu, Lin, Aral, and Brynjolfsson, "Value of Social Network," see note 3.

29. Bill McDermott is quoted by Simon Avery and Sinclair Stewart in "BlackBerry to Use SAP's Programs," *Toronto Globe and Mail,* May 3, 2008, p. B4.

30. Steven Poltrock and Gloria Mark, "Implementation of Data Conferencing in the Boeing Company," in *Implementing Collaboration Technologies in Industry*, ed. Bjørn Erik Munkvold (London: Springer, 2003), p. 157.

31. Peter Cappelli, Laurie Bassi, Harry Katz, David Knoke, Paul Osterman, and Michael Useem, *Change at Work* (New York: Oxford University Press, 1997).

32. David Knoke, *Changing Organizations* (Boulder, CO: Westview Press, 2001).

33. Mark Mortensen, Anita Williams Woolley, and Michael O'Leary, "Conditions Enabling Effective Multiple Team Membership," International Federation for Information Processing Report No. 236 (Laxenburg, Austria: 2007), 215–228.

34. Katherine Chudoba, Mei Lu, Mary Beth Watson-Manheim, and Eleanor Wynn, "How Virtual Are We?" *Information Systems Journal* 15 (2005): 279–306; Mei Lu, Mary Beth Watson-Manheim, Katherine Chudoba, and Eleanor Wynn, "Virtuality and Team Performance," *Journal of Global Information Technology Management* 9, no. 1 (2006): 4–23.

35. Email to coauthor Wellman, March 9, 2009, from an administrator who wishes to remain anonymous.

36. The Nungesser quote comes from Chris Kirkham, "As Oil Spill Cleanup Workers Toil, Officials' Frustration Mounts," *New Orleans Times-Picayune,* June 12, 2010.

37. The general U.S. data come from Madden and Jones, "Networked Workers," see note 14. See also teleworking statistics by occupation, reported in the Bureau of Labor Statistics, U.S. Department of Labor, "Work-At-Home Patterns by Occupation," *Issues in Labor Statistics*, 2009. Meta-analysis reported in Ravi Gajendran and David Harrison, "The Good, the Bad, and the Unknown about Telecommuting," *Journal of Applied Psychology* 92, no. 6 (2007): 1524–1541.

38. Eleni Stroulia, "email," *On Services, 2.0 and 3D* [blog], February 28, 2011, http:// ssrg.cs.ualberta.ca/blogs/WS-20-3D/2011/02/28/email.

39. Tracy Kennedy, Julie Amoroso, and Barry Wellman, "Can You Take It with You?" in *Mobile Communication*, ed. James Katz (Piscataway, NJ: Transaction, 2011), 191–210.

40. Scott Schieman and Marisa Young, "Is There a Downside to Schedule Control for the Work-Family Interface?" *Journal of Family Issues* 31, no. 10 (2010) 1391–1414; Scott Schieman and Marisa Young, "The Demands of Creative Work," *Social Science Research* 39 (2010): 246–259.

41. Sandy Ward interviewed by Mark Evans, "Homezilla CEO on Managing Remote Workers," *Toronto Globe and Mail*, February 12, 2010, http://www.theglobeandmail .com/news/technology/homezilla-ceo-on-managing-remote-workers/article 1464811.

42. The term "travel to trust" was coined by Charles Grantham and Larry Nichols, *The Digital Workplace* (New York: Wiley, 1993).

43. Sara Kiesler and Jonathon Cummings, "What Do We Know about Proximity and Distance in Work Groups?" in *Distributed Work*, eds. Pamela Hinds and Sara Kiesler (Cambridge, MA: MIT Press, 2002): 57–81.

44. More detailed information can be found in Wenhong Chen and Barry Wellman, "Net and Jet," *Information, Communication and Society* 12, no. 4 (2009): 525–547. A recent review is Ivan Light, "Transnational Entrepreneurs in an English-Speaking World," *Die Erde* 14, no. 1 (2010): 1–16.

45. Jennifer Gibbs, Dina Nekrassova, Svetlana Grushina, and Sally Abdul Wahab, "Reconceptualizing Virtual Teaming from a Constitutive Perspective," *Communication Yearbook* 32 (2008): 187–229; Jennifer Gibbs, "Dialectics in a Global Software Team," *Human Relations* 62, no. 6 (2009): 905–935.

46. Arvind Malhotra and Ann Majchrzak, "Enabling Knowledge Creation in Far-Flung Teams: Best Practices for IT Support and Knowledge Sharing," *Journal of Knowledge Management* 8, no. 4 (2004): 75–88; Gloria Mark and Steven Poltrock, "Diffusion of a Collaborative Technology Across Distance," Proceedings of Group '01 Conference on Supporting Group Work (New York: Association for Computing Machinery Press, 2001), http://www.ics.uci.edu/~gmark/Group01.pdf; Steven Poltrock and George Engelbeck, "Requirements for a Virtual Collocation Environment," *Information & Software Technology* 41, no. 6 (1999): 331–339.

47. Dominic Gates, "Boeing 787: Parts from around World Will Be Swiftly Integrated," *Seattle Times* September 11, 2005, http://seattletimes.nwsource.com/html/ businesstechnology/2002486348_787global11.html.

48. J. Lynn Lunsford, "Fastener Woes to Delay Flight of First Boeing 787 Jets," *Wall Street Journal*, November 5, 2008; Wendy Kaufman, "Woes Mount for Boeing's Much-Awaited Dreamliner," *NPR Morning Edition*, December 9, 2010, http://www.npr .org/2010/12/09/131856033/woes-mount-for-boeing-s-much-awaited-dreamliner; Laura Myers and Kyle Peterson, "Boeing's Game-Changing Dreamliner Gets Green Light." See also "Boeing's Dreamliner Becomes Commercial Reality," Reuters, August 26, 2011, http://www.reuters.com/article/2011/08/26/us-boeing-idUSTRE77P0VS201 10826.

49. The quotation is from Lu, Watson-Manheim, Chudoba, and Wynn, "Virtuality and Team Performance," p. 16, see note 34.

50. Quotation from Irving Wladawsky-Berger, "The Entrepreneurial Society," blog entry, April 20, 2009, http://blog.irvingwb.com/blog/2009/04/the-entrepreneurial -society.html.

Chapter 8

1. Justine Yu and Xiaolin Zhuo coauthored this chapter. Brian Keegan and "Wiki-waye" contributed to the writing.
2. For Peter Maranci's posts, visit his blog *Charlie on the Commuter Rail*, http:// charlieonthecommuterrail.blogspot.com.
3. Pew Internet & American Life Project, "Trend Data," September 1, 2011, http:// www.pewinternet.org/Static-Pages/Trend-Data/Online-Activites-Total.aspx.
4. Peter Maranci, "The Diary of a Simple Man—Separated At Birth?" November 16, 2003, http://bobquasit.livejournal.com/32952.html.
5. Abby Philipp, "Obama Takes Questions on YouTube," January 27, 2011, http:// www.politico.com/news/stories/0111/48324.html.
6. "President Obama Takes Questions from YouTube," January 27, 2011, http://www .huffingtonpost.com/2011/01/27/president-obama-youtube-live-video_n_814955 .html. Jeff Zeleny, "Obama Takes Questions on YouTube," *The Caucus* blog, *New York Times,* February 1, 2010, http://thecaucus.blogs.nytimes.com/2010/02/01/obama -takes-questions-on-youtube
7. David Gauntlett, *Making Is Connecting* (Cambridge: Polity Press, 2011), quotation on p. 5.
8. Sebastien Paquet, "Making Group-Forming Ridiculously Easy," *Seb's Open Research* blog, October 9, 2002, http://radio-weblogs.com/0110772/2002/10/09.html.
9. Douglas Rushkoff, "Rise of the Amateur," May 2010, http://www.mpiweb .org/Magazine/Archive/US/May2010/RiseOfTheAmateur.
10. "Wikipedia: Size Comparisons," March 31, 2011, http://en.wikipedia.org/wiki/ Wikipedia:Size_comparisons. There are also Wikipedias in 278 other languages: http://en.wikipedia.org/wiki/Wikipedia#Language_editions.
11. Aaron Swartz, "Who Writes Wikipedia?" September 4, 2006, http://www.aaronsw .com/weblog/whowriteswikipedia. The most detailed description of Wikipedia's ori-gins and policies is in Joseph Reagle, *Good Faith Collaboration* (Cambridge, MA: MIT Press, 2010).
12. Brian Keegan, Darren Gergle, and Noshir Contractor, "A Multi-theoretical, Multi-level Model of High Tempo Collaboration in an Online Community," paper presented at the International Sunbelt Social Network Conference, St. Pete Beach, FL, February 2011, quotation on p. 1.
13. UTC stands for Universal Coordinated Time in French ("Temps Universel Coor-donné"), which is the worldwide time stamp that Wikipedia uses.
14. We thank communication doctoral student Brian Keegan for preparing figure 8.1 and making it available to us, and Mark Smith for writing the NodeXL software used to analyze the data. The Girvan-Newman algorithm was used to group the

nodes into coherent communities. The figure is available in color at: http://www
.flickr.com/photos/7371117@N05/sets/72157626529369758. For further description,
see Brian Keegan, Darren Gergle, and Noshir Contractor, "Hot off the Wiki," SONIC
Lab, School of Communication, Northwestern University, Evanston, IL, March 2011.
For the animated version, "100 Hours of Wikipedia Activity for Sendai Earthquake,"
April 6, 2011, see http://vimeo.com/21088958.

15. Ronald Breiger, "The Duality of Persons and Groups," *Social Forces* 53 (1974):
181–190.

16. Many Wikipedians use aliases to avoid spillovers between Wikipedia conflicts
and the rest of their lives. Disclaimer: Rainie's Pew Internet Project is nonpartisan
and takes no positions on policy and political matters of any kind.

17. Bill Tancer, "Look Who's Using Wikipedia," *Time*, March 1, 2007, http://www
.time.com/time/business/article/0,8599,1595184,00.html; Lev Grossman, "You—Yes
You—Are TIME's Person of the Year," *Time*, December 25, 2006, http://www.time
.com/time/magazine/article/0,9171,1570810,0.html.

18. Much of the information in this section is drawn from the material gathered
at the Library of Congress's Flickr Project at http://www.loc.gov/rr/print/flickr_pilot
.html; 1,600 color photographs were from its Farm Security Administration/Office of
War Information (FSA/OWI) collection and another 1,500 black-and-white photos
were from its George Grantham Bain News Service Collection. All photos can be
accessed at http://www.flickr.com/photos/library_of_congress.

19. Michelle Springer, Beth Dulabahn, Phil Michel, Barbara Natanson, David Reser,
David Woodward, and Helena Zinkham, "For the Common Good: The Library of
Congress Flickr Pilot Project," October 30, 2008, http://www.loc.gov/rr/print/flickr
_report_final.pdf, quotation on p. iv.

20. Flickr Commons Initiative is accessible at http://www.flickr.com/commons. To
view the different tools that have been created, visit http://www.indicommons.org/
tools.

21. For a timeline of the Middle Eastern revolutions of the first quarter of 2011,
see http://www.guardian.co.uk/world/interactive/2011/mar/22/middle-east-protest
-interactive-timeline. Here are links to some key webpages: April 6 Youth Movement,
http://www.facebook.com/april6youth; We Are All Khaled Said, http://www.face
book.com/elshaheeed.co.uk. The Progressive Youth of Tunisia webpage is no longer
available (September 5, 2011).

22. Mona Eltahawy, "Generation Facebook Creating Egyptians' Political Party of
the Internet," Aspen Institute, 2010, http://www.aspeninstitute.org/policy-work/
communications-society/programs-topic/journalism/arab-us-media-forum/dead
-sea-scrolling/generation-.

23. David Kirkpatrick and David Sanger, "A Tunisian-Egyptian Link That Shook Arab
History," *New York Times*, February 14, 2011, http://www.nytimes.com/2011/02/14/
world/middleeast/14egypt-tunisia-protests.html.

24. Dusan Stojanovic, "Serbian Ousters of Milosevic Make Mark in Egypt," *Newsvine*,
February 22, 2011, http://www.newsvine.com/_news/2011/02/22/6104771-serbian

-ousters-of-milosevic-make-mark-in-egypt; Kirkpatrick and Sanger, "A Tunisian-Egyptian Link," see note 23.

25. Kirkpatrick and Sanger, "A Tunisian-Egyptian Link," note 25.

26. "Egypt Protestor Clash with Police," Aljazeera English, January 25, 2011, http://english.aljazeera.net/news/middleeast/2011/01/201112511362207742.html; Kirkpatrick and Sanger, "A Tunisian-Egyptian Link," see note 25.

27. Wael Ghonim, "Inside the Egyptian Revolution," TED Conference, Cairo, Egypt, March 2011, http://www.ted.com/talks/wael_ghonim_inside_the_egyptian_revolution.html.

28. Kirkpatrick and Sanger, "A Tunisian-Egyptian Link," see note 25.

29. Charles Levinson and Margaret Coker, "The Secret Rally That Sparked an Uprising," Wall Street Journal, February 11, 2011, http://online.wsj.com/article/SB1000142405274870413220457613588235653270 2.html.

30. Mahmoud Salem's quotation from John Sutter, "The faces of Egypt's 'Revolution 2.0,'" CNN, February 21, 2011, http://www.cnn.com/2011/TECH/innovation/02/21/egypt.internet.revolution/index.html.

31. "Defiant Al-Jazeera Asks Egypt Audience for Help," CBS News, January 31, 2011, http://www.cbsnews.com/stories/2011/01/31/world/main7300870.shtml?tag=stack.

32. Doug McAdam, Sidney Tarrow, and Charles Tilly, Dynamics of Contention (Cambridge: Cambridge University Press, 2001). See also Jennifer Earl and Katrina Kimport, Digitally Enabled Social Change. (Cambridge, MA: MIT Press, 2011); Philip Howard, The Digital Origins of Dictatorship and Democracy. (New York: Oxford University Press, 2010); Barry Wellman and Christena Nippert-Eng, eds., The Contentious Internet, special section of Information, Community & Society 13, no. 2 (2010): 151–225.

33. Kirkpatrick and Sanger, "A Tunisian-Egyptian Link," see note 23.

34. Marian Wong / ProPublica, "Where Does the U.S. Money to Egypt Go—And Who Decides How It's Spent?" Seattle PostGlobe, January 31, 2011, http://seattlepostglobe.org/2011/02/01/where-does-the-us-money-to-egypt-goand-who-decides-how-its-spent.

35. See http://www.journalism.org.

36. PEJ New Media Index (NMI) methodology described at http://www.journalism.org/index_report/social_media_aid_haiti_relief_effort, at the bottom of the page. From January through June 2010, PEJ used both Icerocket and Technorati.

37. The empirical material in this content comes from an analysis of a year's worth of News Coverage Index findings and New Media Index findings done by the Pew Research Center's Project for Excellence in Journalism called "New Media, Old Media: How Blogs and Social Media Agendas Relate and Differ from the Traditional Press," May 23, 2010, http://www.journalism.org/analysis_report/new_media_old_media.

38. See details of PEJ's News Coverage Index for this week at http://www.journalism.org/index_report/pej_news_coverage_index_march_30_april_5_2009.

39. See details of PEJ's New Media Index for this week at: http://www.journalism.org/index_report/bloggers_focus_april_fools%E2%80%99_joke_interrogation_techniques_and_outspoken_actress.

40. http://www.thesarahpalinblog.com (this is not Sarah Palin's official blog).

41. See, for instance: Aaron Smith, "The Internet and Campaign 2010," Pew Internet & American Life Project, March 17, 2011, http://pewinternet.org/Reports/2011/The-Internet-and-Campaign-2010.aspx; Susannah Fox, "Peer-to-Peer Healthcare," Pew Internet & American Life Project, February 28, 2011, http://pewinternet.org/Reports/2011/P2PHealthcare.aspx.

42. The video that led to Justin Bieber's discovery: http://www.youtube.com/watch?v=eQOFRZ1wNLw; see also http://www.justinbiebermusic.com.

43. Joan Anderman, "All Together Now: His YouTube Mashups Have Become a Hit," *Boston Globe,* May 10, 2009, http://www.boston.com/ae/music/articles/2009/05/10/all_together_now.

44. Scott Thill, "Kutiman's ThruYou Mashup Turns YouTube into Funk Machine," *Wired,* March 25, 2009, http://www.wired.com/underwire/2009/03/kutimans-pionee.

45. Kutiman [Ophir Kutiel], "The Mother of All Funk Chords," http://www.youtube.com/watch?v=tprMEs-zfQA. See the follow-up project at http://thru-you.com.

46. Sasha Frere-Jones, "Heavy Sifting: An Interview with Kutiman," *New Yorker,* May 30, 2009, http://www.newyorker.com/online/blogs/sashafrerejones/2009/03/heavy-sifting-a-1.html.

47. Anderman, "All Together Now," see note 43.

48. The "Beyond Reality" channel can be accessed at: http://www.youtube.com/user/Madrosed.

49. See "All in the Family," a rich portrait of the mother and daughter duo produced by Patricia G. Lange, http://digitalyouth.ischool.berkeley.edu/book-creativeproduction. Find this sidebar in the main volume by scrolling about a third of the way through the website or searching for "Lange."

50. http://www.youtube.com/user/Madrosed, April 19, 2011.

51. Mary Madden and Aaron Smith, "Reputation Management and Social Media," Pew Internet & American Life Project, May 26, 2010, http://www.pewinternet.org/Reports/2010/Reputation-Management.aspx.

52. Frere-Jones, "Heavy Sifting," see note 46.

53. Brian Erkdale, Namkoong Kang, Timothy Fung, and David Perlmutter, "Why Blog? (Then and Now)," *New Media & Society* 12, no. 2 (2009): 217–234, quotation on p. 227.

54. Mizuko Ito, Sonja Baumer, Matteo Bittanti, danah boyd, Rachel Cody, Becky Herr-Stephenson, Heather A. Horst, Patricia G. Lange, Dilan Mahendran, Katynka Z. Martinez, C. J. Pascoe, Dan Perkel, Laura Robinson, Christo Sims, and Lisa Tripp, *Hanging Out, Messing Around, and Geeking Out: Kids Living and Learning with New Media* (Cambridge, MA: MIT Press, 2009).

55. Much of the material about Parles is drawn from a white paper about the rise of empowered "e-patients" that researcher Dr. Tom Ferguson began and his friends, including Pew Internet's Susannah Fox, completed in 2008 after his death: "E-Patients: How They Can Help Us Heal Health Care," http://e-patients.net/e-Patients_White_Paper.pdf. Lung Cancer Online, http://www.lungcanceronline.org.

56. Erkdale, Kang, Fung, and Perlmutter, "Why Blog?," see note 53.

57. Wael Ghonim, "Inside the Egyptian Revolution," TED Conference, Cairo, March 2011, http://www.ted.com/talks/wael_ghonim_inside_the_egyptian_revolution.html.

58. Erkdale, Kang, Fung, and Perlmutter, "Why Blog?," quotation on p. 226, see note 53. One online news source, ProPublica.com, received the 2011 Pulitzer Prize for national reporting about the Wall Street banking scandals. It is the first such prize for journalists not working for traditional printed newspapers. However, they are not citizen journalists but rather professional journalists using a new form of distribution.

59. Elizabeth Eisenstein, *The Printing Press as an Agent of Change* (Cambridge, UK: Cambridge University Press, 1980).

Chapter 9

1. Justine Abigail Yu coauthored this chapter. Sharanpreet Kelley, Gustavo Mesch, Zeynep Tufecki, and Hal Varian provided useful advice.

2. John Seely Brown and Paul Duguid, *The Social Life of Information* (Boston: Harvard Business School Press, 2000).

3. See http://www.experiencefestival.com/a/Talmud/id/1897212 for a description of Daniel Bomberg's interpretation of the modern Talmud with text and commentaries.

4. Vannevar Bush, "As We May Think," *The Atlantic Monthly,* July 1945, http://www.theatlantic.com/magazine/archive/1945/07/as-we-may-think/3881.

5. See Ted Nelson, "Ted Nelson Specs," http://hyperland.com/mlawLeast.html.

6. Tom Rosenstiel has shared his working "list" and presentation slides on this subject. Also see his recent book, coauthored with Bill Kovach: *Blur* (New York: Bloomsbury, 2010).

7. Hal Varian and Peter Lyman, "How Much Information?" 2003, http://www2.sims.berkeley.edu/research/projects/how-much-info-2003/execsum.htm.

8. Report available at http://www.emc.com/collateral/demos/microsites/idc-digital-universe/iview.htm. In 2009, they found that the amount of information in the Digital Universe grew to 800,000 petabytes.

9. Chris Anderson, *The Long Tail* (New York: Hyperion, 2006).

10. Pablo Boczkowski, *News at Work* (Chicago: University of Chicago Press, 2010).

11. Lee Rainie and John Horrigan, "Election 2006 Online," January 17, 2007, http://www.pewInternet.org/~/media/Files/Reports/2007/PIP_Politics_2006.pdf.pdf.

12. John Horrigan, Kelly Garrett, and Paul Resnick, "The Internet and Democratic Debate," October 27, 2004, http://www.pewInternet.org/Reports/2004/The-Internet-and-Democratic-Debate.aspx.

13. Estimates from an analysis of more than twenty sources of information including books, newspapers, satellite radio, and Internet video. Global Industry Information Center, University of California San Diego, "How Much Information?" December 2009, http://hmi.ucsd.edu/pdf/HMI_2009_ConsumerReport_Dec9_2009.pdf.

14. Jenna Wortham, "Feel Like a Wallflower? Maybe It's Your Facebook Wall," *New York Times,* April 9, 2011, http://www.nytimes.com/2011/04/10/business/10ping .html?ref=technology.

15. Deborah Fallows, "Internet Searchers Are Confident, Satisfied and Trusting But Are Also Unaware and Naïve," January 23, 2005, http://www.pewinternet.org/ Reports/2005/Search-Engine-Users.aspx.

16. Lee Rainie and John Horrigan, "Counting on the Internet: Most Find the Information They Seek, Expect," December 29, 2002, http://www.pewinternet.org/Reports/ 2002/Counting-on-the-Internet-Most-find-the-information-they-seek-expect.aspx.

17. Nicholas Negroponte, *Being Digital* (New York: Vintage, 1995). See also the forthcoming: Joseph Turow, *The Daily You: How the New Advertising Industry is Defining Your Identity and Your Worth* (New Haven: Yale University Press, 2012).

18. Paul Hitlin and Lee Rainie, "Booking Travel Online Soars—And Slows," October 2004, http://www.pewinternet.org/~/media/Files/Reports/2004/PIP_Datamemo _Reputation.pdf.pdf. See also Lee Rainie, "Tagging," January 2007, http://www .pewinternet.org/Reports/2007/Tagging/Report.aspx.

19. Socrates in a conversation with Phaedrus in Plato's *Phaedrus.*

20. Nicholas D. Kristof, "D.I.Y. Foreign-Aid Revolution," *New York Times,* October 20, 2010, http://www.nytimes.com/2010/10/24/magazine/24volunteerism-t.html ?pagewanted=1&ref=nicholasdkristof.

21. Nicholas D. Kristof, "Answering Readers on DIY Aid," *On the Ground,* October 29, 2010, http://kristof.blogs.nytimes.com/2010/10/29/a-postscript-on-diy-aid/?ref =magazine.

22. Kristen Purcell, Lee Rainie, Amy Mitchell, and Tom Rosenstiel, "Understanding the Participatory News Consumer," Pew Internet & American Life Project, March 1, 2010, http://pewInternet.org/Reports/2010/Online-News.aspx.

23. Elihu Katz and Paul Lazarsfeld, *Personal Influence* (Glencoe, IL: Free Press, 1955).

24. Jennifer Kayahara and Barry Wellman, "Searching for Culture—High and Low," *Journal of Computer-Mediated Communication* 12, no. 3 (2007): http://jcmc.indiana .edu/vol12/issue3/kayahara.html.

25. Ann Blair, "Reading Strategies for Coping with Information Overload ca. 1550–1700," *Journal of the History of Ideas* 64, no. 1 (2003): 11–28, quotation on p. 11.

26. Kayahara and Wellman, "Searching for Culture," see note 24; see also David Weinberger, *Too Big to Know* (New York: Basic Books, 2012).

27. Fallows, "Internet Searchers," see note 15.

28. Mark Suster, *Both Sides of the Table,* December 20, 2010, http://www.bothside softhetable.com/2010/12/20/the-power-of-twitter-in-information-discovery.

29. Lee Rainie and Aaron Smith, " The Internet and the Recession," July 2009, http:// www.pewinternet.org/Reports/2009/11-The-Internet-and-the-Recession/3-How-the -internet-and-other-sources-have-helped-people-cope-with-the-recession/1-Ameri cans-have-used-several-sources-of-information-and-advice-in-the-recession.aspx.

30. Mary Madden and Aaron Smith, "Reputation Management and Social Media," Pew Internet & American Life Project, May 26, 2010, http://www.pewinternet.org/Reports/2010/Reputation-Management.aspx.

31. Greg Walton, "China's Golden Shield: Corporations and the Development of Surveillance Technology in the People's Republic of China," http://www.dd-rd.ca/site/_PDF/publications/globalization/CGS_ENG.PDF.

32. Article 14 of the government of China's Measures for Managing Internet Information Services, http://www.chinaculture.org/library/2008-02/06/content_23369.htm.

33. Nart Villeneuve, "Breaching Trust: An analysis of surveillance and security practices on China's TOM-Skype platform," http://www.scribd.com/doc/13712715/Breaching-Trust-An-analysis-of-surveillance-and-security-practices-on-Chinas-TOM Skype-platform.

34. The *Wall Street Journal*'s "What They Know" series reports on the Internet-tracking technologies being used by businesses, http://online.wsj.com/public/page/what-they-know-digital-privacy.html.

35. Evgeny Morozov, *The Net Delusion: The Dark Side of Internet Freedom* (New York: Public Affairs, 2011), p. 97.

36. *Wall Street Journal*, "What They Know," see note 34.

37. Julia Angwin, "The Web's New Gold Mine: Your Secrets," *Wall Street Journal,* July 30, 2010, http://online.wsj.com/article/SB10001424052748703940904575395073512989404.html.

38. Lynn Greiner, "The Perils of Social Networking," *ComputerWorld Canada*, October 18, 2010, http://www.itworldcanada.com/news/the-perils-of-social-net working/141749.

39. PleaseRobMe.com has stopped showing these check-ins after making the point that "if you don't want your information to show up everywhere, don't over-share."

40. The term "coveillance" was invented by Barry Wellman. See Steve Mann, Jason Nolan, and Barry Wellman, "Sousveillance," *Surveillance and Society* 1, no. 3 (2003): 331–355.

41. David Westerman, Brandon Van Der Heide, Katherine Klein, and Joseph Walther, "How Do People Really Seek Information about Others?" *Journal of Computer Mediated Communication* 13 (2008), no. 3: 751–767.

42. Bonnie Ruberg, "10 Signs You've Officially Become a Facebook Stalker," January 5, 2009, http://www.heartlessdoll.com/2009/01/10_signs_youve_officially_become_a_facebook_stalke.php.

43. Facebook.com, "Facebook Stalking," http://www.facebook.com/pages/Face-book -Stalking-Admit-it-you-do-it/147838687575.

44. Madden and Smith, "Reputation Management and Social Media," see note 30.

45. Jennifer Gibbs, Nicole Ellison, and Chih-Hui Lai, "First Comes Love, Then Comes Google," *Communication Research* 38, no. 1 (December 2010): 70–100.

46. The term "sousveillance" was invented by Steve Mann. See Mann, Nolan, and Wellman, "Sousveillance,"see note 40.

47. Wikileaks Secret U.S. Embassy Cables database available at http://wikileaks.org/cablegate.html.

48. Robert Booth and Julian Borger, "US Diplomats Spied on UN Leadership," *The Guardian*, November 28, 2010, http://www.guardian.co.uk/world/2010/nov/28/us-embassy-cables-spying-un; *Wikipedia*, "Wikileaks," 2011, http://en.wikipedia.org/wiki/Wikileaks.

49. The National Institute on Money in State Politics was founded in 1999 and "dedicated to accurate, comprehensive and unbiased documentation and research on campaign finance at the state level," http://www.followthemoney.org; Scott Walker information from http://www.followthemoney.org/database/StateGlance/candidate.phtml?c=116585.

50. The OpenNet Initiative is a collaborative project between the Citizen Lab at the University of Toronto, the Berkman Center for Internet and Society at Harvard University, and the SecDev Group: http://opennet.net.

51. OpenNet Initiative Regional Overviews are available at http://opennet.net/research/regions; Information Warfare Monitory, *Tracking GhostNet: Investigating a Cyber Espionage Network,* March 29, 2009, http://www.scribd.com/doc/13731776/Tracking-GhostNet-Investigating-a-Cyber-Espionage-Network.

52. Sun Microsystems' former CEO Scott McNealy said this as early as 1999. See http://www.wired.com/politics/law/news/1999/01/17538.

53. Zuckerberg, Remarks at the *Crunchie Awards,* January 2010, http://www.computerworld.com/s/article/9143859/Facebook_CEO_Zuckerberg_causes_stir_over_privacy; see also http://crunchies2009.techcrunch.com/about.

54. David Kirkpatrick, *The Facebook Effect* (New York: Simon & Schuster, 2010).

55. Interview with Eric Schmidt, "Google's CEO: "The Laws Are Written By Lobbyists," *The Atlantic*, October 1, 2010, http://www.theatlantic.com/technology/archive/2010/10/googles-ceo-the-laws-are-written-by-lobbyists/63908.

56. Mary Madden and Aaron Smith, "Reputation Management Online," September 2011.

57. danah boyd and Eszter Hargittai, "Facebook Privacy Setting: Who Cares?" *First Monday* 15, no. 8 (2010): http://firstmonday.org/htbin/cgiwrap/bin/ojs/index.php/fm/article/view/3086/2589.

58. Amanda Lenhart and Mary Madden, "Teens, Privacy, and Online Social Networks," April 2007, http://www.pewInternet.org/Reports/2007/Teens-Privacy-and-Online-Social-Networks/1-Summary-of-Findings.aspx.

59. Brady Robarts, "Randoms in My Bedroom: Negotiating Privacy and Unsolicited Contact on Social Network Sites," *PRism* 7, no. 3 (2010), http://www.prismjournal.org/fileadmin/Social_media/Robards.pdf.

60. See note 59.

Chapter 10

1. Unlike other names in this chapter, this one is a pseudonym.

2. Barry Wellman and Milena Gulia, "A Network Is More Than the Sum of Its Ties: The Network Basis of Social Support," in *Networks in the Global Village,* ed. Barry Wellman (Boulder, CO: Westview Press, 1999), 83–118.

3. See Jon Kleinberg, "Authoritative Sources in a Hyperlinked Environment," http://www.cs.cornell.edu/home/kleinber/auth.pdf; http://www.cs.cornell.edu/home/kleinber; http://en.wikipedia.org/wiki/Jon_Kleinberg; also David Easley and Jon Kleinberg, *Networks, Crowds and Markets* (Cambridge: Cambridge University Press, 2010).

4. danah boyd, "Social Media Is Here to Stay. Now What?" Microsoft Research Tech Fest, Redmond, Washington, February 26, 2009, http://www.danah.org/papers/talks/MSRTechFest2009.html. All boyd quotations in this chapter section are from this paper.

5. Mary Madden and Aaron Smith, "Reputation Management and Social Media," Pew Internet & American Life Project, May 26, 2010, http://pewinternet.org/Reports/2010/Reputation-Management.aspx; and unpublished data collected in a survey in May 2011.

6. Daniel Solove, *The Future of Reputation* (Ann Arbor, MI: Caravan, 2007), p. 35.

7. *The Godfather* (Paramount, 1972) screenplay by Mario Puzo and Francis Ford Coppola.

8. Rochelle Côté, Gabriele Plickert, and Barry Wellman, "Does the Golden Rule, Rule?" in *Contexts of Social Capital*, ed. Ray-May Hsung, Nan Lin, and Ronald Breiger (London: Routledge, 2009), 49–71.

9. danah boyd, "Taken Out of Context: American Teen Sociality in Networked Publics," PhD dissertation, School of Information, University of California, fall 2008, p. 172, http://www.danah.org/papers/TakenOutOfContext.pdf.

10. Much of this framework for new literacies was inspired by the "Networked Literacy" lecture of Howard Rheingold, University of Toronto, October 2010; see also Howard Rheingold, *Net Smarts* (Cambridge, MA: MIT Press, 2012); the writings of Pam Berger on her Infosearcher blog, *Learning in the Web 2.0 World*, http://infosearcher.typepad.com/infosearcher/2007/04/learning_in_the.html; and the work of Henry Jenkins, especially "Confronting the Challenges of Participatory Culture: Media Education for the 21st Century," with Katie Clinton, Ravi Purushotma, Alice J. Robinson, and Margaret Weigel, http://digitallearning.macfound.org/atf/cf/%7B7E45C7E0-A3E0-4B89-AC9C-E807E1B0AE4E%7D/JENKINS_WHITE_PAPER.PDF.

Chapter 11

1. Christian Beermann and Tsahi Hayat coauthored this chapter. Jesse Hirsh, Eden Litt, Mitchell Ruda, Theresa Senft, and Sarita Yardi gave good advice.

2. Isaac Asimov, *Foundation* (New York: Gnome Press, 1951).

3. Norman Augustine, "What We Don't Know Does Hurt Us," *Science* 279, no. 5357 (1998): 1640–1641, http://www.sciencemag.org/content/279/5357/1640.full.

4. Neal Stephenson, *Snow Crash* (New York: Bantam, 1982).

5. John Smart, Jamais Cascio, and Jerry Paffendorf, *The Metaverse Roadmap*, Acceleration Studies Foundation, 2007, http://metaverseroadmap.org. The authors state that this is "the first public ten-year forecast and visioning survey of 3D Web technologies, applications, markets, and potential social impacts."

6. Gordon Moore, "Cramming More Components onto Integrated Circuits," *Electronics* 38, no. 8 (April 19, 1965), http://download.intel.com/research/silicon/moorespaper.pdf.

7. Smart, Cascio, and Paffendorf, *The Metaverse Roadmap*, see note 5.

8. Chip Walter, "Kryder's Law," *Scientific American*, August 2005, http://www.scientificamerican.com/article.cfm?id=kryders-law.

9. Smart, Cascio, and Paffendorf, *The Metaverse Roadmap*, see note 5.

10. George Gilder, *Telecosm* (New York: Touchstone, 2000); Jakob Nielsen, "Nielsen's Law of Internet Bandwidth," April 5, 1998, http://www.useit.com/alertbox/980405.html; Cooper's Law: story about its creator Marty Cooper in *The Economist*, "Father of the Cell Phone, June 4, 2009, http://www.economist.com/node/13725793?story_id=13725793.

11. Sarah Jacobsson Purewal, "Dropbox Will Hand Over Your Files to the Feds If Asked," *PCWorld*, April 19, 2011, http://www.pcworld.com/article/225549/dropbox_will_hand_over_your_files_to_the_feds_if_asked.html.

12. Wikipedia, "Technologies in *Minority Report*," Accessed April 20, 2011, from update of March 5, 2011, http://en.wikipedia.org/wiki/Technologies_in_Minority_Report.

13. Mark Weiser, "The Computer of the 21st Century," *Scientific American*, September (1991): 66–75; Adam Greenfield, *Everyware* (Indianapolis: New Riders, 2006).

14. Microsoft, "Welcome to Surface 2.0," 2011, http://www.microsoft.com/surface/en/us/default.aspx.

15. Julian Bleeckner, *A Manifesto for Networked Objects* (Cambridge, MA: MIT Press, 2006), p. 4.

16. John Markoff, "Crashes and Traffic Jams in Military Test of Robotic Vehicles," *New York Times*, November 5, 2007, http://www.nytimes.com/2007/11/05/technology/05robot.htm?adxnnl=1&adxnnlx=1303431109-jXXebZtwobwVBjOs1dOsMw; Byron Spice and Anne Watzman, "Carnegie Mellon Tartan Racing Wins $2 Million DARPA Urban Challenge," Carnegie Mellon University, November 4, 2007, http://www.cmu.edu/news/archive/2007/November/nov4_tartanracingwins.shtml.

17. Diana Scearce, *Connected Citizens*, Knight Foundation, Miami, 2011; Stephen Coleman and Peter Shane, eds. *Connecting Democracy* (Cambridge, MA: MIT Press, 2012).

18. Grigoris Antoniou and Frank van Harmelen, *A Semantic Web Primer*, 2nd ed. (Cambridge, MA: MIT Press, 2008).

19. See the discussion of the semantic web in Janna Anderson and Lee Rainie, "The Fate of the Semantic Web," May 2010, http://pewinternet.org/Reports/2010/Semantic-Web/Overview.aspx?r=1; and Lee Rainie and Janna Anderson, "The Future of the Internet III," December 2008, http://www.pewinternet.org/reports/2008/the-future-of-the-internet-iii.aspx.

20. Robin Yapp, "Brazilian Police to Use 'Robocop-Style' Glasses at World Cup," *London Telegraph*, 12 April, 2011, http://www.telegraph.co.uk/news/worldnews/southamerica/brazil/8446088/Brazilian-police-to-use-Robocop-style-glasses-at-World-Cup.html.

21. Amy-Mae Elliott, "10 Amazing Augmented Reality iPhone Apps," Mashable.com, 2009, http://mashable.com/2009/12/05/augmented-reality-iphone; "40 Best Augmented Reality iPhone Applications," iPhoneness.com, http://www.iphoneness.com/iphone-apps/best-augmented-reality-iphone-applications.

22. Google, "Annotating Google Earth," Google Earth Outreach, http://earth.google.com/outreach/tutorial_annotate.html.

23. Howard Rheingold, *Smart Mobs* (New York: Basic Books, 2002).

24. Smart, Cascio, and Paffendorf, *The Metaverse Roadmap*, p. 9, see note 5.

25. David Gelernter, *Mirror Worlds* (New York: Oxford University Press, 1992).

26. Bonnie Nardi, *My Life as a Night Elf Priest* (Ann Arbor: University of Michigan Press, 2010).

27. Stephenson, *Snow Crash*, pp. 35–36, see note 4.

28. Michael Rymaszewski, *Second Life: The Official Guide* (New York: Sybex, 2008); Smart, Cascio, and Paffendorf, *The Metaverse Roadmap*, p. 7, see note 5.

29. Tateru Nino, "Second Life Statistical Charts," 2011, http://dwellonit.taterunino.net/sl-statistical-charts.

30. Isaac Asimov, *The Naked Sun* (New York: Doubleday, 1957).

31. Josh Constine, "Facebook Announces Friendship Pages That Show Friends' Mutual Content," *Inside Facebook*, 2010, http://www.insidefacebook.com/2010/10/28/friendship-pages-mutual-content.

32. Smart, Cascio, and Paffendorf, *The Metaverse Roadmap*, see note 5.

33. Steve Mann, "Wearable Computing," *IEEE Intelligent Systems* 16, no. 3 (2001): 10–15.

34. Theresa Senft, *Cam Girls* (New York: Peter Lang, 2008), p. 1.

35. Theresa Senft,"*Fame to Fifteen: My Talk for TED*," 2010, http://tsenft.livejournal.com/412814.html.

36. FitBit, "About the Fitbit," Fitbit.com, 2010, http://www.fitbit.com/product; Kate Green, "Self Surveillance," *MIT Technology Review*, September 10, 2008, http://www.technologyreview.com/communications/21361/page1.

37. See also Arnold Roosendaal, "Facebook Tracks and Traces Everyone: Like This!" Tilburg Institute for Law, Technology and Society, Tilburg University, Netherlands, November 30, 2010, http://papers.ssrn.com/sol3/papers.cfm?abstract_id=1717563.

38. Justine Yu, email message to coauthor Wellman, April 19, 2011.

39. Nick Bilton, "Tracking File Found in iPhones," *New York Times,* April 20, 2011, http://www.nytimes.com/2011/04/21/business/21data.html?_r=2.

40. Vannevar Bush, "As We May Think," *The Atlantic Monthly* (July 1945): 101–108. The quotation is from p. 106.

41. Gordon Bell and Jim Gemmell, *Total Recall* (New York: Dutton, 2009). The quotation is from p. 218. For a less positive view of a world permeated by networked microsensors, see Neal Stephenson's science fiction novel, *The Diamond Age* (New York: Bantam, 1995).

42. Ronald Baecker and associates, "Digital Life Histories," TAGlab [Techniques for Aging Gracefully], 2011, http://taglab.utoronto.ca/projects/digital-life-history.

43. Jeffrey Rosen, *The Unwanted Gaze* (New York: Knopf, 2001).

44. Rainey Reitman, "Well-Meaning 'Privacy Bill of Rights' Wouldn't Stop Online Tracking," Electronic Frontier Foundation, April 14, 2011, http://www.eff.org/deep links/2011/04/well-meaning-privacy-bill-rights-could-codify.

45. Ciaran Giles, "Internet 'Right to be Forgotten' Debate Hits Spain," *Associated Press,* April 20, 2011, http://www.boston.com/news/world/europe/articles/2011/04/20/ internet_right_to_be_forgotten_debate_hits_spain.

46. Ronald Deibert, John Palfrey, Rafal Rohozinski, and Jonathan Zittrain, eds., *Access Denied* (Cambridge, MA: MIT Press, 2008).

47. This widely circulated saying is attributed to Stewart Brand, also the editor of the epochal *Whole Earth Catalog.* See R. Polk Wagner, "Information Wants to Be Free," University of Pennsylvania Law School, May 2003, http://www.law.upenn.edu/fac/ pwagner/wagner.control.pdf.

48. Jordan Robinson, "The Day Part of the Internet Died," *Washington Times,* January 28, 2011, http://www.washingtontimes.com/news/2011/jan/28/day-part -internet-died-egypt-goes-dark.

49. *Economist,* "The Future of the Internet: A Virtual Counter-Revolution," September 2, 2010, pp. 75–77, http://www.economist.com/node/16941635?story_id =16941635&fsrc=rss.

50. *Wikipedia,* "List of Websites Blocked in the People's Republic of China," accessed April 20, 2011, from update of April 20, 2011, 0758 UTC, http://en.wikipedia.org/ wiki/List_of_websites_blocked_in_the_People%27s_Republic_of_China#Other _websites.

51. Electronic Frontier Foundation, "AT&T's Role in Dragnet Surveillance of Millions of Its Customers," November 12, 2007, http://www.eff.org/files/filenode/att/ presskit/ATT_onepager.pdf.

52. Ronald Deibert, John Palfrey, Rafal Rohozinski, and Jonathan Zittrain, eds., *Access Controlled* (Cambridge, MA: MIT Press, 2010); *The Economist,* see note 49.

53. Eli Pariser, *The Filter Bubble: What the Internet Is Hiding from You* (London: Penguin, 2011).

54. Dan Rowinski, "Wave of the Future: Trusted Identities in Cyberspace," *Read-WriteWeb,* April 20, 2011, http://www.readwriteweb.com/archives/wave_of_the _future_trusted_identities_in_cyberspac.php.

55. Our heading is inspired by William Mitchell, *Me ++* (Cambridge, MA: MIT Press, 2003).

56. Barry Wellman and Barry Leighton, "Networks, Neighborhoods, and Communities," *Urban Affairs Quarterly* 14 (1979): 363–390; Peter Monge and Noshir Contractor, *Theories of Communication Networks* (New York: Oxford University Press, 2003).

57. Motion picture of H. G. Wells's *Things to Come* (United Artists, 1936); Terry Hayes, George Miller, and Brian Hannant, *Mad Max 2: The Road Warrior* (Warner Bros., 1981); Jame Cameron, Gale Anne Hurd, and William Wisher Jr., *The Terminator* (Orion, 1984); George Romero, *George Romero's Land of the Dead* (Universal, 2005); Gary Whitta, *Book of Eli* (Warner Bros., 2010).

58. Walter B. Miller, *A Canticle for Leibowitz* (Philadelphia: Lippincott, 1960); Doris Lessing, *Memoirs of a Survivor* (London: Octagon, 1974); Margaret Atwood, *The Year of the Flood* (London: Bloomsbury, 2009).

59. Lewis Mumford summarizes the change to walled towns in *The City in History* (New York: Harcourt, Brace, 1961). The classic article on the conditions for localism is: Charles Tilly, "Do Communities Act?" *Sociological Inquiry* 43 (1973): 209–240. A good account of gang control of their neighborhoods is in Sudhir Venkatesh, *Gang Leader for a Day* (New York: Penguin, 2008).

60. Jeff Jarvis, "Friends Forever," *BuzzMachine*, November 28, 2007, http://www.buzzmachine.com/2007/11/28/friends-forever-the-advantages-of-publicness.

Index